Topics in
Current Physics

13

Topics in Current Physics Founded by Helmut K. V. Lotsch

Computer Processing of Electron Microscope Images

Edited by P. W. Hawkes

With Contributions by
J. Frank   P. W. Hawkes   R. Hegerl   W. Hoppe
M. S. Isaacson   D. Kopf   J. E. Mellema
W. O. Saxton   M. Utlaut   R. H. Wade

With 116 Figures

Springer-Verlag Berlin Heidelberg New York 1980

Peter W. Hawkes, M. A., Ph. D.

Laboratoire d'Optique Electronique du CNRS, B. P. 4347
F-31055 Toulouse Cedex, France

ISBN-13:978-3-642-81383-2   e-ISBN-13:978-3-642-81381-8
DOI: 10.1007/978-3-642-81381-8

Library of Congress Cataloging in Publication Data. Main entry under title: Computer processing of electron microscope images. (Topics in current physics; v. 13). Bibliography: p. Includes index. 1. Electron microscopy — Data processing. 2. Image processing. I. Hawkes, P. W. II. Frank, Joachim, 1940-. III. Series. QH212.E4C63 535'.3325'02854 79-19555

2153/3130-543210

Preface

Towards the end of the 1960s, a number of quite different circumstances combined
to launch a period of intense activity in the digital processing of electron micrographs. First, many years of work on correcting the resolution-limiting aberrations
of electron microscope objectives had shown that these optical impediments to very
high resolution could indeed be overcome, but only at the cost of immense experimental difficulty; thanks largely to the theoretical work of K.-J. Hanszen and his
colleagues and to the experimental work of F. Thon, the notions of transfer functions were beginning to supplant or complement the concepts of geometrical optics
in electron optical thinking; and finally, large fast computers, capable of manipu-
lating big image matrices in a reasonable time, were widely accessible. Thus the
idea that recorded electron microscope images could be improved in some way or
rendered more informative by subsequent computer processing gradually gained ground.
At first, most effort was concentrated on three-dimensional reconstruction, particularly of specimens with natural symmetry that could be exploited, and on linear
operations on weakly scattering specimens (Chap.1). In 1973, however, R.W. Gerchberg
and W.O. Saxton described an iterative algorithm that in principle yielded the
phase and amplitude of the electron wave emerging from a strongly scattering specimen. Other procedures rapidly followed and a large amount of work has been, and
continues to be, devoted to the so-called phase problem, retrieval of the phase
distribution of the electron wave given that only its intensity can be recorded.
An account of this forms Chap.2.

Meanwhile, the practical problems of image handling—alignment of related images,
numerical superposition to increase the signal-to-noise ratio and many others —were
not neglected. The role of correlation functions in this connection is so important
that an entire chapter of this volume (Chap.5) is devoted to them.

With the success of the early attempts to reconstruct highly symmetric objects,
it was natural that specimens with lower symmetry should be studied (Chap.3) and
that attempts should also be made to reconstruct the three-dimensional structure
of specimens with little or no natural symmetry, particularly those liable to be
damaged by the electron beam itself. Methods of solving this last problem have been
gradually developed over the years by W. Hoppe and colleagues, and the results of
their efforts, and others in this difficult domain, are set out in Chap.4.

The remaining chapters are concerned with two other aspects of electron image processing. In Chap.6, a brief account is given of electron holography, for although digital methods have not yet gained much ground here, owing to the relatively small areas that can be manipulated, it seems likely that some of the experimental techniques would gain considerably from hybridization with the computer: the work of Wahl (referred to in Chap. 6) is a convincing example of this. The last chapter is concerned with on-line processing of the image formed in a scanning transmission electron microscope (STEM), an instrument that lends itself particularly well to such treatment since the image is formed point by point, or rather pixel by pixel, and is directly available as an electrical signal. Many kinds of on-line image processing can thus be envisaged, and the possibilities become even more extensive when we recall that several different kinds of information may be available for each picture element.

Electron image processing by computer is a young subject; nevertheless, it seemed appropriate to collect these accounts of the progress that has been made as the recent literature suggests that the subject is entering a phase of consolidation: processed images of real specimens are appearing at a slowly increasing rhythm while the flood of theoretical papers seems to be abating —to some extent at least. The next few years will show to what extent computer processing of electron images can help to overcome the electron microscopist's problems, particularly where radiation-sensitive specimens are concerned, by means of the techniques described in this volume.

In order to avoid tedious repetition, references to the series of international and European conferences on electron microscopy are identified simply by locality and date. Full bibliographic details are given in the Appendix.

Toulouse, France
November, 1979

P.W. Hawkes

Contents

List of Contributors

FRANK, JOACHIM

    State of New York Department of Health,
Division of Laboratories and Research, Empire State Plaza,
Albany, NY 12201, USA

HAWKES, PETER WILLIAM

    Laboratoire d'Optique Electronique du CNRS, B.P. 4347
F-31055 Toulouse Cedex, France

HEGERL, REINER

    Abteilung für Strukturforschung I, Max-Planck-Institut für Biochemie,
D-8033 Martinsried bei München, Fed. Rep. of Germany

HOPPE, WALTER

    Abteilung für Strukturforschung I, Max-Planck-Institut für Biochemie,
D-8033 Martinsried bei München, Fed. Rep. of Germany

ISAACSON, MICHAEL SAUL

    School of Applied and Engineering Physics,
Clark Hall, Cornell University,
Ithaca, NY 14853, USA

KOPF, DAVID AARON

    Enrico Fermi Institute, University of Chicago,
Chicago, IL 60637, USA

MELLEMA, JAN EGBERT

    Biochemisch Laboratorium, Rijksuniversiteit Leiden, Wassenaarseweg 64,
NL-2300 RA Leiden, The Netherlands

SAXTON, W. OWEN

    High Resolution Electron Microscope, Old Cavendish Laboratory, Free School Lane,
Cambridge CB2 3RQ, Great Britain

UTLAUT, MARK WILLIAM

Enrica Fermi Institute, University of Chicago,
Chicago, IL 60637, USA

WADE, RICHARD HARRY

Département de Recherche Fondamentale, CENG, 85X,
F-38041 Grenoble Cedex, France

1. Image Processing Based on the Linear Theory of Image Formation

P. W. Hawkes

With 12 Figures

For a limited class of specimens, there is a linear relation between the contrast variation in the bright-field electron image and the phase and amplitude variation of the object. For dark-field imaging, a linear relation is satisfied only if very restrictive conditions are satisfied, so restrictive that image processing operations that exploit linearity are of relatively little interest. More than one mechanism is at work in the formation of the bright-field image in a conventional transmission electron microscope. Electrons that are scattered through large angles within the specimen strike the objective aperture and their absence from the image plane creates contrast that is not unlike amplitude contrast in the light microscope. Scattered electrons that are not intercepted by the objective aperture all reach the image plane but the additional deflection due to spherical aberration and defocus combines with the specimen scattering to provide some image contrast. The action of the spherical aberration and defocus is analogous to that of a phase plate of variable thickness in the radial direction and this type of contrast is therefore phase contrast. Unfortunately, owing to the nonuniformity of the phase shifts, the corresponding image detail does not necessarily represent the object structure faithfully. Contrast reversals may occur and some detail will be entirely suppressed. Nevertheless, for specimens that scatter weakly, bright-field image contrast and complex specimen transparency are connected by a linear relation; it is this relation and the various techniques that have been proposed for exploiting it that form the subject of this chapter. Generally speaking, the relation between a dark-field image and the specimen structure is considerably more complicated, and the types of artefact that may arise, although easy to predict, are much more difficult to eliminate; this is true of all the dark-field modes in use. For modest resolutions, image contrast and object structure are related linearly but it seems unlikely that computer image processing will have much to offer in this domain.

1.1 Transfer Functions

The motion of electrons through the lenses of an electron microscope is conveniently described in terms of the (time-independent) wave function $\psi(\underline{x},z)$, the z axis coinciding with the optic axis of the instrument. A detailed account of the background theory is to be found in GLASER's work [1.1,2]; here we set out from the result that the wave function in an arbitrary plane z, downstream from the object plane $z = z_0$, is given in terms of $\psi(\underline{x}_0,z_0)$ by the following formula

$$\psi(\underline{x},z) = \frac{1}{i\lambda h(z)} \iint \psi(\underline{x}_0,z_0) \exp\left\{\frac{i\pi}{\lambda h(z)}[g(z)\underline{x}_0 \cdot \underline{x}_0\right.$$

$$\left. - 2\underline{x} \cdot \underline{x}_0 + h'(z)\underline{x} \cdot \underline{x}]\right\} d^2\underline{x}_0 \quad . \tag{1.1}$$

Here and elsewhere, we denote the cartesian coordinates (x,y) by a single vector $\underline{x}$; the functions g(z) and h(z) are solutions of the paraxial equations of motion satisfying the boundary conditions $g(z_0) = h'(z_0) = 1$, $g'(z_0) = h(z_0) = 0$. In the plane(s) $z = z_d$ in which g(z) vanishes, so that $g(z_d) = 0$, we find that

$$i\lambda h(z_d)\psi(\underline{x},z_d) = E_d \iint \psi(\underline{x}_0,z_0) \exp\left[-\frac{2\pi i}{\lambda h(z_d)} \underline{x}_d \cdot \underline{x}_0\right] d^2\underline{x}_0 \quad , \tag{1.2}$$

where

$$E_d = \exp\left[\frac{i\pi h'(z_d)\underline{x}_d \cdot \underline{x}_d}{\lambda h(z_d)}\right] \quad . \tag{1.3}$$

The wave function in the plane $z = z_d$ is thus proportional to the Fourier transform of that in the object plane, $z = z_0$; $z = z_d$ and any planes conjugate to it are hence the planes in which the diffraction pattern is formed. [It is worth pointing out that the result obtained by KOMRSKA and LENC [1.3], to the effect that " the Fraunhofer pattern is always situated in the image focal plane", does not conflict with this since their boundary conditions place the effective electron source at infinity, whereas the form of (1.1) permits any source position.]

The wave function in the image plane is conveniently obtained in two stages, in order to incorporate the effects due to the aperture and aberrations. The wave function at the image plane, $z = z_i$, is related to ψ in some (arbitrary) intermediate plane, $z = z_a$, by an equation analogous to (1.1)

$$\psi(\underline{x}_i,z_i) = \frac{1}{i\lambda H(z_i)} \iint \psi(\underline{x}_a,z_a) \exp\left\{\frac{i\pi}{\lambda H(z_i)}[G(z_i)\underline{x}_a \cdot \underline{x}_a\right.$$

$$\left. - 2\underline{x}_a \cdot \underline{x}_i + H'(z_i)\underline{x}_i \cdot \underline{x}_i]\right\} d^2\underline{x}_a \tag{1.4}$$

in which $G(z)$ and $H(z)$ satisfy the same boundary conditions as g and h but in the plane $z = z_a$ instead of $z = z_0$. Substituting for $\psi(\underline{x}_a, z_a)$ in (1.4) with the aid of (1.1), we find

$$\psi(\underline{x}_i, z_i)E_i = \frac{1}{M} \iint K(\underline{x}_i; \underline{x}_0)\psi(\underline{x}_0, z_0)E_0 d^2\underline{x}_0 \qquad (1.5)$$

in which M is the magnification,

$$M = g(z_i) \quad . \qquad (1.6)$$

The function $K(\underline{x}_i; \underline{x}_0)$ is of the form

$$K(\underline{x}_i; \underline{x}_0) = \frac{1}{\lambda^2 h^2(z_a)} \iint A(\underline{x}_a) \exp\left[-\frac{2\pi i}{\lambda h(z_a)}(\underline{x}_0 - \underline{x}_i/M) \cdot \underline{x}_a\right] d^2\underline{x}_a \quad . \qquad (1.7)$$

The multipliers E_i and E_0 are quadratic phase factors,

$$E_i = \exp\left(\frac{i\pi}{\lambda M} \frac{g_a h'_i - g'_i h_a}{h_a} \underline{x}_i \cdot \underline{x}_i\right) \qquad (1.8a)$$

$$E_0 = \exp\left(\frac{i\pi}{\lambda h_a} \underline{x}_0 \cdot \underline{x}_0\right) \qquad (1.8b)$$

in which g_a denotes $g(z_a)$ and similarly for the other subscripted functions. Provided that $z_a \approx z_d$, h_a is equal to the objective focal length and we therefore write

$$h_a = f \qquad (1.9)$$

henceforth. The aperture function $A(\underline{x}_a)$ is equal to zero where the objective aperture intercepts electrons and to unity in the opening, in the paraxial approximation. The objective aberrations may be included in this function as a complex factor, or phase shift [1.1,4], and we therefore write

$$A(\underline{x}_a) = a(\underline{x}_a) \exp[-i\gamma(\underline{x}_a)] \quad , \qquad (1.10)$$

where

$$\gamma(\underline{x}_a) = \frac{2\pi}{\lambda}\left[\frac{1}{4} C_s\left(\frac{\underline{x}_a \cdot \underline{x}_a}{f^2}\right)^2 - \frac{1}{2} \Delta \frac{\underline{x}_a \cdot \underline{x}_a}{f^2}\right.$$
$$\left. + \frac{1}{2} C_a \frac{x_a^2 - y_a^2}{f^2}\right] \quad . \qquad (1.11)$$

We have retained only the spherical aberration C_s, the defocus Δ, and the astigmatism C_a in γ. Provided that the other aberrations are negligible, the function K contains x_i and x_o only as the combination $x_i/M - x_o$ with the result that the right-hand side of (1.5) is a convolution. Systems for which the phase shifts associated with aberrations are independent of the object coordinates — for which, that is, the third-order astigmatism, field curvature, coma, and distortion are small — are said to be isoplanatic. The present discussion is restricted to isoplanatic systems for, although the theory can be extended to cover more complicated situations, it does not lead to invertible filters of the type we shall encounter below, for the isoplanatic case [1.5-13].

The sign of the phase shift in $\exp(-i\gamma)$ in (1.10) is to some extent arbitrary, as explained in [1.14] (an alternative explanation is to be found in [1.15]) but the convention whereby we use $\exp(-i\gamma)$ rather than $\exp(i\gamma)$ is equivalent to working with the standard form of the Schrödinger equation rather than with the complex conjugate form that governs ψ^*. In practice, therefore, the form $\exp(-i\gamma)$ is mandatory [1.16].

For isoplanatic systems, then, (1.5) has the form

$$\psi(x_i, z_i) E_i = \frac{1}{M} \iint K(x_i/M - x_o)\ (x_o, z_o) E_o d^2x_o \ . \tag{1.12}$$

Writing

$$\tilde{\psi}_o(\underline{p}) = \iint E_o \psi(x_o, z_o)\ \exp(-2\pi i \underline{p} \cdot x_o) d^2x_o \tag{1.13}$$

$$\tilde{K}(\underline{p}) = \iint K(\underline{x})\ \exp(-2\pi i \underline{p} \cdot \underline{x}) d^2x \tag{1.14}$$

$$\tilde{\psi}_i(\underline{p}) = \iint E_i \psi(Mx_i, z_i)\ \exp(-2\pi i \underline{p} \cdot x_i) d^2x_i$$

$$= \frac{1}{M^2} \iint E_i \psi(x_i, z_i)\ \exp(-2\pi i \underline{p} \cdot x_i/M) d^2x_i \ , \tag{1.15}$$

(1.12) transforms to

$$\tilde{\psi}_i(\underline{p}) = \frac{1}{M} \tilde{K}(\underline{p}) \tilde{\psi}_o(\underline{p}) \tag{1.16}$$

This extremely important equation tells us that the electron microscope, characterized by K or A, acts as a linear scalar filter: the spatial frequency spectrum of the complex image wave function at any frequency $\underline{p}$ is given by the product of K and the component of the spectrum of the complex object wave function at the same frequency.

In order to relate the object wave function to the physical properties of the specimen, we model the latter in terms of a complex transparency $S(\underline{x}_0)$ such that the emergent wave $\psi_0(\underline{x}_0)$ and the incident wave $\bar{\psi}_0(\underline{x}_0)$ are related as follows:

$$\psi_0(\underline{x}_0) = S(\underline{x}_0)\bar{\psi}_0(\underline{x}_0) \quad . \tag{1.17}$$

More realistically, we might attempt to make some allowance for inelastic scattering in the specimen by writing, for example,

$$\psi_0(\underline{x}_0,\lambda') = S(\underline{x}_0,\lambda,\lambda')\bar{\psi}_0(\underline{x}_0,\lambda) \quad , \tag{1.18}$$

where λ denotes the wavelength of the incident beam and λ' is greater than or equal to λ. It proves more natural to use frequency than wavelength, however, because of the direct connection between frequency and energy[1] ($E = \hbar\omega$); we defer discussion of this point to Sect.1.2.

Expanding the specimen transparency $S(\underline{x}_0)$ in the form

$$S(\underline{x}_0) = 1 - S_r(\underline{x}_0) + iS_i(\underline{x}_0) \quad , \tag{1.19}$$

we obtain a general expression, which coincides with that obtained in the weak-phase, weak-amplitude approximation. For, if we write

$$S = (1 - s) \exp(i\varphi) \tag{1.20}$$

and assume that $s \ll 1$ and $\varphi \ll 1$, we have

$$S \approx (1 - s)(1 + i\varphi)$$

$$\approx 1 - s + i\varphi \quad . \tag{1.21}$$

Substituting (1.21) into (1.16), we find

$$\tilde{\psi}_i(\underline{p}) = \frac{1}{M} \tilde{K}(\underline{p})[\delta(\underline{p}) - \tilde{s}(\underline{p}) + i\tilde{\varphi}(\underline{p})] \quad , \tag{1.22}$$

and from (1.7) and (1.14), we know that

$$A(\lambda f\underline{p}) = \tilde{K}(\underline{p}) \quad . \tag{1.23}$$

---

1 $\hbar$ = h/2π (normalized Planck's constant)

Inverting the Fourier transform of (1.22), it is not difficult to show that $j_i = M^2\psi_i\psi_i^*$, which is proportional to the current density at the image, is given by

$$j_i(M\underline{x}_i) = M^2\psi_i(M\underline{x}_i)\psi_i^*(M\underline{x}_i)$$

$$= 1 - 2 \iint a\tilde{s} \cos\gamma \exp(2\pi i\underline{p} \cdot \underline{x}_i)d^2\underline{p}$$

$$+ 2 \iint a\tilde{\varphi} \sin\gamma \exp(2\pi i\underline{p} \cdot \underline{x}_i)d^2\underline{p} \quad . \tag{1.24}$$

If finally we denote the image contrast by C,

$$j_i - 1 = C \quad , \tag{1.25}$$

we see that the spatial frequency spectrum of the image contrast is a linear combination of the spatial frequency spectra of s and φ, modulated by the amplitude and phase contrast transfer functions $-2 \cos\gamma$ and $2 \sin\gamma$, respectively (Fig.1.1),

$$\tilde{C} = - 2a\tilde{s} \cos\gamma + 2a\tilde{\varphi} \sin\gamma \quad . \tag{1.26}$$

The foregoing analysis corresponds to the simplest case possible: a bright-field image formed with perfectly coherent illumination incident perpendicular to the specimen, assumed to lie in a plane perpendicular to the optic axis z. Other cases of practical interest can be analyzed straightforwardly with only a slight modification of the reasoning. By suitably defining the function a(p), the effect of using eccentric or noncircular apertures can be studied [in the analysis given above, it has been assumed that $a(\underline{p}) = a(-\underline{p})$]. By retaining the incident wave function, ψ_o^- of (1.17), we can study the effect of, for example, tilted incident illumination or more generally, the consequences of using a deliberately structured illuminating beam as HOPPE has proposed for ptychography [1.17-20]. For further discussion of half-plane and other nonstandard apertures, see [1.21-42]. Extensive bibliographies are to be found in the review articles of HANSZEN [1.43], SAXTON [1.44], and MISELL [1.45]. Image formation with tilted illumination is analyzed in the language of transfer functions in [1.46-64]. General accounts of transfer theory are given in [1.8,65-72]. The idealization that the illumination consists of a perfectly monochromatic beam of electrons emerging from a point source is normally a moderately good approximation, certainly legitimate for gaining an understanding of the principles of electron microscope transfer theory. In practice, however, we need to make allowance for both the finite energy spread of the incident beam and the finite source size. This can be achieved in various ways, as we explain in Sect.1.2, but the essential result is that the simple linear structure of (1.26)

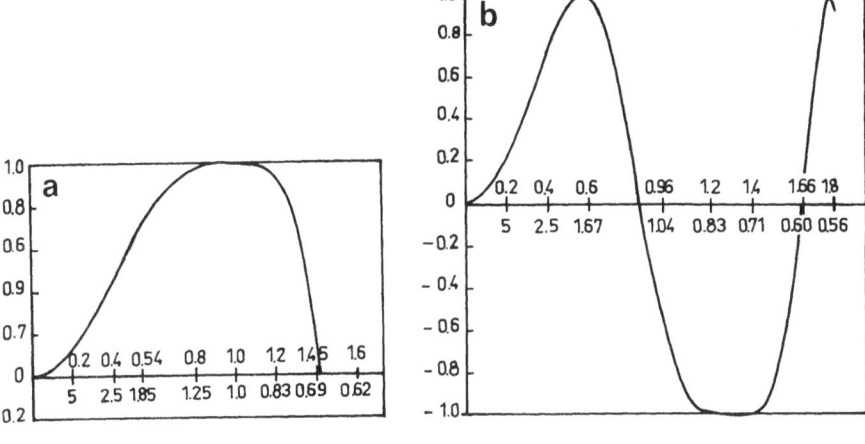

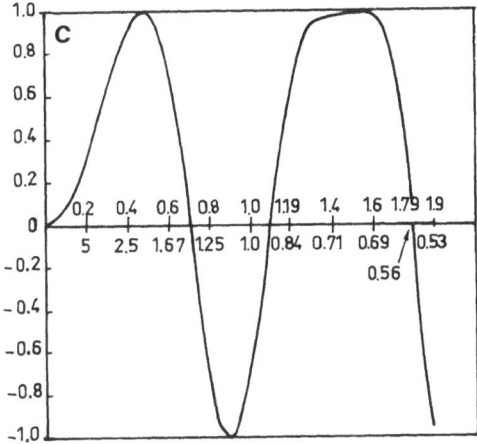

Fig.1.1a-c. The phase contrast transfer function (-sinγ) plotted against
spatial frequency (upper abscissa
scale) or its reciprocal (lower scale).
The latter are scaled with respect to
$C_s\lambda^3)^{1/4}$. The figures show three values
of defocus. (a)$\Delta = (C_s\lambda)^{1/2}$, (b)
$\Delta = (3C_s\lambda)^{1/2}$, (c) $\Delta = (5C_s\lambda)^{1/2}$

is always valid for weakly scattering specimens. Furthermore, in reasonably good
microscope operating conditions, the phase and amplitude transfer functions are
the same as their coherent counterparts, 2 sinγ and -2 cosγ, modulated by (multiplicative) functions characterizing the temporal and spatial partial coherence
separately. The linear relation (1.26) has generated numerous theoretical proposals
and a few experimental attempts to extract the amplitude and phase distributions,
s and φ, from micrographs or sets of micrographs, focal series in particular. It
also underlies the various techniques that are in use for extracting the values of
C_s and Δ from electron micrographs of thin amorphous phase specimens; for the latter
$s \approx 0$ and $\tilde{\varphi}$ is taken to be reasonably uniform over a range of values of spatial
frequency or better, the behavior of $\tilde{\varphi}$ can be obtained from scattering data. These
applications are further discussed in Sect.1.3.

In dark-field conditions, the situation is very different for it is easy to show
that the image contrast is not related linearly to the parameters characterizing the
specimen transparency, whether or not the latter scatters weakly. HANSZEN and ADE

[1.43,66,73-78] have shown that optical artefacts must inevitably be expected in dark-field operation (irrespective of the dark-field mode adopted) and HOCH has made computer simulations of a typical structure in each of the various dark-field modes [1.79]. In particular, he has shown that for the incoherent dark-field mode generated by a thick hollow cone (that is, the beam occupies the volume between two cones with a common vertex but different semiangles), linear transfer is a moderately good approximation only at relatively low resolution. The literature of dark-field image formation is extensive: we refer to the papers by HOCH and to [1.44,55,73,78, 80] for references.

1.2 Transfer Functions with Partially Coherent Illumination

It has long been known that a linear transfer theory can be established, relating image contrast to some suitable specimen property, when the illumination is partially coherent and the specimen scatters weakly [1.81-84]. The proof that the partially coherent transfer functions can be written as modulated versions of the corresponding coherent forms is more recent however: the case of temporal partial coherence was first analyzed by HANSZEN and TREPTE in 1970 [1.85,86] and that of spatial partial coherence by FRANK [1.87] in 1973. The legitimacy of separating the two types of modulation was questioned by McFARLANE (see FRANK [1.88]) but this doubt has since been satisfactorily resolved by WADE and FRANK [1.89] and by McFARLANE himself [1.90]. All this analysis is based on the mutual intensity (J in [1.4]), so that the source is in effect assumed quasi-monochromatic. An alternative approach, which allows us to introduce the energy distribution of the incident electrons in a very convenient way, is used by HAWKES [1.91-94]; here, the cross-spectral density W which is the temporal Fourier transform of the mutual coherence function (Γ in [1.4]) is used. Like Γ, the cross-spectral density W is a function of a pair of sets of position coordinates but the remaining argument is not time but frequency ω. In view of the relation between energy and frequency, $E = \hbar\omega$, we may regard this argument of W as a direct measure of incident electron energy, which explains the interest of this function in electron imaging studies.

We now show how a linear transfer theory arises for weakly scattering specimens, after which we recapitulate the findings of HANSZEN and TREPTE and of FRANK; we then comment briefly on bright-field hollow-cone illumination, an essentially partially coherent mode.

The study of partial coherence is the study of certain second-order correlations between the electron wave function at pairs of points. In particular, if $\Psi(\underline{x},t)$ denotes the (time-dependent) wave function in some plane, then the mutual coherence function $\Gamma(\underline{x},\bar{\underline{x}};\tau)$ is defined as the following time or ensemble average:

$$\Gamma(\underline{x},\bar{\underline{x}};\tau) = \langle \Psi(\underline{x},t + \tau) \Psi^*(\bar{\underline{x}},t) \rangle \quad . \tag{1.27}$$

For quasi-monochromatic sources, we write

$$\Gamma(\underline{x},\bar{\underline{x}};\tau) = J(\underline{x},\bar{\underline{x}}) \exp(-i\bar{\omega}\tau) \quad , \tag{1.28}$$

where $\bar{\omega}$ is the mean frequency of the source, and study the propagation not of Γ but of the mutual intensity, $J(\underline{x},\bar{\underline{x}})$. If, however, we wish to study the effect of the frequency, and hence energy, distribution of the source, it is more convenient to use the Fourier transform of Γ with respect to time, the cross-spectral density W:

$$\Gamma(\underline{x},\bar{\underline{x}};\tau) = \int W(\underline{x},\bar{\underline{x}};\omega) \exp(-i\omega\tau)d\omega \quad . \tag{1.29}$$

The mutual intensities in the object and image planes, suffices o and i, respectively, are connected by the following well-known relation [Ref.1.4, Eq. (10.5.31a); Ref.1.68, Eq. (3.1)]

$$J(\underline{x}_i,\bar{\underline{x}}_i) = \int J(\underline{x}_o,\bar{\underline{x}}_o)\bar{K}(\underline{x}_o,\underline{x}_i)\bar{K}^*(\bar{\underline{x}}_o,\bar{\underline{x}}_i)d^2\underline{x}_o d^2\bar{\underline{x}}_o \quad . \tag{1.30}$$

The function $\bar{K}$ is related to the function K used in (1.7) as follows

$$\bar{K}(\underline{x}_o,\underline{x}_i) = \frac{E(\underline{x}_o)}{ME(\underline{x}_i)} K(\underline{x}_i/M - \underline{x}_o) \tag{1.31}$$

so that (1.30) becomes

$$M^2 E(\underline{x}_i,\bar{\underline{x}}_i)J(\underline{x}_i,\bar{\underline{x}}_i) = \int E(\underline{x}_o,\bar{\underline{x}}_o)K(\underline{x}_i/M - \underline{x}_o)K^*(\bar{\underline{x}}_i/M - \bar{\underline{x}}_o)d^2\underline{x}_o d^2\bar{\underline{x}}_o \quad , \tag{1.32}$$

in which the quadratic phase factors have been written compactly,

$$E(\bar{\underline{x}}_o,\underline{x}_o) = \exp\left[\frac{i\pi r'}{\lambda} (\bar{\underline{x}}_o \cdot \bar{\underline{x}}_o - \underline{x}_o \cdot \underline{x}_o)\right] \tag{1.33a}$$

$$E(\bar{\underline{x}}_i,\underline{x}_i) = \exp\left[\frac{i\pi r}{\lambda M} (\bar{\underline{x}}_i \cdot \bar{\underline{x}}_i - \underline{x}_i \cdot \underline{x}_i)\right] \tag{1.33b}$$

and r, r' are geometrical constants. The mutual intensity below the specimen $J_o(\underline{x}_o, \bar{\underline{x}}_o)$ is related to the incident mutual intensity $J^-(\underline{x}_o,\bar{\underline{x}}_o)$ by the simple expression

$$J_o(\underline{x}_o,\bar{\underline{x}}_o) = S(\underline{x}_o)S^*(\bar{\underline{x}}_o)J^-(\underline{x}_o,\bar{\underline{x}}_o) \tag{1.34}$$

and we can often assume that $J^-(\underline{x}_o,\underline{\bar{x}}_o)$ is of the form $J^-(|\underline{x}_o - \underline{\bar{x}}_o|)$. Substituting (1.34) into (1.32), we find that the image intensity $j(\underline{x}_i)$, given by $J(\underline{x}_i,\underline{x}_i)$, is related to the specimen transparency as follows:

$$j(\underline{x}_i) = \frac{1}{M^2} \int S(\underline{x}_o)S^*(\underline{\bar{x}}_o)J^-(|\underline{x}_o - \underline{\bar{x}}_o|)$$

$$\times K(\underline{x}_i/M - \underline{x}_o)K^*(\underline{x}_i/M - \underline{\bar{x}}_o)E(\underline{x}_o,\underline{\bar{x}}_o)d^2\underline{x}_o d^2\underline{\bar{x}}_o \quad . \tag{1.35}$$

Introducing the Fourier transforms of the various functions occurring in this expression, we find

$$\tilde{j}_i(\underline{p}) = \int \tilde{S}(\underline{\bar{p}} + \underline{p}M)\tilde{S}^*(\underline{\bar{p}})\tilde{T}(\underline{\bar{p}} + \underline{p}M,\underline{\bar{p}})d\underline{\bar{p}} \quad , \tag{1.36}$$

where

$$\tilde{S}(\underline{p}) = \int E(\underline{x}_o)S(\underline{x}_o)\exp(-2\pi i\underline{p} \cdot \underline{x}_o)d^2\underline{x}_o \tag{1.37}$$

and

$$\tilde{T}(\underline{p},\underline{\bar{p}}) = \int J^-(|\underline{\bar{x}} - \underline{x}|)K(\underline{x})K^*(\underline{\bar{x}})$$

$$\times \exp[2\pi i(\underline{\bar{p}} \cdot \underline{\bar{x}} - \underline{p} \cdot \underline{x})]d^2\underline{x}d^2\underline{\bar{x}} \quad . \tag{1.38}$$

The function $\tilde{T}$, which is known as the transmission cross-coefficient, may also be written as

$$\tilde{T}(\underline{p},\underline{\bar{p}}) = \int \tilde{J}^-(\underline{v})\tilde{K}(\underline{p} + \underline{v})\tilde{K}^*(\underline{\bar{p}} + \underline{v})d\underline{v} \quad . \tag{1.39}$$

The most important point to note about (1.36), relating the spatial frequency spectrum of the image to the specimen transparency, is that although it represents a linear filter, telling us how to combine the microscope properties, characterized by $\tilde{T}$, with the specimen function, in which two values of S are involved, it is not a scalar filter but a vector filter: each image spatial frequency consists of a linear combination of contributions from all the specimen spatial frequency spectrum. This complexity renders (1.36) of comparatively little practical interest.

If, however, we replace the general specimen transparency S by the weak scattering approximation, (1.21), we can show that

$$\tilde{j}(\underline{p}/M) - \tilde{T}(\underline{p},0)\delta(\underline{p})$$

$$= \tilde{s}(\underline{p})T_s(\underline{p}) + \tilde{\varphi}(\underline{p})T_\varphi(\underline{p}) \quad , \tag{1.40}$$

in which

$$T_s(\underline{p}) = - [\tilde{T}(\underline{p},0) + \tilde{T}(0, -\underline{p})] \qquad (1.41)$$

$$T_\varphi(\underline{p}) = i[\tilde{T}(\underline{p},0) - \tilde{T}(0, -\underline{p})] \quad . \qquad (1.42)$$

Equation (1.40) is exactly analogous to (1.26), the functions T_s and T_φ replacing the fully coherent transfer functions, $-2 \cos\gamma$ and $2 \sin\gamma$, respectively. A form of this result was established by MENZEL in 1958 [1.81] and it is further discussed in [1.82-84]; a much more detailed version of the foregoing derivation is to be found in [1.68].

We now consider the individual manifestations of partial coherence that are of interest in electron optics: finite source size and finite energy spread of the incident beam. We neglect here the fact that the emergent beam will have a considerably wider energy spread than the incident beam, owing to inelastic scattering; this point is discussed at length in [1.94], and provides a good illustration of the superiority of the cross-spectral density over the mutual intensity when frequency and hence energy is an important parameter. The legitimacy of treating spatial and temporal partial coherence separately is not obvious and has been explored in depth in light optics; a recent discussion with references to earlier work is to be found in [1.95]. In electron optics, a careful examination of the approximations made shows that the two effects can in general safely be treated independently.

We first examine the effect of finite source size, following the analysis of FRANK [1.87] closely. Into the formulae for T_s and T_φ [(1.41) and (1.42)], we substitute for $\tilde{K}$, recalling that the latter is simply related to the aperture function, (1.23), and we neglect the finite aperture size. We find

$$T_s(\underline{p}) = -2 \int \tilde{J}^-(\underline{v}) \cos[\gamma(\underline{p} + \underline{v}) - \gamma(\underline{v})] d\underline{v} \qquad (1.43)$$

$$T_\varphi(\underline{p}) = 2 \int \tilde{J}^-(\underline{v}) \sin[\gamma(\underline{p} + \underline{v}) - \gamma(\underline{v})] d\underline{v} \quad . \qquad (1.44)$$

Expanding $\gamma(\underline{p} + \underline{v})$ about $\underline{p}$, and neglecting $\gamma(\underline{v})$ in comparison, the argument of the sine and cosine terms becomes $\gamma(\underline{p}) + \underline{v} \cdot \text{grad } \gamma(\underline{p})$. Rearranging (1.43) and (1.44), and recalling that the source has been assumed to be symmetric, we find

$$T_s(\underline{p}) = -2 \cos\gamma(\underline{p})E(\underline{p}) \qquad (1.45)$$

$$T_\varphi(\underline{p}) = 2 \sin\gamma(\underline{p})E(\underline{p}) \quad , \qquad (1.46)$$

where

$$E(\underline{p}) = \int \tilde{J}^-(\underline{v})\cos[\underline{v} \cdot grad\gamma(\underline{p})]d\underline{v}$$

$$= J^-\left[\frac{1}{2\pi} grad\gamma(\underline{p})\right] \quad . \tag{1.47}$$

Thus, for ranges of $\underline{p}$ within which $\gamma(p)$ is constant or slowly varying, the coherent transfer functions will be relatively little affected.

Equations (1.45) and (1.46) show that, to a good approximation, the effect of finite source size can be represented as a modulation of the coherent transfer functions. Specific examples are discussed by FRANK [1.87] (and recapitulated in detail in [1.68], where references to related work are to be found). FRANK also considers the effect of the error due to neglect of higher order terms in the expansion of $\gamma(\underline{p} + \underline{v})$ and shows that the small shifts in the zeros of the coherent transfer functions can be accounted for, as well as the attenuation characterized by the modulation functions T_s and T_φ (Fig.1.2).

We now turn to the effect of finite energy spread. It is here that the use of the cross-spectral density gives a clearer insight into the mechanism involved than the mutual intensity for as we have seen, the latter presupposes that the source is quasi-monochromatic, with a particular, narrow frequency distribution. Despite this, the usual treatment of finite energy spread involves adding (incoherently) the contributions to the image corresponding to the range of energies in the incident beam. If we reconsider the cross-spectral density W we see that the intensity, given in terms of the mutual coherence function Γ by $\Gamma(P,P;0)$, is now

$$\Gamma(P,P;0) = \int W(P,P;\omega)d\omega \quad . \tag{1.48}$$

It is thus perfectly legitimate to add the contributions to $W(P,P;\omega)$ corresponding to different values of ω (and hence of energy) in the incident beam. Fortunately, the results are essentially the same, as can be seen if we compare the formulae of [Ref.1.94,Sect.2] with [Ref.1.68,Eq.(3.58)] or (1.51-53) below.

Let us therefore associate a probability distribution, $H(f)$, $\int H(f)df = 1$, with the current density at each point in the image; the argument f is some convenient measure of the energy distribution in the incident beam, the variation in defocus for example. We thus have

$$\frac{dj}{j} = H(f)df \quad , \tag{1.49}$$

Fig.1.2a-f. The effect of finite source size (partial spatial coherence). The phase contrast transfer function is modulated as shown by an envelope function. (a) and (d) $\Delta = (C_s\lambda)^{\frac{1}{2}}$, (b) and (e) $\Delta = (3C_s\lambda)^{\frac{1}{2}}$, (c) and (f) $\Delta = (5C_s\lambda)^{\frac{1}{2}}$. The source is represented by a Gaussian distribution, the width of which for (a) - (c) is half that of (d) - (f) [1.42]. (By courtesy of Wissenschaftliche Verlagsgesellschaft)

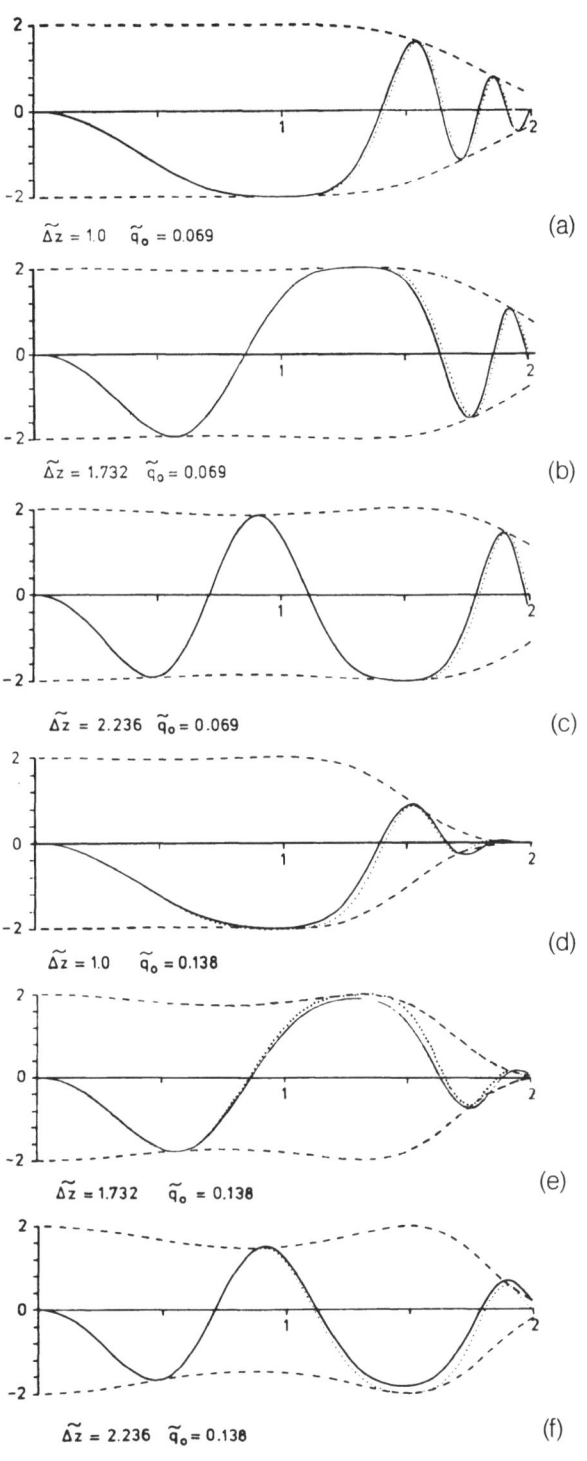

Fig.1.2 a-f.
Caption see opposite page

$\widetilde{\Delta z} = 1.0 \quad \widetilde{q}_o = 0.069$ (a)

$\widetilde{\Delta z} = 1.732 \quad \widetilde{q}_o = 0.069$ (b)

$\widetilde{\Delta z} = 2.236 \quad \widetilde{q}_o = 0.069$ (c)

$\widetilde{\Delta z} = 1.0 \quad \widetilde{q}_o = 0.138$ (d)

$\widetilde{\Delta z} = 1.732 \quad \widetilde{q}_o = 0.138$ (e)

$\widetilde{\Delta z} = 2.236 \quad \widetilde{q}_o = 0.138$ (f) where as usual, j denotes the image current density; for negligible source size, we have

$$j(\underline{x}) \propto 1 - 2 \int a\tilde{s} \cos\gamma \exp(2\pi i\underline{p} \cdot \underline{x}) d\underline{p}$$
$$+ 2 \int a\tilde{\varphi} \sin\gamma \exp(2\pi i\underline{p} \cdot \underline{x}) d\underline{p} \quad . \tag{1.50}$$

Using (1.49) and integrating over f, we obtain

$$j \propto 1 + \int a\tilde{s} T_s'(\underline{p}) \exp(2\pi i\underline{p} \cdot \underline{x}) d\underline{p}$$
$$+ \int a\tilde{\varphi} T_\varphi'(\underline{p}) \exp(2\pi i\underline{p} \cdot \underline{x}) d\underline{p} \quad , \tag{1.51}$$

in which the functions T_s' and T_φ' are given by

$$T_s'(\underline{p}) = -2 \int \cos\gamma(\underline{p},f) H(f) df \tag{1.52}$$

$$T_\varphi'(\underline{p}) = 2 \int \sin\gamma(\underline{p},f) H(f) df \quad . \tag{1.53}$$

If f is chosen to be a measure of defocus variation, we write

$$\Delta = \Delta_0 + f \quad , \tag{1.54}$$

and recalling that

$$\gamma(\underline{p},f) = \frac{\pi}{2} C_s \lambda^3 p^4 - \pi\lambda\Delta p^2 \qquad p = |\underline{p}| \quad , \tag{1.55}$$

we see that

$$T_s'(\underline{p}) = -2 \cos\gamma(\underline{p},\Delta_0) \int H(f) \cos(\pi\lambda f p^2) df$$
$$-2 \sin\gamma(\underline{p},\Delta_0) \int H(f) \sin(\pi\lambda f p^2) df \tag{1.56}$$

$$T_\varphi'(\underline{p}) = 2 \sin\gamma(\underline{p},\Delta_0) \int H(f) \cos(\pi\lambda f p^2) df$$
$$-2 \cos\gamma(\underline{p},\Delta_0) \int H(f) \sin(\pi\lambda f p^2) df \quad . \tag{1.57}$$

We have included Δ_0 explicitly in γ, but $\sin\gamma(\underline{p},\Delta_0)$ and $\cos\gamma(\underline{p},\Delta_0)$ are of course just the coherent transfer functions corresponding to defocus Δ_0. If the distribution H(f) is even, the second terms on the right-hand sides of T_s' and T_φ' vanish and we obtain the simpler expressions

$$T'_s = -2 \cos\gamma(\underline{p},\Delta_0) \int H(f)\cos(\pi\lambda fp^2)df \qquad (1.58)$$

$$T'_\varphi = 2 \sin\gamma(\underline{p},\Delta_0) \int H(f)\cos(\pi\lambda fp^2)df \quad . \qquad (1.59)$$

The modulating function is thus the cosine Fourier transform of H(f), with the appropriate change of scale in the argument. A variety of distributions H(f) are considered by HANSZEN and TREPTE [1.85,86], upon whose work the foregoing analysis is based; their results are recapitulated in considerable detail in [1.68], where references to related work are also listed (Fig.1.3).

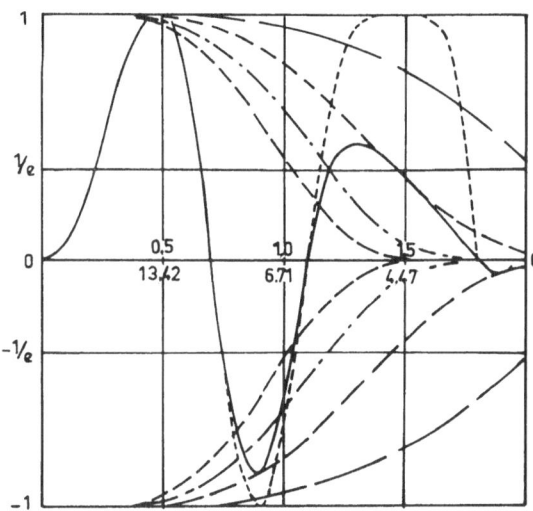

Fig.1.3. The effect of finite energy spread (finite temporal coherence). The phase contrast transfer function (-sinγ) for Δ = (5C_sλ)^½ (short dashes) and the attenuation due to a Gaussian energy spread (solid curve). The envelopes shown correspond to various halfwidths of the Gaussian spread. The abscissa is graduated in generalized spatial frequency p $(C_s\lambda^3)^{1/4}$ above and spacing (1/p) below for a typical objective (C_s = 4 mm, 100 kV) [1.86]. (By courtesy of Wissenschaftliche Verlagsgesellschaft)

If we attempt to combine the analysis of finite source size with that of finite energy spread, the case of practical interest in which both spatially and temporally the illumination is partially coherent, we find that even in a first approximation, it is not obvious that the simple product representation is valid. This point, first raised by McFARLANE (see [1.88]), has now been satisfactorily resolved by McFARLANE [1.90] and WADE and FRANK [1.89]; the product representation, in which the effect of spatial and temporal coherence is obtained by multiplying the coherent transfer functions by the modulating functions described above, is valid everywhere except at high spatial frequencies for which the envelope function attenuates the transfer of information to such an extent that it is effectively suppressed altogether.

Hitherto, we have been considering the effects of departures from perfect coherence on the ordinary bright-field image with conventional illumination. There are, however, other illumination modes that are intrinsically partially spatially coherent, one of which is attracting considerable interest in connection with

ultrahigh resolution work. This is the bright-field mode using a hollow, conical
illuminating beam; such a type of illumination is intrinsically partially spatially
coherent because the contributions from different azimuths of the conical sheaf of
electrons cannot be regarded as part of a single coherent wave structure [1.96].
The consequence of this is that although the first zero of the phase contrast
transfer function can occur at as much as twice the value of spatial frequency
corresponding to an untilted full beam —a result true in general of tilted illumination of which hollow conical illumination is a generalization —the transfer
is always attenuated; it is indeed approximately halved if the resolution is
doubled. Since problems of noise are always particularly acute at high resolution,
this amount of attenuation is a very high price to pay for the increased resolution.
A considerable theoretical and experimental effort is being devoted to this mode
of illumination, in order to establish whether, and in what conditions, it does
yield useful information that could not be obtained otherwise [1.8,53,59,97-101].

Finally, we note that the effect of partial coherence on tilted illumination
has been examined in great detail by WADE and JENKINS [1.64], who establish an
envelope representation for this situation.

1.3 Practical Exploitation of the Linear Relationship

The linear relation between image contrast and the modulus and phase of the complex
specimen function, when the object is a weak scatterer, offers a straightforward
method of establishing the microscope operating characteristics, C_s and Δ. It can
also be used to set the objective stigmator correctly if the microscope image can
be computer-processed on-line. Most interesting of all, it should also be possible
to filter two or more members of a focal series in such a way as to reconstruct
phase and amplitude images from which the deleterious effects of C_s and Δ have been
eliminated as far as possible.

In this section, we consider these various applications in turn.

1.3.1 Measurement of the Microscope Operating Characteristics

Although the history of electron transfer theory goes back to the early 1960s, it
was not until 1966 that THON published the first (optical) Fourier transforms of
micrographs of thin carbon films, showing the ring pattern representing the phase
contrast transfer function, 2 sinγ, along any radius. THON had, however, realized
two years earlier [1.102,103] that the explanation given some years before by LENZ
[1.104,105] of the variation with defocus of electron speckle, as we now call the
grainy image of an amorphous specimen, although qualitatively right, did not give

correct quantitative predictions. His first ring patterns [1.106,107] were obtained, using a light-optical diffractometer at the suggestion of HOPPE, to confirm that the theory accurately predicts the variation of the speckle pattern with defocus if the combined effect of spherical aberration and defocus is taken into account, using the expression for the associated phase shift introduced into electron optics by SCHERZER [1.108]. It was soon realized that the geometry of these ring patterns could be used to obtain the absolute values of the defocus and spherical aberration, by least-squares fitting of the measured radial distribution to the theoretical curve for example. FRANK et al. [1.109] developed a flexible computer program capable of furnishing the astigmatism as well as C_s and Δ and this has since been refined by SAXTON [1.44,110,111] who has incorporated the attenuation produced by the partial coherence of real sources (cf. [1.112-117]) (Fig.1.4).

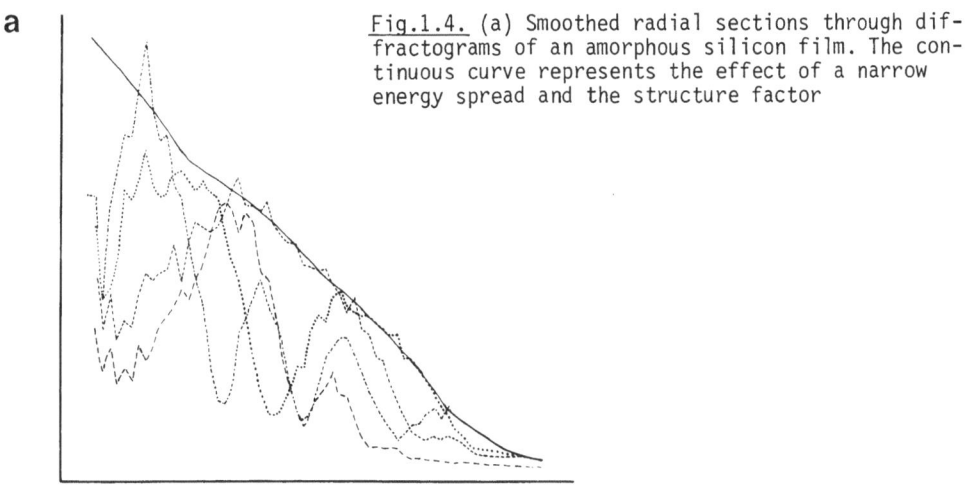

a

<u>Fig.1.4.</u> (a) Smoothed radial sections through diffractograms of an amorphous silicon film. The continuous curve represents the effect of a narrow energy spread and the structure factor

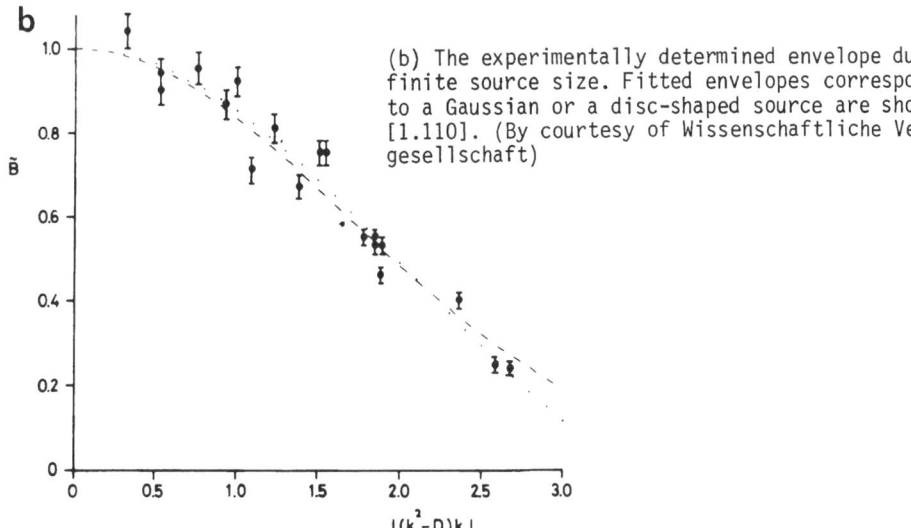

b

(b) The experimentally determined envelope due to finite source size. Fitted envelopes corresponding to a Gaussian or a disc-shaped source are shown [1.110]. (By courtesy of Wissenschaftliche Verlagsgesellschaft)

$\tilde{B}$

$|(k^2-D)k|$

Other methods have been proposed, which are more direct but less accurate than the fitting routine. KRAKOW et al. [1.118] pointed out that tilting the specimen in the microscope effectively gives information about a range of values of defocus simultaneously, which can be exploited optically. Their method has the disadvantage that both the microscope and the diffractometer require some modification. A simpler method that relies on the combined effects of astigmatism and partial coherence has been proposed by BURGE and SCOTT [1.119]. A further method, devised by KRIVANEK [1.120] makes very ingenious use of information already available, that THON [1.106] (cf. [1.121]) had used to derive Δ assuming C_s known. Returning to (1.11), and recalling that $\underline{x}_a = \lambda f \underline{p}$, we have

$$\gamma(\lambda f \underline{p}) = \frac{2\pi}{\lambda} [C_s \lambda^4 (\underline{p} \cdot \underline{p})^2/4 - \Delta\lambda^2 \underline{p} \cdot \underline{p}/2] \tag{1.60}$$

in the absence of astigmatism. The function $\sin\gamma$ will have maxima where

$$C_s \lambda^3 (\underline{p} \cdot \underline{p})^2/4 - \Delta\lambda^2 \underline{p} \cdot \underline{p}/2 = n/4 \quad , \quad n \text{ odd} \tag{1.61}$$

and zeros when the same condition is satisfied with n even. THON used these equations to establish Δ from the ring radii, for a given value of C_s. Alternatively, we may rearrange (1.61) as follows:

$$C_s \lambda^3 \underline{p} \cdot \underline{p} - 2\Delta\lambda^2 = n/\underline{p} \cdot \underline{p} \quad . \tag{1.62}$$

By plotting $n/\underline{p} \cdot \underline{p}$ as a function of $\underline{p} \cdot \underline{p}$, we obtain a family of curves to which the values of p and n obtained from the diffractogram can be matched; this yields straight lines of gradient $C_s \lambda^3$ with intercept $-2\lambda^2 \Delta$ (Fig.1.5).

The envelope representation of the effects of partial coherence can be used to establish values of the source size and wavelength spread, as several authors have shown. We draw particular attention to the work of FRANK et al. [1.88,122], KRIVANEK [1.115], TROYON [1.117], and WADE [1.123] in this connection.

1.3.2 On-Line Processing

For a variety of reasons, it is desirable to adjust the stigmator of the electron microscope rapidly. In particular, if the beam falls on the specimen throughout the period during which the operator is adjusting the stigmator, the corresponding area will have been subjected to a high, perhaps intolerably and always undesirably high, dose of electrons. For fragile specimens, for which radiation damage is a major problem, an automatic method of correcting astigmatism that did not entail prolonged irradiation of the specimen would be very welcome. It might be supposed that if the image is to be processed subsequently in the computer, it would be

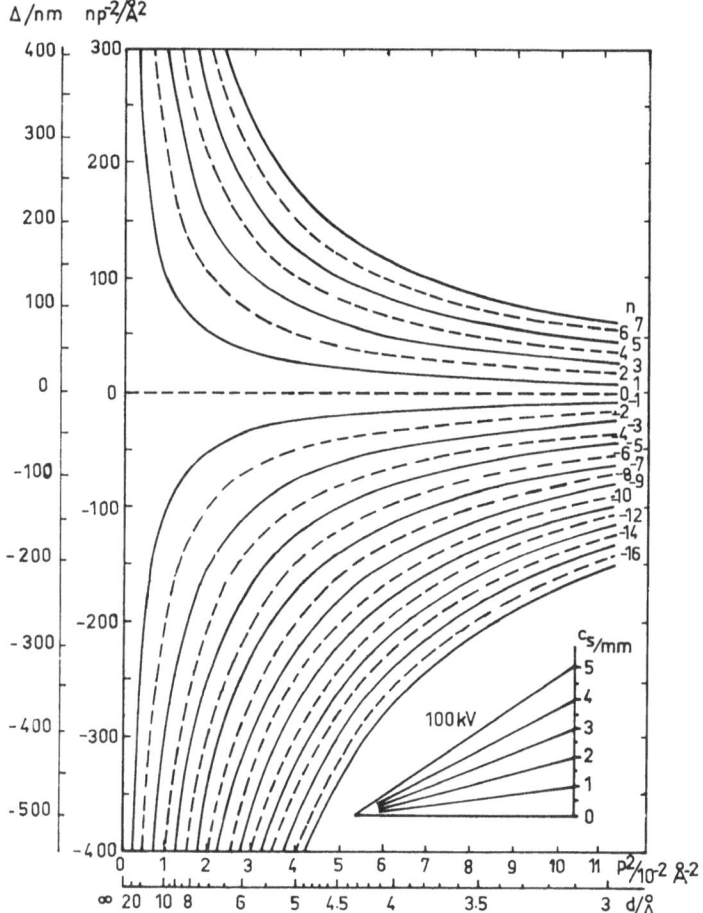

Fig.1.5. A plot of n/p^2 (in $\overset{\circ}{A}{}^2$) against $100\, p^2$ (in $\overset{\circ}{A}{}^{-2}$) for various values of n. The second abscissa scale shows spacing $d(d = 1/p)$ in $\overset{\circ}{A}$. The second ordinate scale shows Δ in nm for 100 kV operation. The pencil of lines inset shows the slope corresponding to various values of C_S [1.120]. (By courtesy of Wissenschaftliche Verlagsgesellschaft)

better to tolerate the astigmatism and eliminate it in the computing stage. Although this is possible, in principle, provided that the astigmatism is not too large, it' is nevertheless better to produce images as free as possible of astigmatism, since the computer routines needed for astigmatic images are appreciably more complicated than those used when it is negligible.

From the practical point of view, it is essential that any automatic technqiue should not introduce undue delay, which in turn requires some means of getting the image from the electron microscope to the automatic correction system directly — without photography, for example, or any other time-consuming intermediate stage. Two types of solution have been proposed: interfaces reading the electron image directly into a computer, and on-line optical analog arrangements. Of the latter, we draw attention to the systems proposed by KÜBLER and WASER [1.124] (patented by K.-H. Müller and F. Thon for the Siemens AG), by BONHOMME et al. [1.125,126] and DUMONT et al. [1.127] and by HERRMANN and KRAHL [1.128-130]. KÜBLER and WASER transfer the electron image by means of a television camera to the control layer of an Eidophor television projector; by illuminating the latter with coherent light, diffractograms can be obtained immediately (Fig.1.6). More recently, DOWNING et al. [1.131] have attempted to use the newly developed General Electric "coherent light valve" for the same purpose, the latter having been developed specifically for operation with coherent light. BONHOMME et al. and DUMONT et al. avoid the intermediate television stage by forming the electron image in the microscope directly on the photoconductive layer of amorphous selenium of a Pockels-effect light valve, Phototitus [1.132]. The magnified image or its Fourier transform is then projected directly onto a screen, and various simple operations, image subtraction for example, can also be performed on-line (Fig.1.7). HERRMANN and KRAHL [1.128-130] too have attempted direct on-line treatment of electron images, without an intermediate television link, using the properties of a thermoplastic layer coupled to a channel-plate image intensifier (Fig.1.8).

We now turn to the various microscope-computer links that have been proposed. One of the first practical designs was described by Glaeser et al. [1.133], subsequently elaborated by KUO and GLAESER [1.134] (Fig.1.9). The following year, WITT et al. [1.135] described a chain leading from the microscope image, via a fiber-optics coupled image intensifier and an SEC vidicon television camera to an analog device. Yet another system was described by GOLDFARB and SIEGEL [1.136-138], and further discussion of the limitations of such schemes is to be found in GOLDFARB [1.136] (Fig.1.10). Meanwhile, HERRMANN et al. (Fig.1.11) produced a series of papers on television image intensifiers and their use for transferring electron images to the computer for processing [1.139-144; cf. 1.145]. See also [1.146-150].

These systems were designed for a variety of reasons, though always with the idea of exposing the specimen to as low a dose of electrons as possible in mind. FRANK [1.151], recognizing that such a direct link should be useful for rapid focusing and stigmator adjustment, showed that the autocorrelation function of a reasonably amorphous region of the specimen gives information about the defocus and the astigmatism of the objective lens simultaneously. Furthermore, since correlation functions average information over a certain area of the specimen, the necessary adjustments can be performed at exposure levels too low for the conventional visual procedure (cf [1.152]).

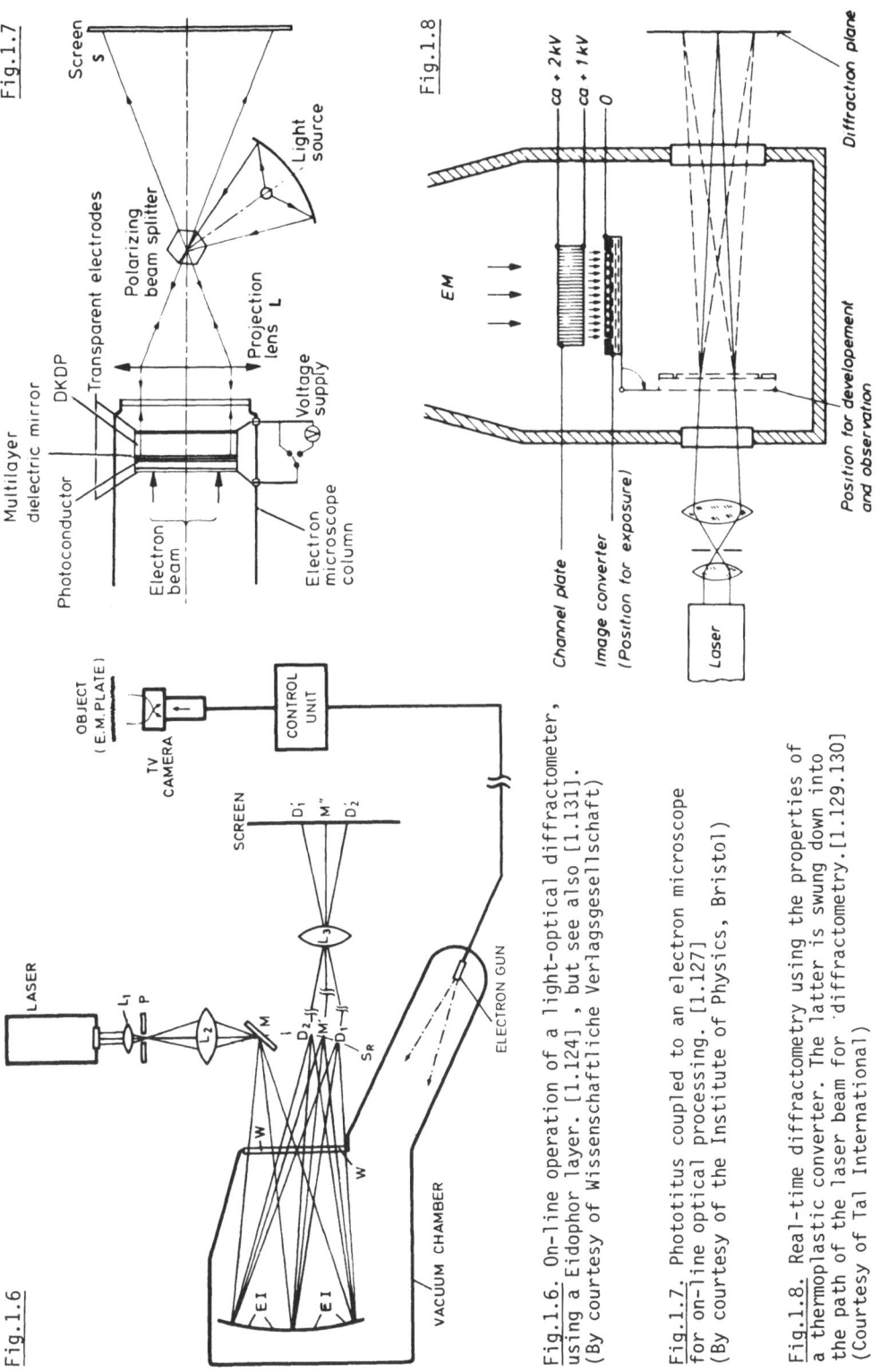

Fig.1.7

Fig.1.8

Fig.1.6

Fig.1.6. On-line operation of a light-optical diffractometer, using a Eidophor layer. [1.124] , but see also [1.131]. (By courtesy of Wissenschaftliche Verlagsgesellschaft)

Fig.1.7. Phototitus coupled to an electron microscope for on-line optical processing. [1.127] (By courtesy of the Institute of Physics, Bristol)

Fig.1.8. Real-time diffractometry using the properties of a thermoplastic converter. The latter is swung down into the path of the laser beam for diffractometry.[1.129.130] (Courtesy of Tal International)

22

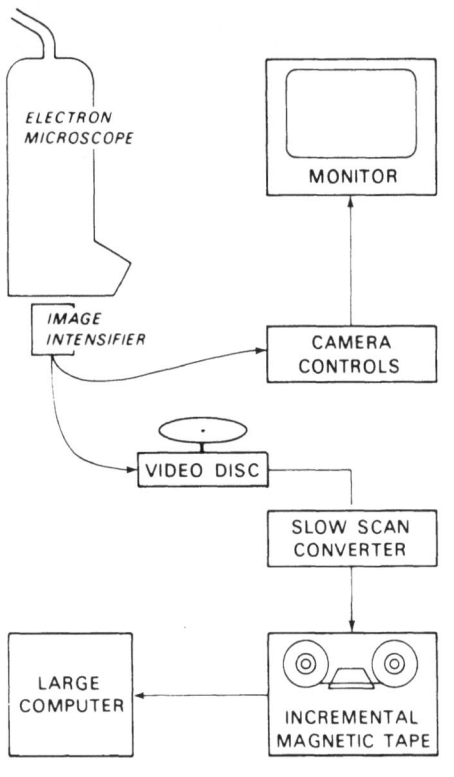

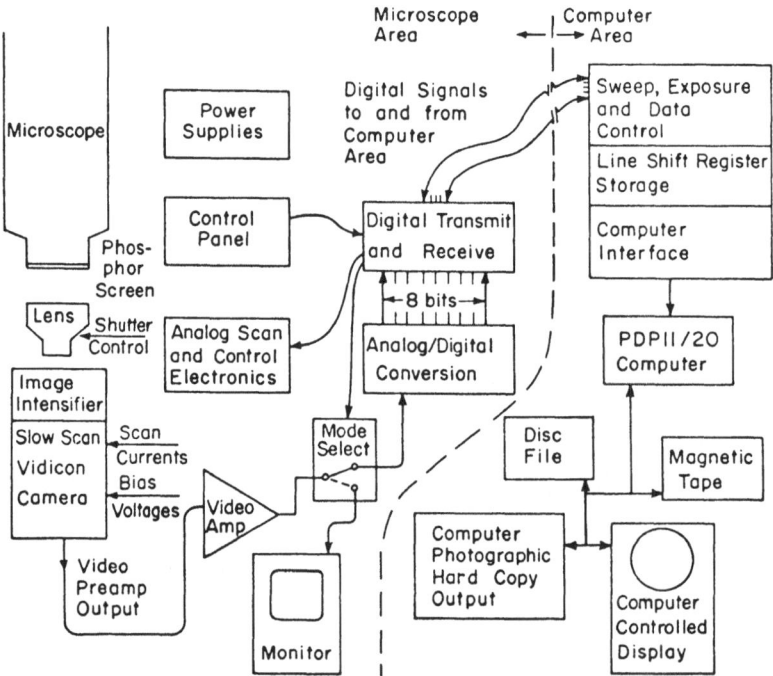

Fig.1.9. A microscope-computer link, using an SEC vidicon TV camera [1.133]. (Courtesy of Claitor, Baton Rouge)

Fig.1.10. Another microscope-computer link, using a Westinghouse high-resolution vidicon camera [1.137]. (Courtesy of Claitor, Baton Rouge)

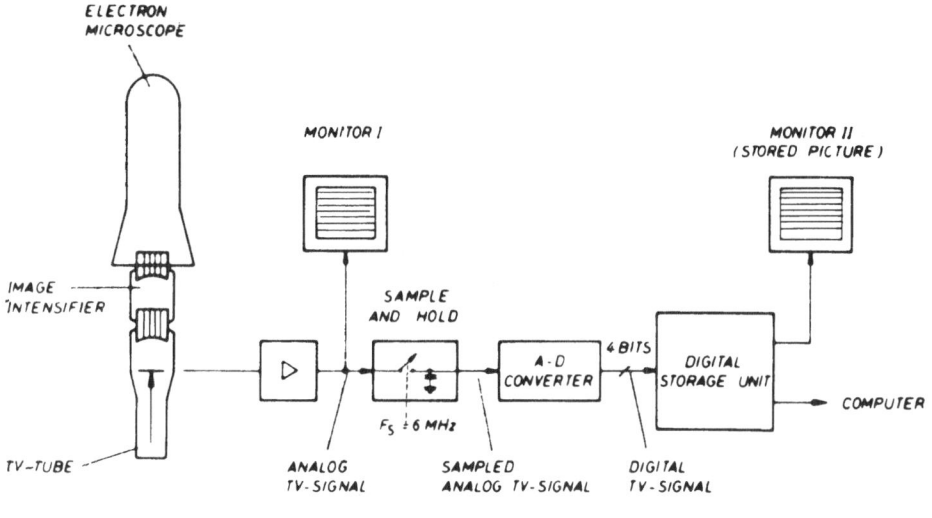

(a)

(b)

Fig.1.11. (a) A microscope-computer link consisting of an image-intensifier coupled to a TV tube, followed by a sample-and-hold circuit, an analog-to-digital converter, and a digital storage unit. (b) The essential components of the digital storage unit, which is interfaced directly to the computer. (After [1.142], by courtesy of Tal International)

1.3.3 Filtering and Reconstruction

We have seen that, for a certain category of specimens, image contrast and the real and imaginary parts of the specimen transparency are linearly related, (1.26). As soon as this result was appreciated, efforts began to extract the modulus and phase of the specimen transparency from the intensity image by inverting the filtering effect introduced by the microscope, characterized by $\sin\gamma$ and $\cos\gamma$. We may summarize the problem as follows: given that for each of N images, the contrast C_j is related to s and φ as in (1.26),

$$\tilde{C}_j = 2\tilde{s} \cos\gamma_j + 2\tilde{\varphi} \sin\gamma_j \quad , \quad j = 1,2, \ldots, N \quad , \tag{1.63}$$

what is the best linear estimate of s and φ? In other words, if we write

$$\tilde{s} = \sum_j a_j \tilde{C}_j \tag{1.64}$$

$$\tilde{\varphi} = \sum_j b_j \tilde{C}_j \quad ,$$

what is the best choice for $\{a_j\}$ and $\{b_j\}$?

Clearly, the answer to this question depends upon how we define "best". If we ignore noise, we obtain the results published by SCHISKE [1.153] in the first paper dealing with this problem in electron optics; Schiske's solution was optimal in the least-mean-squares sense. It is, however, difficult to obtain suitable focal series and the computing requirements rapidly become prohibitive as the number of images is increased. During the next few years, therefore, attempts were made to filter single micrographs, making some allowance for the noise. If we take the specimen to be a pure phase object, so that

$$\tilde{C}_j = 2\tilde{\varphi} \sin\gamma_j \quad , \tag{1.65}$$

we clearly cannot simply invert the equation to obtain $\tilde{\varphi}$ since $\mathrm{cosec}\gamma_j$ is very large in the vicinity of the zeros of $\sin\gamma_j$ so that any errors of measurement in C_j will be hugely increased. If, however, we include an (additive) noise term in (1.65),

$$\tilde{C}_j = 2\tilde{\varphi} \sin\gamma_j + N_j \quad , \tag{1.66}$$

we can again obtain the optimal estimate in the least-mean-squares sense. We obtain the filter structure usually known as the Wiener filter (described in detail below). In the first attempts to process micrographs in this way [1.65,109,154-156], the zeros of the sine function were prevented from enhancing noise in a heuristic way but in 1971, WELTON [1.157] described a computer simulation of an attempt to use the Wiener filter. Further theoretical and experimental papers followed: FRANK [1.158,159] used a reconstruction scheme of the Schiske type and interpreted the phase variations found in terms of the atomic number composition of the specimen; SCHISKE [1.160] extended his earlier results to include noise and HAWKES [1.161] generalized this still further by including more general constraints than the least-mean-squares requirement. WELTON et al. [1.157,162-164] continued with their attempts at Wiener filtering and similar efforts were made in other laboratories [1.165-179; cf 1.180].

We return to (1.26), which we rewrite as follows:

$$\tilde{C}(p,q) = K_s(p,q)\tilde{s}(p,q) + K_\varphi(p,q)\tilde{\varphi}(p,q) + N(p,q) \quad , \tag{1.67}$$

in which

$$K_s = -2 \cos\gamma$$
$$K_\varphi = 2 \sin\gamma \quad . \tag{1.68}$$

We seek filters Q_s and Q_φ, optimal in some well-defined sense, such that good estimates, $\tilde{s}'$ and $\tilde{\varphi}'$, of the specimen functions $\tilde{s}$ and $\tilde{\varphi}$ are given by

$$\tilde{s}' = Q_s\tilde{C}$$
$$\tilde{\varphi}' = Q_\varphi \tilde{C} \quad . \tag{1.69}$$

Suppose first that the specimen is a pure phase object, so that (1.67) simplifies to

$$\tilde{C} = K_\varphi\tilde{\varphi} + N \quad . \tag{1.70}$$

We obtain the least-mean-square error solution by minimizing $\mathscr{E}(|\tilde{\varphi}' - \tilde{\varphi}|^2)$, where $\mathscr{E}$ denotes the expectation value. Substituting for $\tilde{\varphi}'$ from (1.69) and (1.67), we have to minimize.

$$\mathscr{E}[Q_\varphi K_\varphi\tilde{\varphi} + Q_\varphi N - \tilde{\varphi})(Q_\varphi^* K_\varphi^* \tilde{\varphi}^* + Q_\varphi^* N_\varphi^* - \tilde{\varphi}^*)] \quad , \tag{1.71}$$

or

$$Q_\varphi Q_\varphi^* K_\varphi K_\varphi^* \mathscr{E}(\tilde{\varphi}\tilde{\varphi}^*) + Q_\varphi Q_\varphi^* \mathscr{E}(NN^*) + \mathscr{E}(\tilde{\varphi}\tilde{\varphi}^*) \quad , \tag{1.72}$$

in which we have assumed that the noise has zero mean and that signal and noise are not correlated. Setting the derivative of (1.72) with respect to Q_φ^* equal to zero, we obtain

$$Q_\varphi[K_\varphi K_\varphi^* \mathscr{E}(\tilde{\varphi}\tilde{\varphi}^*) + \mathscr{E}(NN^*)] = K_\varphi^* \mathscr{E}(\tilde{\varphi}\tilde{\varphi}^*) \tag{1.73}$$

or finally,

$$Q_\varphi = \frac{K_\varphi^*}{K_\varphi K_\varphi^* + R} \quad , \tag{1.74}$$

where

$$R = \mathcal{E}(NN^*)/\mathcal{E}(\tilde{\varphi}\,\tilde{\varphi}^*) \ . \tag{1.75}$$

Thus the best estimate, in the linear least-mean-square error sense, of the phase variation $\tilde{\varphi}$ is $\tilde{\varphi}' = Q_\varphi \tilde{C}$, where the "Wiener" filter Q_φ is given by (1.74) and (1.75). For small values of $R, Q_\varphi \approx 1/K_\varphi$; this is the simple inverse filter that fails near the zeros of $\sin\gamma$; near these points, however, the presence of R prevents Q_φ from becoming unacceptably large. The term R is the ratio of the power spectra of the noise and the specimen phase and is in practice unknown or difficult to estimate [1.81-183]; an approximate value has therefore usually to be regarded as adequate.

The simple analysis given above may be generalized in various ways, as we have already intimated. Following SCHISKE [1.160], we may retain both the phase and amplitude terms in (1.63) and, assuming that two members of a focal series are available, derive the pair of optimal filters, Q_φ and Q_s, with which estimates $\tilde{\varphi}'$ and $\tilde{s}'$ may be obtained. These filters may be further generalized as shown in HAWKES [1.161], by allowing for the introduction of constraints (see, for example, [1.184-186]).

The earliest practical attempts to restitute images by exploiting these ideas were made by ERICKSON and KLUG [1.65,154,155] in the M.R.C. Laboratory of Molecular Biology in Cambridge, by WELTON et al. [1.157,163,164] at Oak Ridge, and in the research group led by HOPPE in Munich [1.156,187]. At the 1970 EMSA meeting, WELTON described a suite of computer programs and remarked that "A simple filter function has been devised to produce a useful approximation to the maximum-likelihood estimate of the true image. Basically, image and noise intensities are compared at each spatial frequency, and a controlled suppression performed where image information does not exist" [1.157]. At the 1971 EMSA meeting, BALL et al. [1.162] discussed this further and WELTON [1.157] showed simulated images illustrating the effects of a Wiener filter. This work has continued for several years [1.157,163,164]. Meanwhile, at meetings held under the auspices of the Royal Society in London and of the Bunsengesellschaft in Hirschegg, ERICKSON and KLUG [1.155] described their work on catalase, of which a focal series was obtained; the change in the optical diffractogram with defocus was compared with the variation predicted by theory and simulated images were shown, from which the effect of defocus and spherical aberration had been eliminated as far as possible. At the same

---

▶

Fig.1.12a-1. Phase and amplitude reconstruction from a focal series of amorphous germanium, about 10 nm thick. (a) - (d) original images $\Delta/(C_s\lambda)^{1/2}$ = 1.03, 1.28, 1.58, 1.88, where $(C_s\lambda)^{1/2}$ = 81.6 nm, C_s = 1.8 mm, 100 kV. (e) - (f) Restorations of the phase component of the object spectrum and of the object. (g) - (h) Restorations of the amplitude component of the object spectrum and of the object. (i) - (l) Simulated images, corresponding to (a) - (d), calculated from the phase restoration. (Kindly provided by W.O. SAXTON, cf [1.44,177])

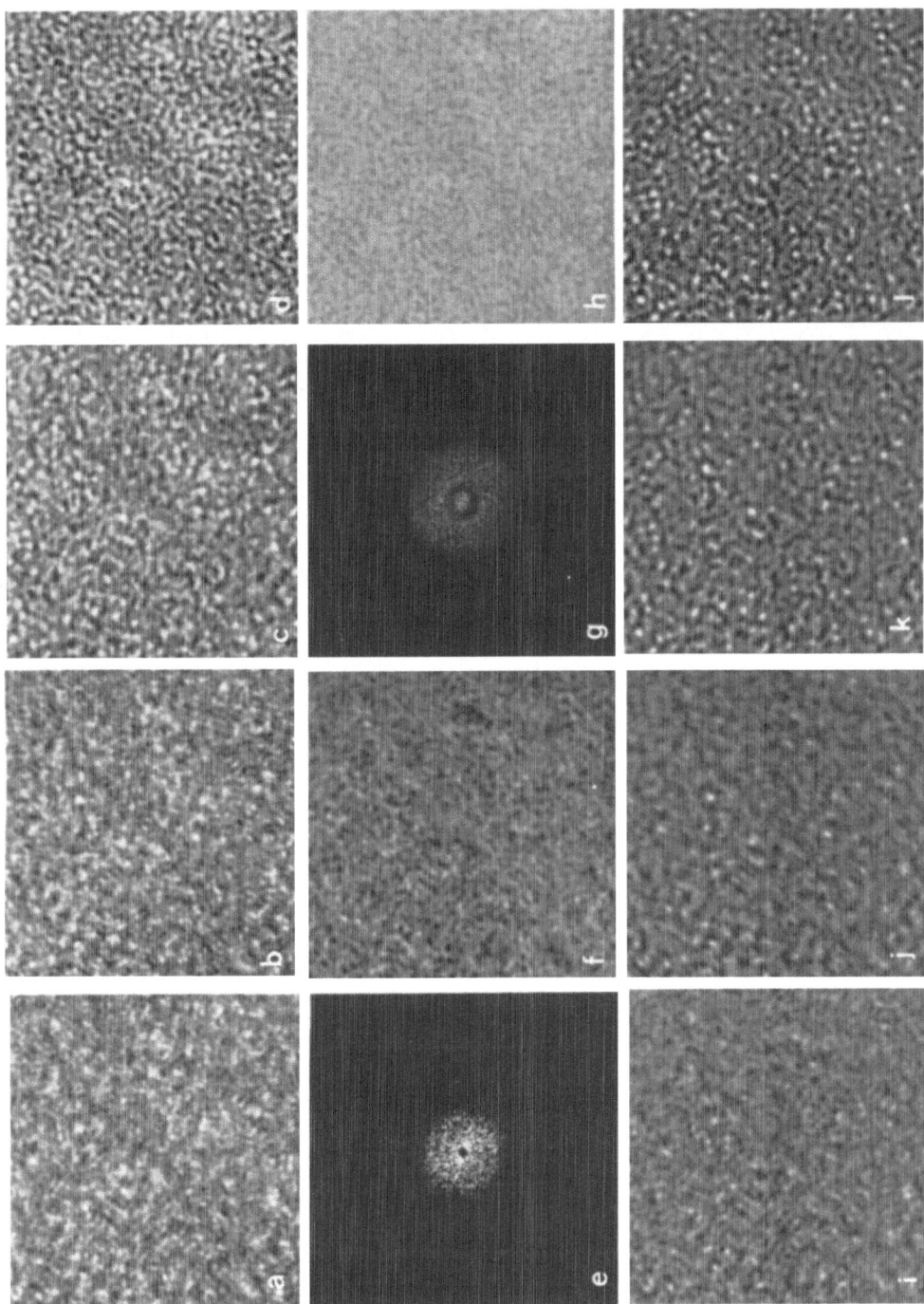

meetings, HOPPE [1.156] and FRANK et al. [1.109] showed simulated micrographs, obtained by filtering the originals with the function,

$$Q_\varphi = \frac{\exp\{a(1-|\sin\gamma|^{-N})\}}{\sin\gamma} \quad;$$

this was chosen somewhat arbitrarily in order to prevent noise amplification around $\sin\gamma = 0$.

In 1972, FRANK [1.158,159] published an account of the first attempts to apply SCHISKE's reconstruction scheme [1.153] to real (rather than simulated) micrographs. The specimen consisted of double-stranded DNA stained with uranyl magnesium acetate, deposited on a thin carbon film ($\sim$ 4 nm thick). Frank's object was to use the atomic number (Z) dependence of the phase part of the specimen transparency to separate the heavy and light atom contributions to the image. In the same year, ANSLEY [1.165] began a study of various filters, using computer simulation, which culminated in a detailed comparison of three filters; the first two are essentially the Wiener filters for amplitude and phase specimens (separately) and the third is a "deterministic" filter, of a somewhat arbitrary design (see too [1.187]).

Digital Wiener filtering has been compared with corresponding optical technique by KÜBLER et al. [1.175,176], as part of a larger effort to use image processing for assessing and minimizing radiation damage.

Interest in filtering individual micrographs or members of focal series has given way more recently to other reconstruction schemes, in particular, those yielding three-dimensional information for comparable radiation loading. These techniques are examined in detail in later chapters. Nevertheless, some quite impressive results have been obtained recently by SAXTON et al. [1.177], studying amorphous germanium and selenium (Fig.1.12).

References

1.1 W. Glaser: *Grundlagen der Elektronenoptik* (Springer, Vienna 1952)

1.2 W. Glaser: "Elektronen- und Ionenoptik, in *Handbuch der Physik*, Vol. XXXIII, ed. by S. Flügge (Springer, Berlin, Göttingen, Heidelberg 1956) pp. 123-395

1.3 J. Komrska, M. Lenc: Manchester (1972) pp. 78-79

1.4 M. Born, E. Wolf: *Principles of Optics* (Pergamon, Oxford and New York 1959; 5th ed., 1975)

1.5 G. Ade: "Der Einfluss der Bildfehler dritter Ordnung auf die elektronen-mikroskopische Abbildung und die Korrektur dieser Fehler durch holographische Rekonstruktion"; Dissertation, Braunschweig (1973); PTB-Bericht A Ph-3, 114 pp (1973)

1.6 G. Ade: Optik *45*, 317-332 (1976)

1.7 G. Ade: Optik *50*, 143-162 (1978)

1.8 K.J. Hanszen, G. Ade: "Problems and Results of the Optical Transfer Theory and of Reconstruction Methods in Electron Microscopy", PTB-Bericht A Ph-5, pp.68 (1974)

1.9   K.J. Hanszen, G. Ade, R. Lauer: Optik *35*, 567-590 (1972)
1.10 P.W. Hawkes: "To What Extent is Isoplanatism a Restriction in Contrast Trans-
      fer Theory?" in *Electron Microscopy and Analysis*, ed. by W.C. Nixon (Insti-
      tute of Physics, London 1971) pp. 230-232
1.11 P.W. Hawkes: Manchester (1972) pp. 398-399
1.12 P.W. Hawkes: Optik *37*, 366-375, 376-384 (1973)
1.13 P. Schiske: "Fourier Methods for the Treatment of Images with Coma and Dis-
      tortion", in *Image Processing and Computer-Aided Design in Electron Optics*,
      ed. by P.W. Hawkes (Academic, London and New York 1973) pp. 54-65
1.14 P.W. Hawkes: Optik *46*, 357-359 (1976)
1.15 M.E.C. Maclachlan: Optik *47*, 363-364 (1977)
1.16 P.W. Hawkes: Optik *48*, 253 (1977)
1.17 R. Hegerl, W. Hoppe: Manchester (1972) pp. 628-629
1.18 W. Hoppe: Acta Cryst. *A25*, 495-501 (1969)
1.19 W. Hoppe: Acta Cryst. *A25*, 508-514 (1969)
1.20 W. Hoppe, G. Strube: Acta Cryst. *A25*, 502-507 (1969)
1.21 W.H.J. Andersen: Manchester (1972) pp. 396-397
1.22 A.G. Cullis, D.M. Maher: Phil. Mag. *30*, 447-451 (1974); Ultramicroscopy *1*,
      97-112 (1975)
1.23 K.H. Downing: *Proc. 30th Annual Meeting EMSA, Los Angeles*, ed. by C.J. Arce-
      neaux (Claitor, Baton Rouge 1972) pp. 562-563
1.24 K.H. Downing: *Proc. 35th Annual Meeting EMSA, Boston*, ed. by G.W. Bailey
      (Claitor, Baton Rouge 1977) pp. 16-17
1.25 K.H. Downing, B.M. Siegel: Optik *38*, 21-28 (1973)
1.26 K.H. Downing, B.M. Siegel: Canberra (1974), Vol. 1, pp. 326-327
1.27 K.H. Downing, B.M. Siegel: Optik *42*, 155-175 (1975)
1.28 K.H. Downing: Ultramicroscopy *4*, 13-31 (1979)
1.29 K.J. Hanszen: Z. Naturforsch. *24a*, 1849 (1969)
1.30 J.R. Harris, J. Kerr: J. Microsc. *108*, 51-59 (1976)
1.31 G.B. Haydon: J. Microsc. *89*, 73-82 (1969)
1.32 G.B. Haydon, R.A. Lemons: J. Microsc. *95*, 483-491 (1972)
1.33 G.B. Haydon, B.C. Hill, R.A. Lemons: *Proc. 29th Annual Meeting EMSA, Boston*,
      ed. by C.J. Arceneaux (Claitor, Baton Rouge 1971) pp. 438-439
1.34 W. Hoppe, R. Langer, F. Thon: Optik *30*, 538-545 (1970)
1.35 S. Nakahara, A.G. Cullis: "Half-Plane and Topographical Contrast in TEM
      Theoretical Considerations", in *Developments in Electron Microscopy and
      Analysis 1977*, ed. by D.L. Misell (Institute of Physics, Bristol 1977)
      pp.263-266
1.36 S. Nakahara, D.M. Maher, A.G. Cullis: Jerusalem (1976), Vol. 1, pp. 85-90
1.37 D.J. Scales: J. Microsc. *102*, 49-58 (1974)
1.38 P. Sieber: Canberra (1974), Vol. 1, pp. 274-275
1.39 J.C.H. Spence: Opt. Acta *21*,835-837 (1974)
1.40 F. Thon: Rome (1968), Vol. 1, pp. 127-128
1.41 F. Thon: "Phase Contrast Electron Microscopy", in *Electron Microscopy in Ma-
      terial Science*, ed. by U. Valdrè (Academic, New York and London 1971) pp.
      570-625
1.42 F. Thon, D. Willasch: Manchester (1972) pp. 650-651
1.43 K.J. Hanszen: Adv. Opt. Electron Microsc. *4*, 1-84 (1971)
1.44 W.O. Saxton: *Computer Techniques for Image Processing in Electron Micros-
      copy* (Academic, New York and London, 1978), supplement 10 to Adv. Electron.
      Electron Phys. 289 pp
1.45 D.L. Misell: Adv. Opt. Electron Microsc. *7*, 185-279 (1978)
1.46 K.H. Downing: Optik *43*, 199-203 (1975)
1.47 W. Goldfarb, W. Krakow, D. Ast, B. Siegel: *Proc. 33rd Annual Meeting EMSA,
      Las Vegas*, ed. by G.W. Bailey (Claitor, Baton Rouge 1975) pp. 186-187
1.48 K.J. Hanszen: Jerusalem (1976), Vol. 1, pp. 95-96
1.49 K.J. Hanszen, G. Ade: "Aspects of Some Image Reconstruction and Holographic
      Methods in Electron Microscopy", PTB-Bericht A Ph-10, pp.38 (1976)
1.50 W. Hoppe, D. Köstler: Jerusalem (1976), Vol. 1, pp. 99-104
1.51 W. Hoppe, D. Köstler, P. Sieber: Z. Naturforsch. *29a*, 1933-1934 (1974)
1.52 W. Hoppe, D. Köstler, D. Typke, N. Hunsmann: Optik *42*, 43-56 (1975)

1.53   W.K. Jenkins, R.H. Wade: "Contrast transfer in the Electron Microscope for Tilted and Conical Bright Field Illumination", in *Developments in Electron Microscopy and Analysis 1977*, ed. by D.L. Misell (Institute of Physics, Bristol 1977) pp.115-118

1.54   A.G. Kiselev, M.B. Sherman: Optik *46*, 55-60 (1976)

1.55   W. Krakow: "Calculation and Observation of Atomic Structure for Tilted Beam Dark-Field Microscopy", in *Developments in Electron Microscopy and Analysis*, ed. by J.A. Venables (Academic, London and New York 1976) pp. 260-264

1.56   W. Krakow: *Proc. 34th Annual Meeting EMSA, Miami Beach*, ed. by G.W. Bailey (Claitor, Baton Rouge 1976) pp. 566-567

1.57   S.C. McFarlane: J. Phys. C: Solid State Phys. *8*, 2819-2836 (1975)

1.58   S.C. McFarlane, W. Cochran: J. Phys. C: Solid State Phys. *8*, 1311-1321 (1975)

1.59   H. Rose: Ultramicroscopy *2*, 251-267 (1977)

1.60   D. Typke, D. Köstler: Optik *45*, 495-498 (1976)

1.61   D. Typke, D. Köstler: Ultramicroscopy *2*, 285-295 (1977)

1.62   R.H. Wade: Phys. Status Solidi (a) *37*, 247-256 (1976)

1.63   R.H. Wade: Optik *45*, 87-91 (1976)

1.64   R.H. Wade, W.K. Jenkins: Optik *50*, 1-17 (1978)

1.65   H.P. Erickson: Adv. Opt. Electron Microsc. *5*, 163-199 (1973)

1.66   K.J. Hanszen: "Contrast Transfer and Image Processing", in *Image Processing and Computer-Aided Design in Electron Optics*, ed. by P.W. Hawkes (Academic, London and New York 1973) pp. 16-53

1.67   P.W. Hawkes: "Introduction to Electron Optical Transfer Theory", in *Image Processing and Computer-Aided Design in Electron Optics*, ed. by P.W. Hawkes (Academic, London and New York 1973) pp. 2-14

1.68   P.W. Hawkes: Adv. Opt. Electron Microsc. *7*, 101-184 (1978)

1.69   F. Lenz: Optik *22*, 270-288 (1965)

1.70   F. Lenz: Lab. Invest. *14*, 808-818 (1965)

1.71   F. Lenz: "Fourier Electron Optics", in *Electron Microscopy and Analysis*, ed by W.C. Nixon (Institute of Physics, London 1971) pp. 224-229

1.72   F. Lenz: "Transfer of Image Information in the Electron Microscope", in *Electron Microscopy in Material Science*, ed. by U. Valdrè (Academic, New York and London 1971) pp. 540-569

1.73   G. Ade: Optik *42*, 199-215 (1975)

1.74   K.J. Hanszen: "The Relevance of Dark Field Illumination in Conventional and Scanning Transmission Electron Microscopy", PTB-Bericht A Ph-7, pp.28 (1974)

1.75   K.J. Hanszen: Optik *46*, 107-123 (1976)

1.76   K.J. Hanszen, G. Ade: Canberra (1974), Vol. 1, pp. 196-197

1.77   K.J. Hanszen, G. Ade: Optik *42*, 1-22 (1975)

1.78   K.J. Hanszen, G. Ade: Optik *44*, 237-249 (1976)

1.79   H. Hoch: Optik *38*, 220-222 (1973); *47*, 65-85 (1977)

1.80   D. Typke, R. Hegerl, W. Hoppe: Canberra (1974), Vol. 1, pp. 198-199

1.81   E. Menzel: Optik *15*, 460-470 (1958): "Die Abbildung von Phasenobjekten in der optischen Übertragungstheorie", in *Optics in Metrology*, ed. by P. Mollet (Pergamon, Oxford 1960) pp. 283-293

1.82   E. Menzel, W. Mirandè, I. Weingärtner: *Fourier-Optik und Holographie* (Springer, Vienna and New York 1973)

1.83   S. Slansky: J. Phys. Radium *20*, 13 S - 14 S (1959); Rev. Opt. *39*, 555-577 (1960); Opt. Acta *9*, 277-294 (1962)

1.84   S. Slansky, A. Maréchal: C.R. Acad. Sci. Paris *250*, 4132-4134 (1960)

1.85   K.J. Hanszen, L. Trepte: Grenoble (1970), Vol. 1, pp. 45-46

1.86   K.J. Hanszen, L. Trepte: Optik *32*, 519-538 (1971)

1.87   J. Frank: Optik *38*, 519-536 (1973)

1.88   J. Frank: Optik *44*, 379-391 (1976)

1.89   R.H. Wade, J. Frank: Optik *49*, 81-92 (1977)

1.90   S.C. McFarlane: Optik, to be published

1.91   P.W. Hawkes: Optik *47*, 453-467 (1977)

1.92   P.W. Hawkes: Optik *49*, 146-161 (1977)

1.93   P.W. Hawkes: "Image Formation in Partially Coherent Illumination Including Inelastic Scattering", in *Developments in Electron Microscopy and Analysis 1977* , ed. by D.L. Misell (Institute of Physics, Bristol 1977) pp. 123-126

1.94   P.W. Hawkes: Optik *50*, 353-370 (1978)

1.95   L. Mandel, E. Wolf: J. Opt. Soc. Am. *66*, 529-535 (1976)

1.96   H.A. Ferwerda: Optik *45*, 411-426 (1976); Jerusalem (1976), Vol. 1, pp. 261-262

1.97   K.J. Hanszen, L. Trepte: Optik *33*, 166-198 (1971)

1.98   W. Krakow: *Proc. 35th Annual Meeting EMSA, Boston*, ed. by G.W. Bailey (Claitor, Baton Rouge 1977) pp. 72-73

1.99   H. Niehrs: Optik *38*, 44-63 (1973)

1.100 H. Rose, J. Fertig: *Proc. 35th Annual Meeting EMSA, Boston*, ed. by G.W. Bailey (Claitor, Baton Rouge 1977) pp. 200-201

1.101 W.O. Saxton, W.K. Jenkins, L.A. Freeman, D.J. Smith: Optik *49*, 505-510 (1978)

1.102 F. Thon: Prague (1964), Vol. 1, pp. 127-128

1.103 F. Thon: Z. Naturforsch. *20a*, 154-155 (1965)

1.104 F. Lenz, W. Scheffels: Z. Naturforsch. *13a*, 226-230 (1958)

1.105 B. von Borries, F. Lenz: Stockholm (1956) pp. 60-64

1.106 F. Thon: Z. Naturforsch. *21a*, 476-478 (1966)

1.107 F. Thon: Kyoto (1966), Vol. 1, pp. 23-24

1.108 O. Scherzer: J. Appl. Phys. *20*, 20-29 (1949)

1.109 J. Frank, P. Bussler, R. Langer, W. Hoppe: Ber. Bunsen-Ges. Phys. Chem. *74*, 1105-1115 (1970); Grenoble (1970), Vol. 1, pp. 17-18

1.110 W.O. Saxton: Optik *49*, 51-62 (1977)

1.111 W.O. Saxton: "Coherence in Bright Field Microscopy of Weak Objects", in *Developments in Electron Microscopy and Analysis 1977*, ed. by D.L. Misell .(Institute of Physics, Bristol 1977) pp. 111-114

1.112 A. Beorchia, P. Bonhomme: Optik *39*, 437-442 (1974)

1.113 R.E. Burge, J.C. Dainty, J. Thom: Jerusalem (1976), Vol. 1, pp. 256-258

1.114 R.E. Burge, J.C. Dainty, J. Thom: "The Spatial Coherence of Electron Beams", in *Developments in Electron Microscopy and Analysis*, ed. by J.A. Venables (Academic, London and New York 1976) pp. 221-224

1.115 O.L. Krivanek: Optik *43*, 361-372 (1975)

1.116 O.L. Krivanek: Jerusalem (1976), Vol. 1, pp. 263-264

1.117 M. Troyon: Optik *49*, 247-251 (1977)

1.118 W. Krakow, K.H. Downing, B.M. Siegel: *Proc. 31st Annual Meeting EMSA, New Orleans*, ed. by C.J. Arceneaux (Claitor, Baton Rouge 1973) pp. 278-279; Optik *40*, 1-13 (1974). Cf Ref. 1.188

1.119 R.E. Burge, R.F. Scott: Optik *43*, 503-507 (1975)

1.120 O.L. Krivanek: Optik *45*, 97-101 (1976)

1.121 F. Thon: Phys. Bl. *23*, 450-458 (1967)

1.122 J. Frank, S.C. McFarlane, K.H. Downing: Optik *52*, 49-60 (1978/9)

1.123 R.H. Wade: Ultramicroscopy *3*, 329-334 (1978)

1.124 O. Kübler, R.Waser: Optik *37*, 425-438 (1973)

1.125 P. Bonhomme, A. Beorchia, B. Meunier: C.R. Acad. Sci. Paris *B 282*, 63-66 (1976)

1.126 P. Bonhomme, A. Beorchia, B. Meunier, F. Dumont, D. Rossier: Optik *45*, 159-167 (1976)

1.127 F. Dumont, D. Rossier, P. Bonhomme, A. Beorchia, B. Meunier: J. Phys. E: Sci. Instrum. *10*, 520-524 (1977)

1.128 K.-H. Herrmann, D. Krahl: Canberra (1974), Vol. 1, pp. 102-103

1.129 K.-H. Herrmann, D. Krahl: Optik *45*, 231-247 (1976)

1.130 K.-H. Herrmann, D. Krahl: Jerusalem (1976), Vol. 1, pp. 312-313

1.131 K.H. Downing, O. Kübler, M. Noble: *Proc. 35th Annual Meeting EMSA, Boston*, ed. by G.W. Bailey (Claitor, Baton Rouge 1977) pp. 76-77

1.132 G. Marie, J. Donjon, J.-P. Hazan: Adv. Image Pickup Display *1*, 225-302 (1974)

1.133 R.M. Glaeser, I. Kuo, T.F. Budinger: *Proc. 29th Annual Meeting EMSA, Boston*, ed. by C.J. Arceneaux (Claitor, Baton Rouge 1971) pp. 466-467

1.134 I. Kuo, R.M. Glaeser: Ultramicroscopy *1*, 53-66 (1975)

1.135 V. Witt, H.P. Englmeier, W. Hoppe: Manchester (1972) pp.632-633

32

1.136 W. Goldfarb: Jerusalem (1976), Vol. 1, pp. 316-317
1.137 W. Goldfarb, B.M. Siegel: *Proc. 31st Annual Meeting EMSA, New Orleans*, ed.
       by C.J. Arceneaux (Claitor, Baton Rouge 1973) pp. 264-265
1.138 W. Goldfarb, B.M. Siegel: *Proc. 33rd Annual Meeting EMSA, Las Vegas*, ed. by
       G.W. Bailey (Claitor, Baton Rouge 1975) pp. 124-125
1.139 R. Brüders, K.-H. Herrmann, D. Krahl, H.-P. Rust: Jerusalem (1976), Vol. 1,
       pp. 318-319
1.140 W. Fröhlich, K.-H. Herrmann, D. Krahl, H. Kuschel, V. Rindfleisch: Grenoble
       (1970), Vol. 1, pp. 337-338
1.141 K.-H. Herrmann, D. Krahl, V. Rindfleisch: Siemens Forsch. u. Entwickl. Ber.
       1, 167-178 (1972)
1.142 K.-H. Herrmann, D. Krahl, H.-P. Rust: Jerusalem (1976), Vol. 1, pp. 320-321
1.143 K.-H. Herrmann, D. Krahl, H.-P. Rust, O. Ulrichs: Optik *44*, 393-412 (1976)
1.144 H.-P. Rust: Canberra (1974), Vol. 1, pp. 92-93
1.145 A.C. van Dorsten, P.H. Broerse, H.F. Premsela: Kyoto (1966), Vol. 1, pp.
       275-276
1.146 H. Alsberg, R.E. Hartman: Adv. Electron. Electron Phys. *40A*, 287-300 (1976)
1.147 F.C. Billingsley: Adv. Opt. Electron Microsc. *4*, 127-159 (1971)
1.148 R.E. Hartman, H. Alsberg, R.S. Hartman, R. Nathan, P. Wendell: Canberra (1974),
       Vol. 1, pp. 96-97
1.149 J. Podbrdský: *Proc. XVth Czechoslovak Conf. Electron Microscopy*, ed. by
       V. Viklický, J. Ludvík (Czechoslovak Academy of Sciences, Prague 1977), Vol.
       B, pp. 607-608
1.150 Yu.M. Voronin, R.Yu. Khaitlina: Izv. Akad. Nauk SSSR (Ser. Fiz) *41*, 871-875
       (1977); [English transl. Bull. Acad. Sci USSR (Phys. Ser.) *41*(5), 25-28 (1977)]
1.151 J. Frank: J. Phys. E: Sci. Instrum. *8*, 582-587 (1975)
1.152 Yu.M. Voronin, I.R. Demenchenok, R.Yu. Khaitlina: Opt. Mekh. Prom. *39*(3),
       7-8 (1972); [English transl. Sov. J. Opt. Technol. *39*, 130-131 (1972)]
1.153 P. Schiske: Rome (1968), Vol. 1, pp. 145-146
1.154 H.P. Erickson, A. Klug: *Proc. 28th Annual Meeting EMSA, Houston*, ed. by C.J.
       Arceneaux (Claitor, Baton Rouge 1970) pp. 248-249
1.155 H.P. Erickson, A. Klug: Ber. Bunsen-Ges. Phys. Chem. *74*, 1129-1137 (1970);
       Phil. Trans. Roy. Soc. Lond. *B261*, 105-118 (1971)
1.156 W. Hoppe: Phil. Trans. Roy. Soc. Lond. *B261*, 71-94 (1971)
1.157 T.A. Welton: *Proc. 27th Annual Meeting EMSA, St. Paul*, ed. by C.J. Arceneaux
       (Claitor, Baton Rouge 1969) pp. 182-183;
       Proc. 28th Annual Meeting EMSA, Houston, ed. by C.J. Arceneaux (Claitor, Baton
       Rouge 1970) pp. 32-33;
       Proc. 29th Annual Meeting EMSA, Boston, ed. by C.J. Arceneaux (Claitor, Baton
       Rouge 1971) pp. 94-95;
       Proc. 30th Annual Meeting EMSA, Los Angeles, ed. by C.J. Arceneaux (Claitor,
       Baton Rouge 1972) pp. 592-593;
       Proc. 32nd Annual Meeting EMSA, St. Louis, ed. by C.J. Arceneaux (Claitor,
       Baton Rouge 1974) pp. 338-339;
       Proc. 33rd Annual Meeting EMSA, Las Vegas, ed. by G.W. Bailey (Claitor, Baton
       Rouge 1975) pp. 196-197;
       Proc. 35th Annual Meeting EMSA, Boston, ed. by G.W. Bailey (Claitor, Baton
       Rouge 1977) pp. 176-177
1.158 J. Frank: Biophys. J. *12*, 484-511 (1972)
1.159 J. Frank: "Use of Anomalous Scattering for Element Discrimination", in *Image
       Processing and Computer-aided Design in Electron Optics*, ed. by P.W. Hawkes
       (Academic, London and New York 1973) pp. 196-211
1.160 P. Schiske: "Image Processing Using Additional Statistical Information About
       the Object", in *Image Processing and Computer-aided Design in Electron Optics*,
       ed. by P.W. Hawkes (Academic, London and New York 1973) pp. 82-90
1.161 P.W. Hawkes: Optik *41*, 64-68 (1974)
1.162 F.L. Ball, W.W. Harris, T.A. Welton: *Proc. 39th Annual Meeting EMSA, Boston*,
       ed. by C.J. Arceneaux (Claitor, Baton Rouge 1971) pp. 88-89

1.163 T.A. Welton, W.W. Harris: Canberra (1974), Vol. 1, pp. 318-319
1.164 T.A. Welton, F.L. Ball, W.W. Harris: *Proc. 31st Annual Meeting EMSA, New Orleans*, ed. by C.J. Arceneaux (Claitor, Baton Rouge 1973) pp. 270-271
1.165 D.A. Ansley: *Proc. 30th Annual Meeting EMSA, Los Angeles*, ed. by C.J. Arceneaux (Claitor, Baton Rouge 1972) pp. 596-597; Optik *40*, 400-422 (1974)
1.166 H.C. Benski: "Restoration of Weak Phase Objects from Electron Microscopy" in *Digital Image Processing and Analysis*, ed. by J.C. Simon, A. Rosenfeld (Noordhoff, Leiden 1977) pp. 479-483
1.167 H.C. Benski, C. Sonrel: Rev. Phys. Appl. *12*, 543-546 (1977)
1.168 P. Bussler, A. Feltynowski, W. Hoppe: Manchester (1972) pp. 626-627
1.169 H.A. Ferwerda, B.J. Hoenders: *Proc. 32nd Annual Meeting EMSA, St. Louis*, ed. by C.J. Arceneaux (Claitor, Baton Rouge 1974) pp. 334-335
1.170 H.A. Ferwerda, B.J. Hoenders: Optik *39*, 317-326 (1974); Opt. Acta *22*, 25-34, 35-36 (1975)
1.171 M. Hahn, J. Seredynski: Canberra (1974), Vol. 1, pp. 234-235
1.172 P.W. Hawkes: Optik *40*, 539-556 (1974)
1.173 B.J. Hoenders: Optik *35*, 116-133 (1972)
1.174 B.J. Hoenders, H.A. Ferwerda: Optik *37*, 542-556 (1973); *38*, 80-94 (1973)
1.175 O. Kübler, M. Hahn, J. Seredynski: Jerusalem (1976), Vol. 1, pp. 306-307
1.176 O. Kübler, M. Hahn, J. Seredynski: Optik *51*, 171-188, 235-256 (1978)
1.177 W.O. Saxton, A. Howie, A. Mistry, A. Pitt: "Fact and Artefact in High Resolution Microscopy", in *Developments in Electron Microscopy and Analysis 1977*, ed. by D.L. Misell (Institute of Physics, Bristol 1977) pp. 119-122
1.178 Yu.M. Voronin, A.V. Mokhnatkin, R.Yu. Khaitlina: Izv. Akad. Nauk SSSR (Ser. Fiz.) *38*, 1382-1384 (1974); [English transl. Bull. Acad. Sci. USSR (Phys. Ser.) *38* (7), 20-22]
1.179 K. Welles, W. Krakow, B.M. Siegel: Canberra (1974), Vol. 1, pp. 320-321
1.180 G.W. Stroke, M. Halioua, F. Thon, D.H. Willasch: Proc. IEEE *65*, 39-62 (1977)
1.181 J. Frank: *Proc. 34th Annual Meeting EMSA, Miami Beach*, ed. by G.W. Bailey (Claitor, Baton Rouge 1976) pp. 478-479
1.182 J. Frank, L. Al-Ali: Nature *256*, 376-379 (1975)
1.183 J. Frank, L. Al-Ali: "The Focus Dependence of the Signal-to Noise Ratio in Electron Micrographs", in *Developments in Electron Microscopy and Analysis*, ed. by J.A. Venables (Academic, London and New York 1976) pp. 229-231
1.184 H.C. Andrews, B.R. Hunt: *Digital Image Restoration* (Prentice-Hall, Englewood Cliffs, N.J. 1977)
1.185 W.K. Pratt: *Digital Image Processing* (Wiley, New York and Chichester 1978)
1.186 A. Rosenfeld, A.C. Kak: *Digital Picture Processing* (Academic, New York and London 1976)
1.187 W. Hoppe, P. Bussler, A. Feltynowski, N. Hunsmann, A. Hirt: "Some Experience with Computerized Image Reconstruction Methods", in *Image Processing and Computer-Aided Design in Electron Optics*, ed. by P.W. Hawkes (Academic, London and New York 1973) pp. 92-126
1.188 P. Schiske: Toronto (1978), Vol. 1, pp. 216-217

2. Recovery of Specimen Information for Strongly Scattering Objects

W. O. Saxton

With 6 Figures

The previous chapter described the highly satisfactory theory of image formation and processing that results when the specimen is sufficiently thin and light to be treated in projection, and produces no more than slight attenuation and phase shifting of the incident electron beam. This limit is an increasingly close approximation as accelerating voltage and resolution increase; it is an approximation at best, however, and is sometimes simply untenable. Accordingly, this chapter is devoted to the much more difficult problem of image correction for stronger objects, when the imaging process is nonlinear. Although more laborious than the analyses applicable to weak objects, the methods reviewed here are still useful if applied to specimens which are too thick or heavy to be *weak*, but not too much so to be treated in projection still.

The allowable thickness decreases as resolution improves, but even so it seems that their principal limitation in high-resolution applications is not the restricted specimen thickness so much as the inability of most of the methods so far put forward to make adequate allowance for the limited coherence properties of the imaging. The possibility of treating these properly is a considerable recommendation for the first group of methods put forward (Sect.2.2), which are based on iterative application of the linear theory. After these, methods relying on twin intensity measurements (image/diffraction, two defocused images, images recorded with complementary halves of the diffraction plane obscured) are described (Sects.2.3,4).

A powerful body of theory devoted to "analytic" images, based on the constraint of a finite diffraction plane aperture, is collected in Sect.2.5; this provides new ways of interpreting many of the problems considered in the chapter, besides suggesting techniques for analyzing some previously intractable problems. Section 2.6 considers "holographic" methods, dependent on the addition or removal of a "reference" beam from the image, and Sect.2.7 "ptychographic" methods in which neighboring diffracted beams are made to overlap in such a way as to reveal their relative phases. Section 2.8 describes a final, new, method based on bright-field/dark-field subtraction, which is simple and noniterative, and allows a proper treatment of the coherence conditions, though requiring four separate images as data.

The practical problems encountered, such as coherence, inelastic scattering, recording noise, alignment, and deconvolution, are given some attention in Sect.2.9, which concludes with some reference to other constraints (those used in X-ray crystallography; the relative likelihood of different object waves) that might be used to supplement those already exploited. Firstly, however, the context within which we propose to operate must be clarified.

2.1 Image Formation and Interpretation

2.1.1 Recapitulation of Coherent Image Formation

A brief summary of the equations describing the propagation of the electron wave through the microscope is necessary first of all. To avoid inessential complications, a single thin imaging lens is assumed and astigmatism is neglected; a fuller description of the general case appears in HAWKES [2.1] (see also Chap.1).

The wave function leaving the object plane is denoted by $\psi_0(r_0)$. With the effective illumination source at infinity, the next plane of interest is the back focal (diffraction, aperture) plane of the lens, in which the wave function in the absence of aberrations would be

$$\psi_a(r_a) = \exp(-\pi i r_a^2/\lambda d) \int \psi_0(r_0)\exp(-2\pi i r_0 \cdot r_a/\lambda f)d^2r_0 \quad ; \tag{2.1}$$

λ denotes the electron wavelength, f the lens focal length, and d the distance between the diffraction and image planes. Here and subsequently, normalizing constants that preserve the total electron flux $(\int|\psi|^2)$ are omitted. Continued propagation to the image plane is described by

$$\psi_i(r_i) = \exp(\pi i r_i^2/\lambda d) \int \psi_a(r_a)\exp[\pi i(r_a^2/\lambda d - 2r_a \cdot r_i/\lambda d)]d^2r_a \quad . \tag{2.2}$$

Apart from the quadratic phase factor, ψ_a is the Fourier transform of ψ_0:

$$\psi_a(r_a) = \tilde{\psi}_0(r_a/\lambda f) \quad . \tag{2.3}$$

The quadratic phase factor cancels with a similar factor in (2.2), reducing the integral to a further Fourier transform; in this way, and apart again from a quadratic phase factor, ψ_i recovers ψ_0:

$$\psi_i(r_i) = \psi_0(-f r_i/d) = \psi_0(-r_i/M) \quad , \tag{2.4}$$

where M is the magnification. It is rarely necessary to consider the quadratic phase factors in practice, and they are commonly omitted (as subsequently here) from expressions for the wave functions.

The separation of the two transforms is useful because the combined effects of the only serious lens aberration (third-order spherical) and instrumental defocus is to modify the phase of the diffraction plane wave function by an aberration function γ. Referring image plane coordinates back to the object plane, the object and image plane wave functions are connected by the simple relation

$$\tilde{\psi}_i(\underline{t}) = \tilde{\psi}_0(\underline{t})\exp[-i\gamma(\underline{t})] \qquad\qquad (2.5)$$

between the transforms; γ takes the form

$$\gamma(\underline{t}) = \pi\lambda\left(\tfrac{1}{2}\, C_s\lambda^2 t^4 - \Delta t^2\right) \quad, \qquad\qquad (2.6)$$

involving the spherical aberration coefficient C_s and the defocus Δ, and $\underline{t} = (p,q) = \underline{r}_a/\lambda f$. $\tilde{\psi}_i(\underline{t})$ is commonly referred to as if it actually were the diffraction plane wave function. The effects of a diffraction plane aperture are easily included via a masking factor $B(\underline{t})$ in (2.5); the effects of the incomplete coherence achieved in practice are considered from time to time as necessary—particularly in Sect.2.9.1.

The most that can be sought from a microscope is the complex wave function leaving the specimen $\psi_0(\underline{r})$, and its extraction from the *intensity* measurements provided by micrographs is the main theme of this chapter; Sect.2.1.2 considers briefly the other half of the problem, namely drawing from $\psi_0(\underline{r})$ reliable conclusions about the specimen structure. Most of the methods here seek to establish the image wave $\psi_i(\underline{r})$ immediately, obtaining $\psi_0(\underline{r})$ from this by the use of (2.5). The deconvolution is straightforward because the "transfer function" $\exp[-i\gamma(\underline{t})]$ has no gaps; this is in marked contrast to the transfer functions that apply to the *derived* problems (Chap.1) of relating image intensity linearly to the modulus and phase variations in $\psi_0(\underline{r})$.

To the limited extent that scanning transmission electron microscopes (STEM) can be used in modes analogous to those of the conventional instrument, what is said is relevant to both instruments; it is with the latter primarily in mind that I write, however.

2.1.2 Interpreting the Specimen Wave

Very loosely, we may say that, provided the specimen is reasonably thin, the phase of $\psi_0(\underline{r})$ is proportional to the projected potential within the specimen (and hence roughly to projected mass), but that its modulus is usually uninterpretable; for thicker specimens even the phase is uninterpretable.

Atoms are nearly transparent to the electron wave: they form potential wells $V(\underline{r},z)$ that shorten the electron wavelength and advance the phase by

$$\phi(\underline{r}) = 2\pi em\lambda (1 + e\Phi/mc^2)/h^2 \int V(\underline{r},z)dz \qquad (2.7)$$

(Φ is the accelerating voltage). In this (WKB) approximation, lateral movement of the electrons within the specimen is neglected, as the deformation of the wave fronts occurs on a scale much larger than the wavelength. If $\phi(\underline{r})$ is small compared with unity, we have the weak phase object limit; the thickness at which that approximation becomes poor depends on the composition of the specimen (via V), being about 15 nm for light disordered specimens (e.g., amorphous carbons, metallic glasses, unstained biological specimens), and 5 nm for heavy disordered specimens (e.g., stained biological specimens; MISELL [2.2,3]), but dropping as low as 1 nm for crystals containing heavy atoms one above another [2.4]; all figures refer to 100 kV electrons.

The "lateral movement" of electrons neglected in (2.7) is of course simply defocusing. The thickness beyond which it cannot be ignored is indicated by the defocusing phase shifts produced in the diffraction plane — cf (2.6). Limiting these to half a radian, say, gives a maximum thickness

$$d \leq \tfrac{1}{2} x^2/\pi\lambda \quad , \qquad (2.8)$$

making clear the strong dependence on resolution through the object period x: at 100 kV, the permissible thickness is over 40 nm at a resolution of 1 nm, but only 4 nm at 0.3 nm resolution. (The fact that both sets of figures — for the breakdown of the weak phase approximation and the projection approximation — improve as the wavelength falls is a compelling argument for the use of higher accelerating voltages in future.)

When the condition (2.8) is seriously violated, the phase ϕ of ψ_0 becomes uninterpretable: image evaluation can be carried out only by the *calculation* of images for comparison with micrographs, on the basis of an assumed specimen structure, for example by the multislice method in which the propagation through the specimen ·is treated by alternate defocusing and phase shifting (e.g., [2.5]). Accordingly, (2.8) sets an ultimate limit on the useful specimen thickness if direct interpretation is to be attempted. We see that the linear methods of Chap.1 will always be adequate for high-resolution study of thin amorphous materials, for example, because the phase shifts remain small long after (2.8) breaks down; on the other hand, there is a significant gap between the breakdown of the weak object approximation and the failure of (2.8) for most other cases, and it is to these that the methods of this chapter are relevant.

Two important cases should be excepted, as being unlikely in most cases to benefit from the methods here: negatively stained biological specimens for which the low-resolution (~ 2 nm) images normally obtainable are already faithful projections of the stain density; and isolated heavy atoms (e.g., used as "markers" in macromolecules), which are better imaged in an incoherent mode, and for which in any case the exact image profile does not matter, position being the only feature of interest.

The reason the modulus of $\psi_0(\underline{r})$ is rarely helpful is that too many effects contribute to it. Variations in the height of a given feature in the specimen affect it, by way of the focus shifts (2.6); failure of (2.8) at the level of individual heavy atoms (failure of the Born approximation) contributes; the loss of some electrons by inelastic scattering processes contributes; and the interception at the diffraction plane aperture of electrons scattered through high angles by heavy atoms, although not affecting the *actual* $|\psi_0|$, inevitably reduces the *reconstructed* $|\psi_0|$ over the scattering centers.

The "charge density" approximation [2.4] would seem at first sight to offer an alternative to image correction altogether, suggesting that at a suitable focus level the image contrast is directly proportional to the projected specimen charge (rather than potential) density. The approximation depends in turn on the weak object approximation however, and is limited to lower than ~ 1 nm resolution besides [Ref.2.6, pp.18,19]. MISELL [Ref.2.3, pp.267,268,270] considers the use of the approximation *after* image correction, which rather misses the point of the approximation, namely its great simplicity within its limits.

2.1.3 Rendering Images Discrete

Virtually all of the methods described in this chapter presuppose that the wave functions or intensity distributions are readily available for mathematical manipulation by a computer. This entails representing the functions by a set of sample values, usually on a square sampling grid. The details of how the "discrete" function approximates the continuous one, particularly in Fourier transform processes, are well known (see, for example, [Ref.2.6, pp.34-43]) and will not be given here. The numerical transform process has an exact inverse, so that large numbers of transforms in both directions may be performed without any significant errors being introduced into the data — i.e., iterative processes involving transforms are quite feasible (and indeed quite rapid).

In one particular case, representation by a set of discrete sample values is in fact exact: this is the case of periodic signals with a finite number of Fourier components, where the sampling lattice is chosen to match the periodicity, so that in the transform plane the sampling points coincide with the reciprocal lattice sites.

In describing processing methods here, I have used discrete or continuous notations simply according to whichever happens temporarily to allow the most compact explanation of the principles of the method; a discrete image is always implied in practice of course.

2.2 Methods Iterating the Linear Theory Solution

The previous chapter was devoted to the limiting case in which the scattered part of the wave is small compared with the unscattered part throughout the microscope. If the image plane wave function is separated into three parts

$$\psi_i(\underline{r}) = 1 + u(\underline{r}) + iv(\underline{r}) \quad , \tag{2.9}$$

being the unscattered wave, the real part of the scattered wave, and the imaginary part, the image intensity becomes

$$I(\underline{r}) = 1 + 2u(\underline{r}) + |u(\underline{r})|^2 + |v(\underline{r})|^2 \quad ; \tag{2.10}$$

the weak scattering (weak object) limit allows the neglect of the squared terms and the direct observation of the real part $u(\underline{r})$ of the scattered wave. Since this is in general only *half* the information needed about the scattered wave, the object cannot be recovered from this alone. However, for any bright-field imaging conditions, a relationship can be established between the transform of u and those of the real and imaginary (amplitude and phase) parts of the specimen transparency $1 + S_r(\underline{r}) + iS_i(\underline{r})$, of the form

$$2\tilde{u}(\underline{t}) = \tilde{T}_r(\underline{t})\tilde{S}_r(\underline{t}) + \tilde{T}_i(\underline{t})\tilde{S}_i(\underline{t}) \quad . \tag{2.11}$$

$\tilde{T}_r$ and $\tilde{T}_i$ are called weak-object amplitude and phase contrast transfer functions, respectively; they vary with such instrumental parameters as illumination direction and defocus, so that a *pair* of images does in general allow full recovery of the object transparency by providing two simultaneous equations of the form of (2.11), though the equations may become singular at particular points or lines in the transform ($\underline{t}$) plane, necessitating a third image to resolve the ambiguity.

If we now move away from the weak scattering limit, we can retain the above equations, though the names *real* and *imaginary part* transfer functions become preferable for $\tilde{T}_r$ and $\tilde{T}_i$. The image contrast, $I(\underline{r}) - 1$, must now be regarded as giving only a first approximation to $2u(\underline{r})$, because the squared terms are no longer insignificant; however, provided those are at least rather smaller than $2u(\underline{r})$, the approximation is reasonably close. A pair of data images thus allows first estimates to be made of S_r and S_i, by solution of (2.11); from these an estimate can be made by direct forward calculation of the scattered wave in both data image planes, allowing the squared terms in (2.10) to be estimated and taken into account as a *better* estimate of $2u(\underline{r})$ is extracted. The process can be repeated once or several times as necessary until it converges, i.e., until the values obtained for S_r and S_i stabilize.

It is not easy to formulate precise conditions (on S_r, S_i, $\tilde{T}_r$, and $\tilde{T}_i$, say) for convergence, but if in a given case the process is not obviously converging after a cycle or two, it should probably be abandoned forthwith. What is *very* important about this type of iterative method is that, unlike most of those considered later in this chapter, it allows a proper treatment of the effects of illumination divergence and focus spread (spatial and temporal coherence). This is modeled (in principle) by averaging the image intensity (2.10) over a number of different illumination directions and focus levels; the average over the linear term $2u(\underline{r})$ carries over to an average over $\tilde{T}_r$ and $\tilde{T}_i$ in (2.11), where it can usually be carried out analytically for reasonable models of the illumination and focus profiles (cf Chap.1). S_r and S_i can thus be recovered from $2u(\underline{r})$ even without coherent imaging conditions; the squared terms can be calculated from these allowing for the imperfect coherence, so the iteration can proceed as before except that calculation of the squared terms is much more time consuming than previously. It is well known that the transfer functions $\tilde{T}_r$ and $\tilde{T}_i$ decay to zero for high spatial frequencies (typically 3 nm^{-1}), this fact limiting the ultimate recovery of information in the microscope; patently, any methods disregarding coherence requirements cannot be used near this resolution with any confidence.

The previous chapter also considered the optimization of the solution of (2.11) in the presence of noise; much of this remains feasible and useful when the process is iterated.

This approach of iterating the linear theory solution has been applied to model problems by a number of authors. MISELL et al. [2.7], and MISELL and GREENAWAY [2.8] apply it to images formed with special apertures obscuring the two halves of the diffraction plane in turn, though they use a different route from (2.11) to recover the squared terms (cf Sect.2.4). They found convergence within a few cycles when the squared terms accounted for as much as 50% of the image intensity variation, which is very encouraging. However, they assume perfect coherence and discard the method in favor of another described in Sect.2.4.2. LANNES [2.9,10] assumes similar data and proposes an iteration that differs in only minor respects which he applies to calculated images of a dislocation dipole; again convergence is reported in a few cycles with substantial image contrast. LANNES notes the possibility of treating coherence properly, though he does not include such a treatment in his trial calculations.

Finally, VAN TOORN et al. [2.11] use the method applied to images at two different focus levels; varying the object strength (up to 3% amplitude variation and
phase variations up to ±0.6 rad) they found never more than 20 cycles necessary for
convergence. They study the onset of failure as the image noise levels increase,
reporting, for their strongest object, failure when the noise level on the measured
intensities exceeds 0.05%; this figure is hard to credit, since useful reconstructions of weaker objects have been made using the simple linear theory, in the presence of noise levels 50 times higher. Of considerable interest, however, is their
proposal that a diffraction pattern be used to reduce the noise sensitivity of the
process: in solving (2.11) at each stage, they constrain $|\tilde{S}_r + i\tilde{S}_i|$ to the known
diffraction plane modulus (assumed noise free; a more mature version of the process
would obviously use a slightly different compromise) and choose a phase which gives
a least-squares fit to the equations. The result is not simply a greatly improved
noise tolerance, but much better convergence too: seven cycles appear adequate in
all cases.

2.3 Methods Requiring No Special Apertures

2.3.1 The Data Used

The two earliest methods proposed for reconstructing the object plane wave function
ψ_o without any restriction on the strength of the object are the simplest experimentally: one relies on two differently defocused images (the "two-defocus" method),
just as are used for the reconstruction of the weak objects to which the linear
theory (Chap.1) applies, and the other on a single image combined with a diffraction
pattern from the same specimen area (the "image/diffraction" method). Recording the
image at more than one focus level is routine practice in any case, so it presents
no problems beyond the normal ones of specimen damage (Sect.2.9.3); the requirement
in the image/diffraction method that the diffraction pattern be recorded illuminating only the specimen area that is subsequently analyzed is very difficult to meet,
however, in general, and restricts the method at the present time to periodic
specimens, for which the relative intensities at the reciprocal lattice sites are
independent of the number of specimen cells illuminated. Recording the diffraction
pattern of a periodic specimen has recently been elegantly automated [2.12].

While it is not an unreasonable supposition that the two differently defocused
images will determine the wave function uniquely (by inference from the linear
theory), the uniqueness of the solution for the image/diffraction method is a very
different question, considered in Sect.2.3.4. In both cases uniqueness appears to
be guaranteed "in general" provided the image intensities are adequately closely
sampled, and provided the diffraction plane wave function (even for the two-defocus

method) is still significant at the edge of the diffraction plane aperture.
Whether this latter proviso is in fact really necessary is still open to discussion: problems in which it does not hold have nevertheless been solved successfully by both methods.

2.3.2 The Iterative Transform Algorithm

The same algorithm can in fact be used to solve both of the problems now posed.
It was proposed by GERCHBERG and SAXTON [2.13,14], with an immediate application to
the image/diffraction problem; and its adaptation for the two-defocus problem was
reported by MISELL [2.15,16].

The algorithm relies on the calculation from a set of sample estimates for a
wave function in one data plane of the corresponding set of samples in the other
data plane. For the image/diffraction problem this requires no more than a single
(forward or reverse) Fourier transformation (cf Sect.2.1); for the two-defocus
method it involves a transformation $\psi_i \rightarrow \tilde{\psi}_i$, multiplication by a phase shifting
factor $\exp(\pm i\lambda\Delta_r t^2)$, where Δ_r is the relative defocus of the two images, and inverse transformation [cf (2.5,6)]. In each cycle of the algorithm, the current
wave function estimate is carried to the other data plane, where its modulus is
corrected to the observed modulus (square root of intensity) at all sample points,
the phases being preserved; the wave function is then returned to the original
plane, where a similar modulus correction is performed. Convergence to a solution
is indicated by the modulus corrections becoming insignificant.

In order to discuss the convergence of this process, we define a suitable
measure of how consistent a given wave function estimate is with the data modulus
(intensity) distributions. Using subscripts 1 and 2 to label the data planes, ψ
to denote the wave functions and m to denote the observed modulus distributions,
a suitable measure of the "error" in an estimate ψ_1 having a modulus m_1 is the sum
square "residue"

$$E = \sum (|\psi_2| - m_2)^2 , \qquad (2.12)$$

the sum being taken over all sample points. This is the sum square difference
between predicted and observed modulus values for the *other* data plane; for the
image/diffraction method it may be quoted as a percentage of the total "flux"
$\sum m_2^2$, and for the two-defocus method (in view of the smaller difference between the
data modulus distributions) as a percentage of $\sum (m_1 - m_2)^2$. It is simple to demonstrate that E cannot increase during application of the algorithm [2.13,17], and
this proof follows.

Given an initial wave function estimate ψ_1, with modulus m_1, suppose corrections
c_2, d_1, and e_2 are added to the estimate in plane 2, plane 1, and plane 2 again,
to correct the modulus as the algorithm proceeds. The error E in the initial wave

function is simply $\sum |c_2|^2$. Considering the correction process in plane 1,

$$|\psi_1 + c_1| = m_1 + |d_1| \leq m_1 + |c_1| \qquad (2.13)$$

(the latter part by the triangle inequality), so that $|d_1| \leq |c_1|$. The error in this next estimate is $\sum |e_2|^2$, and an argument similar to that just given shows that $|e_2| \leq |d_2|$. Accordingly

$$\sum |e_2|^2 \leq \sum |d_2|^2 = \sum |d_1|^2 \leq \sum |c_1|^2 = \sum |c_2|^2 \qquad (2.14)$$

[using the fact that the total flux is the same in the two planes — cf (2.5) in conjunction with Parseval's theorem], which is the result that was to be proved.

This means that the algorithm is completely safe from catastrophic divergence, no matter how far from the root the initial estimate may be, and is reasonably insensitive to observational noise. At the same time, the fact that a stronger statement is not possible means that convergence is often quite slow (cf the examples in Sect.2.3.3). The first few iterations are always the most useful, and little further change takes place after 10 or 20 cycles — apparently regardless of the complexity of the problem. Convergence to a zero E is not guaranteed either: a steady decay to a finite value is not infrequently observed with an unfortunate initial estimate: if we regard E as a function of the unknown ψ_1 phases, there is no reason to suppose it has not *local* minima as well as a zero global minimum, and started near one of these the algorithm will move into it. It is obviously preferable to supply an initial estimate based on approximate analysis of the images using the linear theory, or a low resolution model for the specimen struc-ture (for example), than to use random or zero initial phases, but the latter course *is* viable (cf Sect.2.3.3) with a few exceptional cases [Ref.2.6, pp.111,154].

For low contrast images at least, the error criterion E can be simply and use-fully related to the rms phase error ϕ in the estimate, at least during the later stages of the iteration [Ref.2.6, p.122]: E is simply a fraction ϕ^2 of the total flux $\sum m_2^2$. Thus a phase accuracy of 0.1 rad, for example, requires E < 1%, while E < 0.1% implies a phase accuracy of about 0.03 rad. Equation (2.12) shows that, if the rms fractional modulus defect in plane 2 is f, E is simply a fraction f^2 of the total flux, so that E $\simeq$ 0.1% for example implies that the modulus values are fitted to no better than 3.2%. This in turn indicates the accuracy required in the modulus *data*, for there is no gain in fitting the modulus values more closely than they are known: an rms phase error ϕ requires a fractional error rather less than ϕ in the data moduli. Determination of all intensities to within 10% (moduli to within 5%) thus allows a phase accuracy of 0.1 rad, but a phase accuracy of 0.01 rad would require the intensities to within 1%.

2.3.3 Examples and Practical Applications

Two practical applications of the iterative transform algorithm have been made, by GERCHBERG [2.18] and by CHAPMAN [2.19] (see also [2.20]). GERCHBERG analyzed an image and diffraction pattern of a thick (60 nm), Al-U stained, catalase crystal, obtaining a low-resolution (> 2 nm) solution for the image wave function. At this resolution, object plane phase shifts would roughly follow the projected stain density (by way of specimen potential), and aberrations would have little effect, so that the image plane phase variations should be similar; the image intensity variation is produced mainly by scattering contrast, and follows the same pattern. GERCHBERG obtained an image phase which did indeed closely match the observed intensity variations. His final value for E (1.6%) indicates an rms phase error about 0.13 rad, which is at least well below the peak shifts found (about a radian); on the other hand it indicates that the observed modulus values, which showed no more than an 8-9% variation, were being fitted to no better than 13%, and were therefore contributing little. Accordingly, GERCHBERG repeated the solution using a constant modulus in place of the observed image modulus, i.e., operating between the *object* and diffraction plane, and recovered the object plane phase shifts directly — as anticipated, very similar to the image plane shifts.

The possibility thus suggested of dispensing with an image altogether for pure (strong) phase objects was taken up by CHAPMAN, who analyzed the phase variation (anything from 3-30 radians) across stripe magnetic domains in cobalt and permalloy, using as data simply the low angle diffraction pattern from the films, and exploiting the stripe character of the domains to work in one dimension. He weakened the fitting process slightly, correcting this modulus values at each stage *not* to the (noisy) observed values, but simply to within estimated experimental error of them; he was also able (and found it necessary for convergence) to supply reasonable initial estimates of the phase distributions, of the form $\phi_0 (\sin kx)^{\frac{1}{2}}$, relying on the algorithm merely for refinement. It is unfortunately on the *derivative* of the recovered phase that the magnetic flux density (the real object of interest) depends, which puts stringent requirements on the smoothness of the recovered phase that the algorithm does not wholly meet. Nevertheless, in this context the algorithm has provided a partial answer to a previously wholly intractable direct interpretation problem. CHAPMAN found, for example, that the flux density was almost uniform within each domain in cobalt, with sharp reversals at the walls, in contrast to a much more slowly varying pattern in permalloy.

No practical applications of the algorithm using the two-defocus data have been reported. However, MISELL [2.16] has made a careful study of its behavior in simple one-dimensional problems in bright- and dark-field conditions, including the effects of observational noise, inelastic background intensity, and slight misalignment between the data images. He finds it broadly reliable, except that it is noise sensitive in bright-field conditions; since his maximum image-plane phase shifts are only 0.05 rad in those circumstances, the noise sensitivity is hardly surprising. While MISELL accordingly recommends the method primarily for dark-field images, my own trials [Ref.2.6, pp.154-162] in fact suggest the contrary, the solution being found much more reliably in bright-field conditions than in dark, and some spurious solutions with a *low*, if not zero, E being found in dark-field conditions besides; the question is therefore still open. LANNES [2.9] reports successful solution using the method on the image of a dislocation dipole in both bright-field and dark-field conditions, in the course of comparing it with the method described in Sect.2.2.

Reference [2.6] also studies the image/diffraction method applied to a number of one- and two-dimensional problems in bright- and dark-field conditions. It appears to be increasingly difficult to *find* solutions as the complexity of the image grows (i.e., E does not decay so far), particularly in bright-field conditions; spurious solutions are usually (but not quite always) identifiable by a high final value of E. This deterioration as the size of the problem grows does not arise with the two-defocus method: this is probably because defocusing is a largely local process, so that two separate problems set side by side do not interfere with each other in the two-defocus method in the way they obviously do with diffraction plane data. HUISER et al. [2.21] report the performance of the image/diffraction method in small one-dimensional trials with an image phase variation between 0 and π rad; they consider these unsatisfactory because the phase recovered is not smooth, does not exactly match the actual phase distribution, and varies somewhat with the (random) initial phase estimate. However, except when their final error criterion (not precisely defined) clearly indicates convergence failure, the solutions are all within $\pi/8$ of the true phase, and a useful solution is found to a problem in which an alternative analysis of the data (see Sect.2.3.6) fails.

Figure 2.1 illustrates the image/diffraction method applied to a model problem in dark-field conditions, and Figs.2.2 and 2.3 the two-defocus method applied in bright-field conditions; the results are certainly encouraging. Table 2.1 gives the computer program used to obtain the result in Fig.2.2: this is written in "Semper", a high-level language developed for image processing at Cambridge [2.23] and may be comprehensible in outline even without any prior knowledge of the language.

2.3.4 Uniqueness

The existence of a solution to the problem posed in Sect.2.3.1 is obvious if noise is discounted, though it is likely that no *exact* solutions exist in the presence of noise. Whether the solutions are unique is an important question, however. The image/diffraction problem is discussed first; [Ref.2.6, pp.78-93] gives a fuller account.

Fig.2.1   Fig.2.2

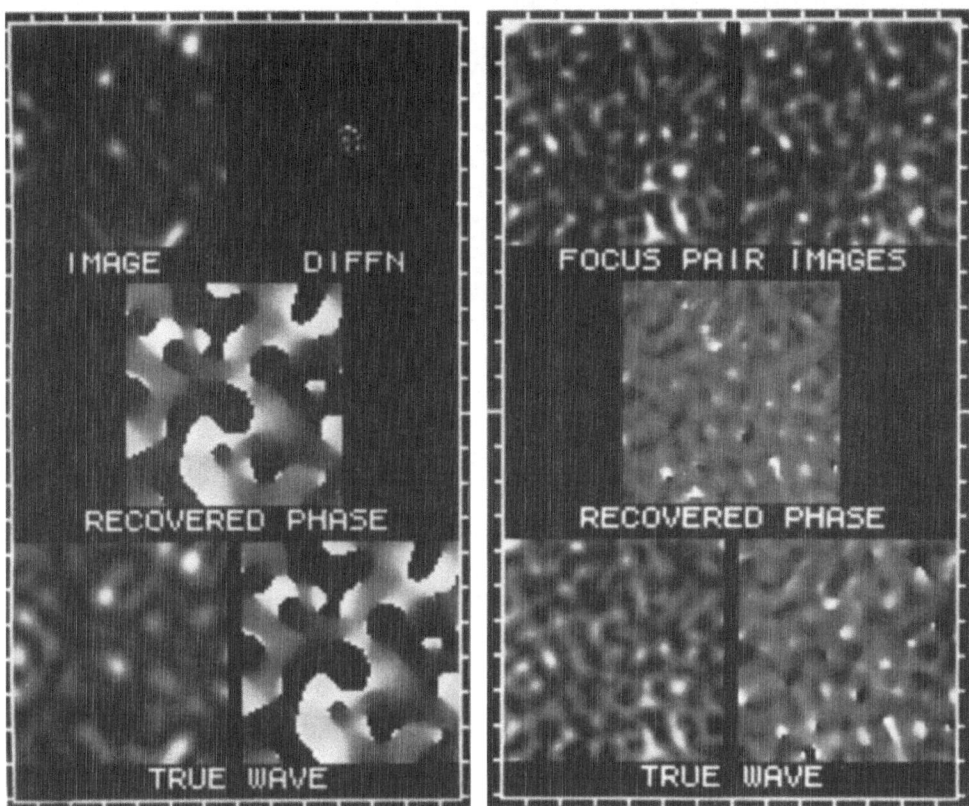

Fig.2.1. Recovery of the complex wave function from image and diffraction pattern: a worked example. *Bottom*: a hypothetical image **wave** function with 80 nonzero Fourier components, for central stop dark-field conditions, shown as modulus (left) and phase (right) —the latter covering the full range -π to π. *Top*: the corresponding image and diffraction plane intensities, serving as test data. *Center*: the phase recovered from the test data after 15 cycles of the iteration, commencing with phases up to ±π/2 from the true phase. (No solution was found to this problem from a completely random starting point)

Fig.2.2. Recovery of the complex wave function from differently defocused images: a worked example. *Bottom*: a hypothetical image wave function with 315 nonzero Fourier components, for bright-field conditions, shown as modulus (left) and phase (right)- the latter covering the range -2 to 2 rad. *Top*: the corresponding intensities in focus (left) and defocused (right) so as to shift the phase of the extreme Fourier components by 1.2 rad; both exhibit 100% contrast. *Center*: the phase recovered from the intensities after 8 cycles from zero initial phases

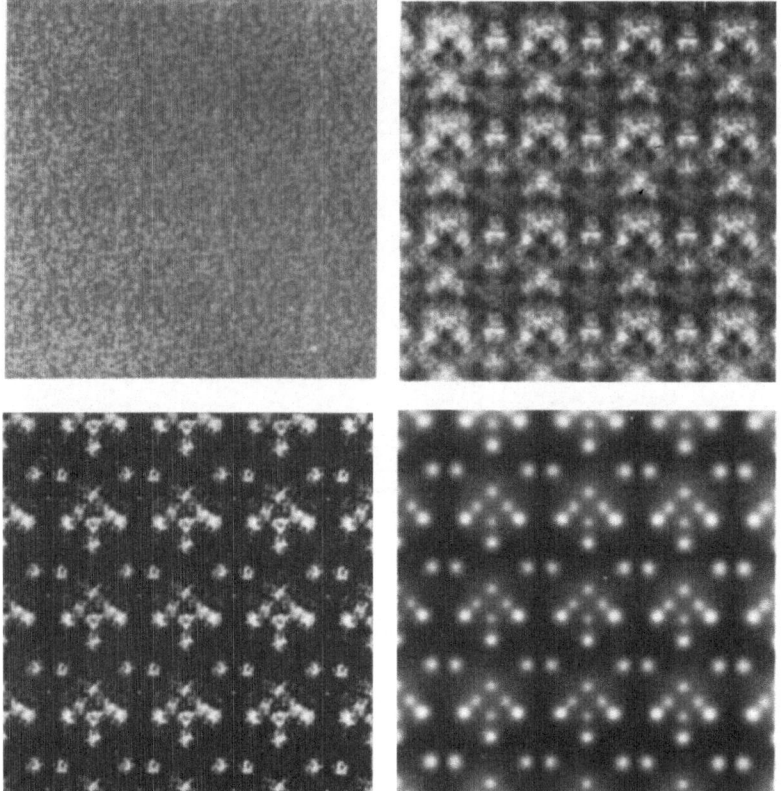

Fig.2.3. Recovery of the complex wave function from differently defocused images: a worked example with noise. The hypothesized image wave function has a constant modulus, and a phase distribution as shown bottom right, covering the range 0 to 2 rad; images at two focus levels are shown top left and top right (both as modulus rather than intensity, the true values being distorted by up to ±20% at random to simulated recording noise). The phase distribution recovered after 18 cycles from zero initial phases is shown bottom left [2.6]. (By permission of Academic Press)

Absolute phase is meaningless in any case, so all solutions are subject to an additive constant. Simple two- or fourfold ambiguities arise if the intensity distributions are centrosymmetric or mirrorsymmetric (e.g., centrosymmetry in one plane also permits the complex conjugate of any wave function solution in the other plane). Some extremely ingenious examples permitting a large degree of ambiguity have been constructed by SCHISKE [2.24], but it is not suggested that such cases (dependent on a great deal of internal symmetry) are likely in practice; the situation is perhaps analogous to the problem of "homometric structures" in X-ray crystallography, where although the number of possible homometric sets of structures grows with the complexity of the problem, the chance that a particular set of data permits ambiguous interpretation declines (see, for example, LIPSON and COCHRAN [Ref.2.25, p.139]).

Table 2.1. A "Semper" program to recover the complex wave from two images at different focus levels. (The character "C" introduces explanatory comments)

```
C AT OUTSET -
C   M1,M2 ARE FOCUS PAIR IMAGE INTENSITIES
C   E1,E2 ARE DEFOCUS/REFOCUS PHASE SHIFT FACTORS

C OBTAIN IMAGE MODULI
  ROOT M1; ROOT M2

C SET WAVE TO INITIAL ESTIMATE (UNITY)
  SELECT WAVE; INSERT 1,0
  CYCLE=0

RECYCLE:  C CORRECT MODULUS IN PLANE 1
             MODULUS WAVE TO TEMP
             SELECT WAVE; MULTIPLY M1; DIVIDE TEMP
          C AND DEFOCUS
             FOURIER; MULTIPLY E1; IMAGE
          C CORRECT MODULUS IN PLANE 2
             MODULUS WAVE TO TEMP
             SELECT WAVE; MULTIPLY M2; DIVIDE TEMP
          C AND REFOCUS
             FOURIER; MULTIPLY E2; IMAGE

C REPEAT PROCESS 10 TIMES
  CYCLE=CYCLE+1; UNLESS CYCLE>10 JUMP RECYCLE

C COMPLETE - DISPLAY FINAL WAVE
...
```

The image/diffraction method, however, fails completely when the image phase variation is *weak* (e.g., when the specimen is weak), so that the term $|v(\underline{r})|^2$ in (2.10) can be neglected: except possibly when the image intensity is very low at some points, a twofold choice becomes possible for the wave function at all points throughout half the diffraction plane [Ref.2.6, p.82]. The situation is still worse if the object plane is substituted for the image plane, as was suggested in the previous section: the scattered wave at the object is small and purely imaginary, and assigning it a completely arbitrary phase in the diffraction plane (provided it is conjugate-antisymmetric) will preserve this fact, and therefore the object plane modulus. We thus see that the image/diffraction method *depends* on the squared terms in the image intensity, which is no doubt why it is more reliable in dark-field than in bright-field conditions.

SCHISKE [2.24] has examined the problem posed by specifically discrete or periodic data, where a discrete Fourier transform is assumed between the data in the two planes, using the Galois theory of extension fields, and he finds that uniqueness is *not* to be expected in general, but *is* to be expected when only the central half of the diffraction plane values are nonzero. This is a reasonable requirement: the transform of the image intensity is twice as wide as the diffraction plane wave function (being its autocorrelation), so it merely implies adequate sampling (in the sense of the sampling theorem) of the image intensity; to phase a 0.2 nm diffracted beam, the image must be recorded at 0.1 nm intervals. Since SCHISKE is also able to demonstrate the linear independence of the image intensity values, treated as equations in the unknown diffraction plane phases, the restriction on the number of unknowns means the equations are in fact overdetermined. A conclusion similar to Schiske's is derived in Sect.2.3.5.

HUISER et al. [2.21] consider the problem under the assumption that the diffraction plane wave is continuous and differentiable within the aperture, and non-vanishing at its ends (or possibly perimeter; the work is one dimensional). While the case is not of practical relevance to electron microscopy, where as we have seen the method is limited to periodic specimens, their results are nevertheless interesting: treating $\tilde{\psi}_i(p)$ and $\tilde{\psi}_i^*(-p)$ as formally independent (the former being the diffraction plane wave function as usual), they develop a set of four coupled integral equations for the *difference* between two supposed solutions, and are able to state conditions under which the set has no solution, i.e., the original problem has a unique solution. It is sufficient if the diffraction plane intensities differ at the two ends of the aperture (p = ±a), and if

$$\mathrm{Im}\{\tilde{\psi}_i^*(a)\tilde{\psi}_i^*(-a)\tilde{\psi}_i(p)\tilde{\psi}_i(-p)\} \neq 0 \quad , \tag{2.15}$$

though the *necessary* conditions are somewhat weaker.

Whatever the theoretical uniqueness of the solution given arbitrarily accurate modulus determination, the fact that FOWLE [2.26] is able to give a method for constructing two different wave functions *approximately* consistent with arbitrary intensity distributions in image and diffraction planes (subject of course to flux conservation between the planes), and suggests via worked examples that the approximation becomes *closer* as the complexity of the problem grows, is very disturbing. His method is based on approximating the Fourier integral by the stationary-phase method and requires that there be one stationary point only, so that the phases of the functions he constructs grow rapidly and monotonically towards the outside of the functions in a manner quite unrealistic for an electron image wave function; however, a numerical algorithm not suitably constrained (as the algorithm described is not) might well find such a phase distribution. Admittedly, FOWLE's method cannot be extended into two dimensions [Ref.2.6, p.86].

Turning now to the two-defocus method, the first point to be made is that this does *not* fail in the weak specimen case, unlike the image/diffraction method: in that case, solution is possible using the linear theory of Chap.1. We may accordingly expect better performance from this method in bright-field conditions. The theoretical uniqueness problem has been examined for periodic specimens by SCHISKE [2.27], and for differentiable diffraction plane wave functions by DRENTH et al. [2.28].

SCHISKE assumes a finite diffraction plane aperture, giving well-defined outside diffraction orders, and the uniqueness of the solution follows at once from the method of solution he develops, which although differently presented is essentially the "direct" method outlined in Sect.2.3.6. The set of papers by DRENTH et al. parallels closely the set [2.21] from the same group on the image/diffraction problem, and is the more relevant to electron microscopy because the two-defocus method is not restricted to the periodic case. Under reasonable conditions on the diffraction plane wave, they find by integral equation methods similar to those used in [2.21] a unique solution provided the diffraction plane wave does not vanish at either end of the aperture, and provided the relative defocus of the two pictures is limited by the condition (2.29) deduced in Sect.2.3.6.

An interestingly different approach is given by ROBINSON [2.29] at the end of an article reviewing the uniqueness problem for both sets of data. He defines a function similar to the radar ambiguity function,

$$\chi(p',x) = \int \tilde{\psi}_i^*(p - p')\tilde{\psi}_i(p)\exp(2\pi ixp)dp \quad , \tag{2.16}$$

and notes that $\chi(0,x)$ and $\chi(p',0)$ follow at once from $|\tilde{\psi}_i(p)|^2$ and $|\psi_i(x)|^2$, while

$$\int \chi(p',x)dx = \tilde{\psi}_i^*(-p')\tilde{\psi}(0) \quad , \tag{2.17}$$

so that a projection of χ would yield the diffraction plane wave function. The point of interest is that recalculation of the function with $\tilde{\psi}_i$ replaced by $\tilde{\psi}_i \exp(-\pi i\lambda\Delta p^2)$ — cf (2.6) —yields $\chi(p',x - \Delta p')\exp(\pi i\lambda\Delta p'^2)$, so that a defocused image can be used to derive a section through χ at an angle determined by the defocus Δ. The use of a *sufficient number* of defocused images would thus appear to determine χ, and hence ψ_i, completely.

2.3.5 Periodic Images and Complex Zeros

A useful simple treatment is possible of the image/diffraction problem in the case of periodic images with finite diffraction plane apertures (i.e., *band-limited* periodic functions, with a finite number of nonzero Fourier components only); this also looks forward to some more general results given later. If the image wave function takes the form (in one dimension)

$$\psi_i(x) = \sum_{j=-n}^{n} c_j \exp(2\pi ijx) \quad , \tag{2.18}$$

we can remove a factor $\exp(2\pi inx)$ to leave a polynomial in $w = \exp(2\pi ix)$ [2.30,31],

$$\psi_i(x) = \bar{\psi}_i(w) = w^{-n} \sum_{0}^{2n} \bar{c}_j w^j \quad , \qquad \bar{c}_j = c_{j-n} \quad . \tag{2.19}$$

Choosing an origin such that $\psi_i(0) \neq 0$, this is easily recast as a product over the 2n (in general complex) zeros w_i of the polynomial,

$$\bar{\psi}_i(w) = \psi_i(0)w^{-n} \prod_0^{2n} [(w - w_i)/(1 - w_i)] \quad . \tag{2.20}$$

Thus, the image wave function is almost completely characterized by the positions of its "complex zeros". We can use (2.20) to construct other wave functions with the same number of Fourier components simply by moving the zeros around. However, the zero manipulations that preserve the image intensity $|\psi_i|^2$ are strictly limited. This is because $|\psi_i(x)|^2$, having no more than twice the number of components in ψ_i itself, has a similar polynomial representation, with 4n zeros rather than 2n, and the polynomial with the value $|\bar{\psi}_i(w)|^2$ on $|w| = 1$ (x real) is $\bar{\psi}_i(w)\bar{\psi}_i^*(1/w^*)$, therefore having zeros in pairs $(w_i, 1/w_i^*)$. One zero of each pair must appear in any new function $\bar{\psi}_i$ we construct, so the only allowable manipulations are "flipping" operations in which a zero w_i is moved to $1/w_i^*$.

This means that the image intensity alone in this case restricts the possible phase values to a set of 2^{2n} distinct arrangements — an impossibly large number still admittedly. What becomes possible now, however, is to explore the effect on $|c_j|$ of flipping particular zeros: it has been shown [Ref.2.6, pp.89-91] that moving w_i to $1/w_i^*$ multiplies $|c_n|$, and divides $|c_{-n}|$, by $|w_i|$. Since $|c_n|$ and $|c_{-n}|$ are the peripheral diffraction order moduli, and $|w_i| \neq 1$ (or the flipping is trivial), we conclude that flipping individual zeros changes the diffraction plane intensity, and that the image/diffraction problem has a unique solution in general, provided the intensity is sampled adequately ($\Delta x \leq 1/4n$): the existence of a subset of zeros for which $\prod |w_i| = 1$ is obviously a coincidence.

These conclusions are borne out by trials on small problems [Ref.2.6, pp.144-147] with a different algorithm suited only to this particular task: although multiple solutions were found several times when there was no restriction on the number of nonzero diffraction orders, no spurious solutions could be found at all when the number was halved.

Extension of the ideas here to two dimensions is *not* trivial, and is considered in Sect.2.5.5; for the present no more will be said than that they are considerably *more* powerful in two dimensions than in one.

It should perhaps be stressed that the existence of a unique solution does not mean we can necessarily *find* it: it is clear from the earlier sections that diffi- culties do arise in this respect, and the next section considers some other methods of analysis.

2.3.6 Other Methods of Analysis

There are three main directions to consider: variations on the iterative transform algorithm, the "direct" method, and the "steepest descent" method.

GASSMANN [2.32] has proposed a variation of the iterative transform algorithm, applied to the image/diffraction problem. He considers the modulus correction process in one plane as a filtering process applied to the wave function in the other plane, and seeks to optimize the filter "in a Wiener sense". Interesting as the idea is, it is difficult to reconcile it with the normal theory of Wiener filtering, as the crucial average over an ensemble of current estimates is not fully applied [Ref.2.32, Eq.(7a), or Ref.2.33, Eqs. (3a') and (3b')], resulting in a filter depending not simply on the statistical (i.e., ensemble average) properties of the estimate, but on specific properties (namely its modulus). Gassmann's results suggest multiplication of the transform not simply by a modulus correction factor, but also by a factor that approaches 1 as the modulus corrections become smaller, but is initially well below 1, particularly at low-intensity points in the wave function. This means a reduction in the total flux, which seems likely to have serious consequences for the *next* correction stage; on the other hand, the initial suppression of the low-intensity points may be beneficial, allowing phasing of the larger orders first, as in normal crystallographic practice. This variation remains untested, however; some other simple variations (under- or overcorrection, re-laxation) have been explored [Ref.2.6, pp.112-115] without any useful result.

The "direct" method is attractively simple. Let us denote the Fourier components (diffraction plane sample values) of an image wave function by c_j, as in (2.18); the image intensity may be transformed to yield their correlation,

$$d_j = \sum_\ell c^*_{\ell-j} c_\ell \quad . \tag{2.21}$$

(For simplicity only, the method is explained in one-dimensional terms.) We consider the image/diffraction problem first, in which we seek c_j given $|c_j|$ and d_j. We fix a phase origin by choosing c_n real, say; c_{-n} then follows from

$$d_{2n} = c^*_{-n} c_n \quad . \tag{2.22}$$

The next correlation value introduces two more coefficients

$$d_{2n-1} = c^*_{-(n-1)} c_n + c^*_{-n} c_{n-1} \quad . \tag{2.23}$$

Since the modulus of the two terms is known, a twofold choice is possible in general for these, and hence for c_{n-1} and $c_{-(n-1)}$, since c_n and c_{-n} are now known. Each new correlation value introduces two fresh coefficients; exceptionally there may be only one solution for them, but two or none will be the normal number. We must therefore explore every branch in turn until it encounters an insoluble equation or the equation involving d_n, which introduces c_0 for the first time and completes the solution: a suitable program is easy to write in a recursive language. However, the time required to explore all the branches is quite prohibitive, as it doubles roughly with every fresh diffraction order included! DALLAS [2.34] does not appear to be aware of this difficulty, but gives a good discussion of the exceptional circumstances that may interrupt the solution "tree" (see also [Ref.2.6, pp.95-96]). One complication is the possibility that for some j,

$$c_{-n}^* c_n + c_{-n}^* c_j = 0 \quad , \tag{2.24}$$

for in that case the equation involving d_{n+j} gives no information on the phases of c_j and c_{-j}. Excluding this possibility covers the condition (2.15), required on quite different grounds by HUISER et al. [2.21], who in fact also propose this method of solution, with the important difference that they envisage treating sample values of a continuous wave function, allowing them to apply a smoothness criterion to select one of the two choices at every stage, making the process viable, and they present a number of one-dimensional model problems solved by the method, including cases where the smoothness criterion makes the wrong choice. BATES [2.35] also proposes a variant of the direct method with a "smoothness" constraint that effectively forces three out of every four image sample points to be equal.

Applied to the two-defocus problem, the direct method ceases to suffer from solution branching. We now have two correlations,

$$d_j = \sum_{\ell} c_{\ell-j}^* c_\ell$$
$$d_j' = \sum_{\ell} c_{\ell-j}'^* c_\ell' \tag{2.25}$$

(primes denoting quantities at the higher focus value), with d_j and d_j' given, and the knowledge that

$$c_j' = c_j T_j = c_j \exp(-\pi i \lambda \Delta p_j^2) \tag{2.26}$$

— cf (2.6) — where Δ is the relative defocus of the two images and p_j the spatial frequency of the jth diffraction order; again we seek the c_j values. We begin as before at the outside

$$d_{2n} = c_{-n}^* c_n \quad . \tag{2.27}$$

We fix a phase origin by taking c_n real; moreover we now take it equal to 1, since we do not know its true modulus, and let us suppose this is r times the true modulus: the factor carries through the solution and is determined at the end. Equation (2.27) now gives a value for c_{-n}, too *small* by a factor r. The next pair of correlation values

$$d_{2n-1} = c^*_{-(n-1)} c_n + c^*_{-n} c_{n-1}$$

$$(2.28)$$

$$d'_{2n-1} = c^*_{-(n-1)} c_n T^*_{-(n-1)} T_n + c^*_{-n} c_{n-1} T^*_{-n} T_{n-1}$$

now form a pair of simultaneous equations for c_{n-1} and $c^*_{-(n-1)}$, allowing solution for these; c_{n-1} is too large, and $c^*_{-(n-1)}$ too small, by a factor r. We proceed inwards in this way until the equations in d_n and d'_n, which introduce c_0 and c^*_0: r is revealed by the ratio of the values obtained if this pair is solved as if c_0 and c^*_0 were independent. It is easy to show from (2.26) that the condition

$$\Delta < 1/\lambda p_n^2$$

$$(2.29)$$

is *sufficient* to prevent the coefficient determinant vanishing in (2.28); this re-stricts the maximum phase shifts associated with the relative defocus to $< \pi$ rad. This rather modest focus difference means that the two image intensities are quite similar, implying relatively high noise sensitivity; the condition is not *neces-sary*, given fortunate sampling, and the iterative transform algorithm at least has solved problems with greater focus differences.

Applied to the two-defocus problem, the direct method was first put forward by SCHISKE [2.27], who proposed the use of a third image at a further focus level to resolve the exceptional singular cases; DALLAS [2.36] and DRENTH et al. [2.28] sug-gest its use with sample values from a continuous wave function, the latter with a thorough analysis and several one-dimensional problems solved by the method.

While the simplicity of the direct method must certainly recommend it, it has two drawbacks. The obvious problem is its noise sensitivity: since it works serially inwards from the periphery of the diffraction plane, and each new value found de-pends on the values established earlier, errors due to noise in the data will ac-cumulate steadily. This is particularly so because the process begins with the outer-most values in the diffraction plane, where the data will be the least accurate; my own trials diverge disastrously if the outermost $|c_j|$ values (in the image/dif-fraction problem) are halved, for example. The d_j values for $0 \le j < n$, which have not yet been used, should in principle allow a reduction of the noise sensitivity, but no suggestions have been yet made as to how they could be used for the purpose. The second problem is that, even *without* the solution branching of the image/dif-fraction application, computing time rises with the *square* of the number of data points — because each fresh d_j equation examined involves the calculation of a par-tial correlation sum over the values so far established. This means the direct method will compare badly with iterative transform methods on large problems: FERWERDA [2.37] is incorrect in describing it as competitive in this respect. Reference [2.21] records the same time being required for both methods even in a

100-point one-dimensional problem. We must conclude then that the direct method will be useful only when a) the number of Fourier components contributing to the image is not too great, b) observational noise levels are low, and c) the diffraction plane has a well-defined cutoff, up to which the wave function remains significant.

Lastly, methods of the "steepest descent" (Newton-Raphson) type are considered. We suppose, as in Sect.2.3.2, observed modulus distributions m_1 and m_2 in the two planes, and a wave function estimate $\psi_1 = m_1\exp(i\phi)$ in the first; the corresponding estimate in the other plane is given by

$$\psi_2 = T\{\psi_1\} \qquad (2.30)$$

in which T denotes a transformation or defocusing operator, according to which problem is being addressed. We view the sum square residue E defined in (2.12) as a function of the phase ϕ of ψ_1, and seek a minimum of E by direct consideration of the local shape of the (hyper-)surface: we evaluate the gradient $\underline{\nabla}E = \partial E/\partial\phi$, and move ϕ in the opposite direction by subtracting

$$E\underline{\nabla}E/|\underline{\nabla}E|^2 \qquad (2.31)$$

(ϕ and $\partial E/\partial\phi$ will be vectors with N components, deriving from N sample points used to represent ψ_1). It is not difficult to show [Ref.2.6, pp.98-102,153] that the gradient takes the simple form

$$-2\ \mathrm{Im}\{\psi_1^* T^{-1}(m_2\psi_2/|\psi_2|)\} \qquad (2.32)$$

so that it can be calculated quite quickly, the computation involved being effectively that involved in a single cycle of the iterative transform algorithm. Accordingly, the repeated calculation of (2.32), and phase adjustment by (2.31), provides a promising way of converging on a zero in the residue function E. Reference [2.6], pp.96-105,123-129,153 considers convergence acceleration near the root, and a variation needed when E has a minimum only rather than a root [so that (2.31) becomes singular], and presents a number of simple problems solved from random starting points by the method. Its convergence *is* rather faster than the iterative transform method, but its stability is poorer —particularly in two-dimensional problems —and it seems most usefully applied to improve a reasonable estimate already produced by another route (e.g., by the iterative transform method).

Using only the gradient $\partial E/\partial\phi$ is a rather weak form of the steepest descent method: it would be more normal to consider the residue *vector*

$$f = |\psi_2| - m_2 \qquad (2.33)$$

instead of its squared modulus E, and use the gradient of each component of this to develop a second-order convergent iteration based on matrix inversion —the above method is only first-order convergent. However, computing time rises very rapidly with the size of the problem in this method (as the *cube* of the number of sample points), and it is not therefore feasible to use it. DRENTH et al. [2.28] found similar objections to a further method of this type (their "Newton-Kantorovich" method), in which an integral equation was developed for the (assumed small) difference between the actual wave function and the estimate ψ_1. GONSALVES [2.38] on the other hand reports successful use of a standard (Fletcher-Powell) algorithm of this type, to minimize a residue very like E by varying a *small number* of parameters involved in a polynomial approximation for the wave function phase.

2.3.7 Conclusions

Summarizing the results achieved so far with image/diffraction and two-defocus data, we may say that both sets yield a unique solution in general, but that the uniqueness sometimes depends on very accurate intensity measurement; that none of the methods of analysis so far developed can be relied on always to *find* the solution, though they will usually establish its broad features; that the computation involved is substantially greater than that involved in restorations based on the linear theory (though *iterated* application of the linear theory, as in Sect.2.2, forms an intermediate case); and that the image/diffraction data are viable only for *periodic* specimens, *strong* or imaged in *dark-field* conditions, though the two-defocus data appear useful in all circumstances.

Experience with restoring focal series of micrographs within the linear theory [2.39,40] strongly suggests, however, that practical noise levels necessitate more than an exactly determined set of data for a reliable object reconstruction. To some extent then, discussion as to uniqueness for a given set of data may be beside the point, and the real direction of future effort should be to seek a method of analyzing numerically, without excessively long computation, intensity data provided in *several* planes simultaneously. SAXTON [Ref.2.6, pp.162-167] has reported an adaptation of the iterative transform algorithm that uses three data planes—a bright-field image, a dark-field image, and a diffraction pattern—the uniqueness problem disappears there completely, but there still appear to be difficulties in *finding* the solution sometimes. The use of a diffraction pattern in conjunction with two defocused bright-field images, albeit only in the context of an iterated application of the linear theory [2.11](cf Sect.2.2) is another example of progress in this direction.

This area is also reviewed at length by MISELL [2.3], while FERWERDA et al. [2.41] give a very useful summary of their own group's work on the problem: see also FERWERDA [2.42]. HOENDERS [2.43] reviews the next stage of the reconstruction process, namely deducing the specimen structure from the wave function recovered.

2.4 Methods Using Half-Plane Apertures

2.4.1 Hilbert Transforms

This section describes how the formation of images with one half of the diffraction plane obscured by a suitably shaped aperture can be used to establish the full complex wave leading the specimen. Two such images are required in general, with complementary halves of the diffraction plane obscured, but with the primary (un-scattered) beam itself being passed in both cases; the method works when the specimen is strong, but *not too strong*.

Firstly, we establish a relation between the real and imaginary parts of an image plane wave function recorded with a diffraction plane aperture B(p,q) which is opaque for p < 0, except for a small indentation to accommodate the primary beam. We separate the real and imaginary parts of the scattered wave function ψ_s in the image plane, writing

$$\psi_i(\underline{r}) = 1 + \psi_s(\underline{r}) = 1 + u(\underline{r}) + iv(\underline{r}) \tag{2.34}$$

as in (2.9). Since $\tilde{\psi}_i(\underline{t}) = \tilde{\psi}_i(p,q) = 0$ for p < 0, we have

$$\tilde{v} = i\tilde{u} \tag{2.35}$$

over the half-plane p < 0. Since, however, u and v are real, we can use the con-jugate symmetry of their transforms to obtain $\tilde{v}$ over the other half-plane too: in general,

$$\tilde{v} = -i \, \text{sgn}(p)\tilde{u} \quad . \tag{2.36}$$

Now the inverse Fourier transform of $-i \, \text{sgn}(p)$ is $\delta(y)/\pi x$ (the forward transform is easily evaluated by contour integration); we rephrase (2.36) as a convolution

$$v(\underline{r}) = -\frac{1}{\pi} \int \frac{u(x',y)}{x'-x} \, dx' \quad , \tag{2.37}$$

so obtaining the well-known Hilbert transform integral (dispersion relation) between the real and imaginary parts of a function whose transform vanishes over one half-plane. The stroke denotes a Cauchy principal part: the integral is never evaluated directly, however, as the convolution is much more efficiently calculated by means of the filtering operation (2.36) applied to the transform of u [2.44].

The immediate relevance of this relation to electron microscopy is that u(r) is directly observable, as half the image contrast, in the weak scattering limit (cf Sect.2.2). Thus, the full scattered wave may be found for each half of the dif-fraction plane in turn, after which phase shifting to compensate for aberrations (2.5) allows complete recovery of the object plane wave function. However, this by itself is not really new, even if differently presented [2.44,45]. Imaging of weak objects with half-plane apertures is described by straightforward transfer functions, namely (for p < 0 obscured)

$$T_r(\underline{t}) = \exp[-i \ \mathrm{sgn}(p)\gamma(\underline{t})]A(\underline{t})$$

(2.38)

$$T_i(\underline{t}) = i \ \mathrm{sgn}(p)\exp[-i \ \mathrm{sgn}(p)\gamma(\underline{t})]A(\underline{t}) \quad ,$$

in which $A(\underline{t})$ denotes an envelope factor, identical with that used in the normal two-sided case [cf (1.45,46,58,59], describing the effects of imperfect coherence, and $\gamma(\underline{t})$ is the aberration function of (2.6)[1]; accordingly the normal linear theory can be used to recover the object by way of (2.11). There is in fact a close connection with holography, discussed further in Sect.2.6.1, and it was with an analysis more of that nature that HOPPE et al. [2.46] (see also MISELL [2.47]) proposed the method for reconstructing complex objects. DOWNING and SIEGEL [2.48] report such a reconstruction, with full discussion of the practical problems of the approach (the most serious of which is the introduction of further phase shifts into the diffraction plane wave by the accumulation of charge on the edge of the aperture near the beam); see MISELL [Ref.2.3, pp.233-235] for comments on their results.

The proposal by MISELL et al. [2.7] to apply the process *iteratively*, however, extended its scope at once beyond weak objects, and studies arising from this proposal have already been mentioned in Sect.2.2 [2.8-10]. In coherent conditions, however, iteration can be bypassed by an ingenious process described next.

2.4.2 Logarithmic Hilbert Transforms

The Hilbert transform relationship (2.37) depended solely on the fact that the transform of the scattered wave $\psi_s = u + iv$ vanished over the half-plane p < 0. Now if this is true of ψ_s it is also true of $\ln \psi_i = \ln(1 + \psi_s)$, provided that $|\psi_s| < 1$ everywhere, i.e., provided the scattered wave is never stronger than the primary at any point in the image; this is because the logarithm may then be expanded

$$\ln(1 + \psi_s) = \psi_s - \psi_s^2/2 + \psi_s^3/3 \ \cdots \quad ,$$

(2.39)

and the transform of each term vanishes for p < 0, because the convolution of two one-sided functions remains one sided [2.49]. Accordingly we can apply (2.37) to the real and imaginary parts of $\ln(1 + \psi_s)$ to obtain

---

1 These do not seem to have been stated previously in full, but are easily established.

$$\arg[1 + \psi_s(r)] = -\frac{1}{\pi} \int \frac{\ln|1+\psi_s(x',y)|}{x'-x} dx' \quad , \tag{2.40}$$

which allows the image wave phase to be deduced at once from its observed modulus (and hence intensity).

This method was introduced to electron microscopy by BURGE et al. [2.50] (see also [2.51] for a review), who followed an ingenious argument due to PEŘINA but corrected a serious fallacy in the original argument. MISELL and GREENAWAY [2.8] report some one-dimensional trial calculations. The requirement $|\psi_s| < 1$ is less restrictive than it might seem at first because half the wave scattered at the specimen is intercepted at the half-plane aperture, reducing the typical value of $|\psi_s|$. It will be clear, on the other hand, from (2.39) that the transform of $\ln(1 + \psi_s)$ is much wider than that of ψ_s (strictly, infinitely wide), so that closer sampling of the image intensity is needed when (2.40) is evaluated [by a transform plane filter again, like (2.36)].

2.4.3 Real Aperture Shapes

The fact that half-plane apertures must include an additional indentation to pass the primary beam means that the relation (2.36) cannot be applied to the very low spatial frequencies, though it does not destroy its validity elsewhere. Rather more serious is the probable loss of information on the line dividing the two half-planes on which (2.36) is not defined: the best course is probably to allow the aperture to extend slightly beyond $p = 0$ and accept $\tilde{v} = 0$ on the line itself. The problems are more serious for the logarithmic form, since the convolutions in (2.39) will spread into the $p > 0$ half-plane from the central indentation, destroying the strict validity of the Hilbert transform (2.40), while a loss of information on $p = 0$ will render the image wave subject to an unknown unimodular factor depending on y, the value of which affects $\tilde{\psi}_s$ everywhere, not simply on $p = 0$ [Ref.2.6, pp.68-71].

The seriousness (or otherwise) of the loss of information is illustrated in Fig.2.4, which shows the result of using the logarithmic transform with two half-plane apertured images of the same (model) specimen area as was treated earlier by the iterative transform methods (Sect.2.3.3). The only real solution to the information loss is to use more than two images. HOPPE [2.52] shows how four successive displacements of a *circular* aperture (or of the illumination direction) can be used to phase a wide area of the diffraction plane; he does not consider strong objects, but of course the logarithmic transform could be applied to each image in turn to extend the method to such objects.

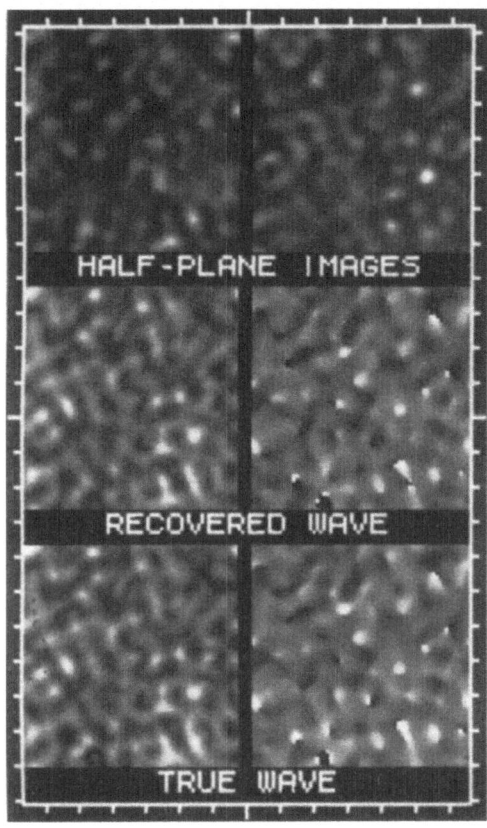

Fig.2.4. Recovery of the complex wave
function from half-plane apertured
images: a worked example. *Bottom*: the
hypothetical image wave function used
previously in Fig.2.2, once more as modulus (left) and phase (right). *Top*:
image intensities when the left- (left)
and right- (right) hand half-planes (plus
the dividing line in both cases) are obscured; both exhibit 100% contrast.
Center: the approximate wave recovered
from the two half-plane images, as modulus (left) and phase (right). The accuracy of the solution is the more sa-
tisfying since the $|\psi_s| < 1$ condition of
Sect.2.4.2 is violated by the wave function hypothesized

2.4.4 Dark-Field Conditions

The logarithmic transform (2.40) fails completely in dark-field conditions, i.e.,
if the half-plane apertures also intercept the unscattered beam so that the image
wave is simply ψ_s rather than $1 + \psi_s$: the integral does not even converge in general since the logarithm will tend to infinity (as ψ_s tends to zero) at both ends
of the x' axis. A similar integral can be used to construct one of *many possible*
wave functions consistent with the observed intensity, but a proper account of the
ambiguities involved must wait until the next section: they are associated with
complex zeros of the type discussed briefly in Sect.2.3.5. MISELL and GREENAWAY
[2.8] illustrate the breakdown with model calculations; LANNES [2.10] gives an
adaptation of the iterative method that finds at least one possible solution.

2.5 Analytic Wave Functions and Complex Zeros

2.5.1 Zero Distributions and Zero Flipping

This section takes up the ideas hinted at in Sect.2.3.5 as to the crucial role
played by the "complex zeros" of an image wave function formed with a finite diffraction plane aperture: these provide new ways of analyzing some of the methods
outlined previously as well as others to appear later. Most of the material presented is covered more fully by SAXTON [Ref.2.6, pp.44-77] and GREENAWAY [2.53].
The discussion is initially one dimensional.

An image wave function $\psi_i(x)$ formed with a finite diffraction aperture takes
the form

$$\psi_i(x) = \int_{-a}^{a} \tilde{\psi}_i(p)\exp(2\pi ixp)dp \quad . \tag{2.41}$$

The wave function is *analytic* [loosely, because it is a finite superposition of
analytic functions $\exp(2\pi ixp)$]; it can therefore be continued off the real axis
as a function of $z = x + iy$; it does not diverge as $|z|$ tends to infinity any
faster than $\exp(2\pi a|z|)$ [loosely, because $\exp(\pm 2\pi iaz)$ are the fastest varying com-
ponents present]; its behavior as $|y|$ tends to infinity is determined by $\tilde{\psi}_i$ near
$p = \pm a$ [because of the factor $\exp(\pm 2\pi ay)$ in the integrand]; as $|x|$ tends to in-
finity, $|\psi_i|$ tends to zero, its precise behavior depending only on $\tilde{\psi}_i$ near $p = \pm a$
and near any discontinuities [because when the $\exp(2\pi ixp)$ factor varies rapidly
enough, only the incomplete cycles at the discontinuities do not average to zero].
Paley and Wiener established in 1934 that *any* analytic function growing no faster
than $\exp(2\pi a|z|)$ had a Fourier transform over the real axis restricted to the
interval (-a,a).

Much as a polynomial would be, ψ_i is almost entirely determined by the positions
of its zeros z_i: Hadamard's factorization theorem gives

$$\psi_i(z) = c_1 \exp(c_2 z) \prod_i [(1 - z/z_i)\exp(-z/z_i)] \quad . \tag{2.42}$$

The number of zeros is always infinite; their density falls faster than $1/|z|$,
however, and for large $|z|$ they are evenly distributed within a finite distance
from the real axis; the product (2.42) converges absolutely, allowing us to consider infinite subsets of its factors freely.

The transform of the image intensity is restricted to the interval (-2a,2a), so
it too may be continued off the real axis as the analytic function $\psi_i(z)\psi_i^*(z^*)$;
this has zeros in conjugate pairs (z_i,z_i^*); knowledge of $|\psi_i(x)|^2$ thus determines
$\psi_i(x)$ itself up to a two-fold choice at every zero pair [and the constants c_1 and
c_2 in (2.42)]. In fact, as WALTHER [2.54] pointed out, either choice may be made
for each pair without affecting the width of the transform of $\psi_1(x)$: "flipping"

a zero in (2.42) from z_i^* to z_i is effected by an additional factor (a "Blaschke" factor)

$$(z - z_i)/(z - z_i^*) \tag{2.43}$$

which does not affect the limiting behavior of $\psi_i(z)$ as $|z| \to \infty$.

There is no ambiguity associated with *real* zeros (for which $z_i = z_i^*$), but in general there will be an infinite number of complex zeros. However, most of these are too far away to matter: if $|z_i| \to \infty$, (2.43) $\to$ constant, besides which we have already seen that the outermost zeros relate only to the edges or discontinuities of $\tilde{\psi}_i$. In practice an image requiring N sample points for reasonably close approximation (i.e., for which the product of its own and its transform's widths is N) will have only N zeros that matter. Section 2.3.5 demonstrates this clearly for the special case of periodic functions; note that with the mapping $w = \exp(2\pi i z)$ applied there, exchanging w_i and $1/w_i^*$ is equivalent to exchanging z_i and z_i^*.

Locating the zeros of the intensity is best carried out by obtaining its Fourier transform $\tilde{I}(p)$ initially, and then calculating the continuation $I(z)$ by repeating the transform integral

$$I(z) = \int_{-a}^{a} \tilde{I}(p)\exp(-2\pi yp)\exp(2\pi ixp)dp \tag{2.44}$$

for one y value after another; there is little danger of missing zeros through unlucky sampling of the z plane, because the moduli of analytic functions do not have minima except at zeros, so that any local minimum in the sample values of $|I(z)|$ indicates an adjacent zero. Figure 2.5 shows a perspective view of an intensity continuation made in this way, illustrating how clearly the zeros are defined, and how many are actually significant in practice.

The conclusion is that the restriction on the diffraction plane width alone reduces the number of wave functions consistent with the image intensity to 2^N only, all of which can be constructed directly from the observed intensity. The force of the result may be dependent in practice on rather accurate measurement of the image intensity, or alternatively on the use of a diffraction plane aperture intercepting the wave function while it is still relatively large, so that the finite-width constraint is genuinely effective.

2.5.2 An Example

It will probably clarify the concept of zero flipping to consider a specific example, simple enough to be handled algebraically. Suppose that, on forming an image with the central three beams diffracted by a periodic specimen, we observe an intensity

$$I(x) = 125 - 100 \cos(2\pi x) - 100 \sin(2\pi x) + 40 \sin(4\pi x) \quad . \tag{2.45}$$

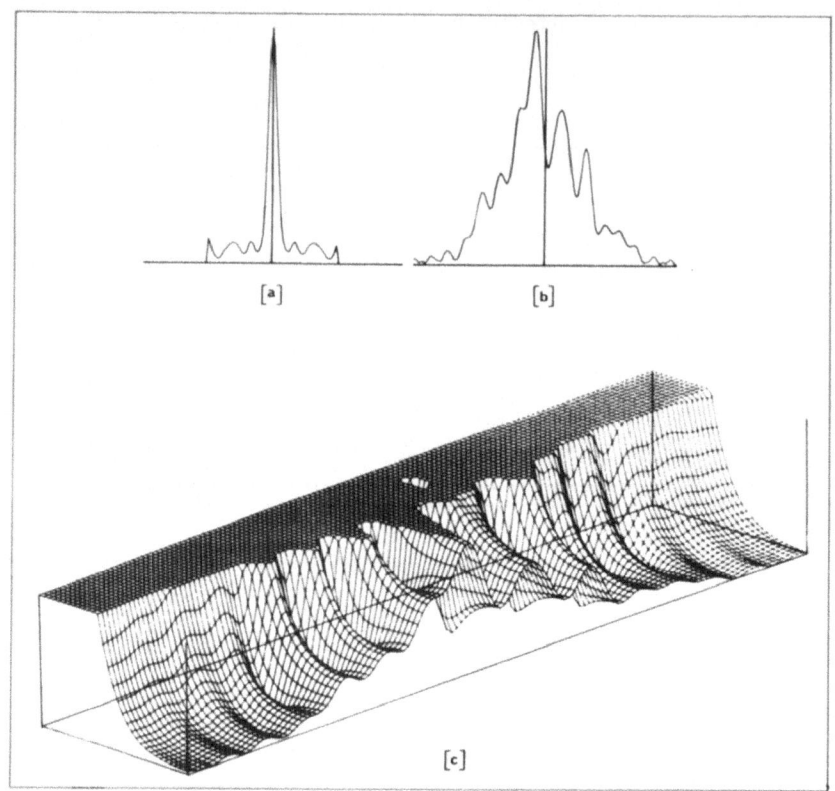

Fig.2.5a-c. Analytic continuation of an image intensity off the real axis: (a) and
(b) are corresponding diffraction and image plane intensities, and (c) part of the
squared modulus of the continuation of (b) into the upper half-plane, beginning
from the real axis. (For clarity, since the zeros are so near the real axis, the
imaginary axis is shown expanded 3 times)

In terms of $w = \exp(2\pi ix)$ —cf Sect.2.3.5—this is

$$[20i - 50(1 + i)w + 125w^2 - 50(1 - i)w^3 - 20iw^4]/w^2 \quad , \tag{2.46}$$

the polynomial having two pairs of roots for w, namely 2, ½ and 2i, ½i. We
can construct one possible wave function by taking the roots 2,2i and using (2.20):

$$\psi_{2,2i} = 5(w - 2)(w - 2i)/[w(1 - 2)(1 - 2i)] \tag{2.47}$$

$$= (8 - 4i)/w - (2 - 6i) - (1 + 2i)2 \tag{2.48}$$

$$= [-2 + 7 \cos(2\pi x) - 2 \sin(2\pi x)] + [6 - 6 \cos(2\pi x) - 9 \sin(2\pi x)]i \quad ,\tag{2.49}$$

which has diffraction plane intensities 80, 40, and 5, from (2.48). We have fixed a phase origin by taking $\psi_i(0)$ real; its modulus, 5, is simply $\sqrt{I(0)}$. Alternatively, choosing the roots ½, 2i gives

$$\psi_{\frac{1}{2},2i} = 5(w - \frac{1}{2})(w - 2i)/[w(1 - \frac{1}{2})(1 - 2i)] \tag{2.50}$$

$$= -(4 - 2i)/w + (7 - 6i) + (2 + 4i)w \tag{2.51}$$

$$= [7 - 2\cos(2\pi x) - 2\sin(2\pi x)] + [-6 + 6\cos(2\pi x) + 6\sin(2\pi x)]i \; , \tag{2.52}$$

with diffraction plane intensities 20, 85, and 20. It may be easily verified that both waves have a squared modulus matching (2.45). The two remaining possibilities are the complex conjugates of these—flipping *all* a function's zeros conjugates the function—with diffraction intensities 5, 40, 80 and 20, 85, 20 again; apart from the conjugate ambiguity in the wave function $\psi_{\frac{1}{2},2i}$ therefore, image and diffraction plane intensity measurements would clearly distinguish the various possible wave functions.

The zeros, 2,½ in w correspond to $z = n \mp i(\ln 2)/2\pi$, and the zeros 2i, ½i to $z = n + \frac{1}{4} \mp i(\ln 2)/2\pi$, so that the zeros of $I(z)$ occur in complex conjugate pairs, four to each cycle. We can generate an *aperiodic* example corresponding to the above by "spreading" each of the delta functions comprising $\tilde{\psi}_i(p)$ into abutting rectangle functions: this corresponds to multiplying $\psi_i(x)$ by a "sinc" function $\sin(\pi x)/\pi x$, and $I(z)$ therefore by $\sin^2(\pi z)/(\pi z)^2$. The sinc factor has real zeros only, which introduce no further ambiguity, so we are again concerned with selecting one zero from each of the two pairs occurring in each cycle of $I(z)$. What is different is that whereas the prior information that the wave function was periodic allowed us to consider one cycle only, the zeros in other cycles being forced to conform to preserve the periodicity, we can now flip zeros in different cycles independently; the ambiguity, if we are given no more information than the image intensity and the finite diffraction plane width, is therefore much higher, even if we note that only the zeros in the few cycles near the origin have any significant effect on $\psi_i(x)$. If the diffraction plane intensity is known to consist of three rectangles, however, the ambiguity is resolved as before, though it is more tedious to demonstrate this.

2.5.3 Immediate Applications

An application of the zero flipping theory to answering the uniqueness of the image/diffraction problem of Sect.2.3 has been included in the previous section. In fact, what has so far been established by this means is disappointing, and it is probably reasonable to hope for stronger results in future. SAXTON [Ref.2.6, pp.87-88] shows that a zero z_i^* of $\psi_i(z)$ may be moved to z_i without affecting the diffraction plane intensity only if the transform of $\psi_i(z)/(z - z_i^*)$ has a constant modulus, which is obviously a coincidence in general; however, the proof does not extend readily to the simultaneous flipping of *sets* of zeros. HOENDERS [2.55] makes the additional assumption of a finite *object* field leading to an *analytic* diffraction plane wave too; he is able to demonstrate that flipping any number of the image plane zeros results in a diffraction plane wave that diverges when continued outside the aperture, only the true solution tending to zero at both ends of the real axis.

BATES, who has pioneered practical attempts to locate and manipulate zeros, has made a number of interesting suggestions for their use. In restoring a real positive function of finite spatial extent from the modulus of its transform, he demonstrates (using a small problem with a limited number of complex zeros) how each of the possible solutions can be constructed using the ideas above, and most eliminated because they contain negative regions [2.56]; however, he accepts that constructing all possible solutions when there are many complex zeros is a prohibitively lengthy process. BATES and NAPIER [2.57] show how to identify and correct *errors* in the recorded transform phase: if the (accurately recorded) intensity and the inaccurate complex wave are both continued off the real axis, the zeros of the latter, although not quite coincident with zeros of the former, will at least indicate *which* of each conjugate pair of intensity zeros should be selected to construct an accurate complex wave. (The same suggestion is made in BATES [2.35].) Considering the deconvolution of an image $\psi_i(x)$ formed by a finite width point spread function $T(x)$ applied to a similarly limited object field $\psi_o(x)$, according to

$$\tilde{\psi}_i(p) = \tilde{\psi}_o(p)\tilde{T}(p) \quad , \tag{2.53}$$

BATES et al. [2.58] point out that since the complex zeros of $\tilde{\psi}_i$ should include those of $\tilde{\psi}_o$ *and* $\tilde{T}$, the latter of which is known, some of the noise in the recorded ψ_i can be removed by forcing back into position zeros in $\tilde{\psi}_i$ that obviously derive from $\tilde{T}$ but are displaced because of the recording noise; they also show that the deconvolution may be carried out correctly when only $|\tilde{T}|$, rather than $\tilde{T}$, is known, since the zeros of $\tilde{\psi}_i$ will identify which of each pair of conjugate zeros of $|\tilde{T}|^2$ is present in $\tilde{T}$, and present some convincing worked examples in one and two dimensions. These examples should amply demonstrate the power of the theory.

As an alternative to forming an infinite product (2.42) over half the zeros of the intensity (!), an integral of the Hilbert type can be used to construct *one* wave function consistent with a given intensity—namely the wave function for which all the zeros are in the *lower* half-plane. This is established by considering the integral

$$\int \ln \psi_i(z)/[z'^2(z' - x)]dz' \quad , \tag{2.54}$$

taken over a contour consisting of the real axis and an upper half-plane semicircle "at infinity". Since $|\psi_i(z')|$ grows no faster than $\exp(2\pi a|z'|)$ as $|z'|$ tends to infinity, and $\arg[\psi_i(z')]$ is restricted to the interval $(-\frac{1}{2}\pi, \frac{1}{2}\pi)$ because of the absence of upper half-plane zeros, the integral over the semicircle vanishes, so that the integral over the real axis is determined by the residues at $z' = x$ and 0:

$$\int \frac{\ln \psi_i(x')}{x'^2(x'-x)} dx' = \frac{\pi i \ln \psi_i(x)}{x^2} - \frac{\pi i \ln \psi_i(0)}{x^2} - \frac{\pi i \psi_i'(0)}{x \psi_i(0)} \quad . \tag{2.55}$$

Taking the real part of this gives the phase of the required wave function

$$\arg[\psi_i(x)] = -\frac{x^2}{\pi} \int \frac{\ln|\psi_i(x')|}{x'^2(x'-x)} dx' + c_1 + c_2 x \quad , \tag{2.56}$$

in which for simplicity the last two terms of (2.55) have been replaced by $c_1 + c_2 x$ [2.53,59]. Several variants of (2.56) exist—see SAXTON [Ref.2.6, pp.73-77], HOEN-DERS [2.55], or TOLL [2.60], for example; note that the integral is still a convolution, and can be evaluated as a transform product like (2.36). The wave function produced from the intensity in this way is called the "minimal phase", or "Hilbert phase" wave function, and is not usually directly useful; it gives a diffraction plane intensity relatively concentrated near $p = -a$. In general, Blaschke factors (2.43) must be applied to the wave function to raise into the upper half-plane the zeros that ought to be there; the question of course is which zeros they are, and this will be reconsidered in Sect.2.5.4.

The relation (2.56) applies equally to image wave functions formed with a half-plane aperture in the diffraction plane: variation of the constant c_2 interconverts the one-sided and two-sided cases. We can now understand the level of ambiguity applying to half-plane apertured images in dark-field conditions (Sect.2.4.4). In bright-field conditions, however, the case is different because the addition of the strong background term to the image wave function eliminates zeros from the upper half-plane, giving a unique solution: some further simplification is possible, leading back to (2.40). A generalization of this idea of removing zeros by addition of some other larger function will be considered in Sect.2.6.

As an alternative to *eliminating* the upper half-plane zeros, ROSS et al. [2.61] investigate how the zeros may be moved sufficiently far from the origin for the minimal phase to be an adequate approximation. They demonstrate, with examples, how an image wave function not initially recoverable by Hilbert transform techniques can become so if the diffraction plane is suitably apodized, e.g., multiplied by a sinc or Gaussian; this is extremely interesting work, though they no longer feel their discussion of partial coherence as a means of apodizing is helpful [2.62].

2.5.4 Reformulation of Zero Flipping

When zeros cannot be eliminated or moved away as above, it becomes necessary to consider more carefully how they might actually be located, i.e., how further information might be used to establish which of each pair of conjugate zeros in $I(z)$ is in fact present in $\psi_i(z)$. For this purpose, a reformulation due to GREENAWAY [2.53] (see also BURGE et al. [2.59]) of the effect of the Blaschke factors (2.43) is useful.

The Blaschke product $\Pi(z - z_i)/(z - z_i^*)$ taken over a subset of the N significant zeros z_i^* in the lower half-plane of the intensity continuation $I(z)$ can be easily recast in the form

$$
1 + \left[\prod_{i=1}^{N} \left(1 + k_i \frac{z_i^* - z_i}{z - z_i^*} \right) - 1 \right] \quad , \tag{2.57}
$$

in which the k_i values are 0 or 1 according to which zeros are to be included in the upper half-plane of $\psi_i(z)$. The expression in square brackets vanishes at infinity and is patently analytic except for a number of poles z_i^*; accordingly it is simply the sum of the principal parts at its poles. This gives for the Blaschke product

$$
1 + \sum_i k_i \frac{z_i^* - z_i}{z - z_i^*} \prod_{j \neq i} \left(1 + k_j \frac{z_j^* - z_j}{z_i - z_j^*} \right) \quad , \tag{2.58}
$$

or, introducing c_i and returning z to the real axis,

$$
1 + \sum_i c_i (z_i^* - z_i)/(x - z_i^*) \quad . \tag{2.59}
$$

(GREENAWAY in fact uses a different proof of this identity). Direct evaluation of $|c_i|$ shows that it *exceeds unity*, except that if k_i is zero, so is c_i: this is very significant for the subsequent determination of the c_i values from noisy data. Defining finally

$$
H_i(x) = H_0(x)(z_i^* - z_i)/(x - z_i^*) \quad , \tag{2.60}
$$

in which $H_0(x)$ is the minimal phase wave function derived from (2.56), we find that multiplying $H_0(x)$ by the Blaschke product can be re-expressed as a simple sum

$$
\psi_i(x) = H_0(x) + \sum_i c_i H_i(x) \tag{2.61}
$$

over known functions. Not the least convenient feature of this is the simplicity of its Fourier transform

$$
\tilde{\psi}_i(p) = \tilde{H}_0(p) + \sum_i c_i \tilde{H}_i(p) \quad . \tag{2.62}
$$

GREENAWAY [2.53] applies this to a variety of the phase determination problems discussed in this chapter; we shall refer to it again in Sect.2.6. For the present it is illustrated by its application to the two-defocus problem of Sect.2.3. We add a second subscript 1 or 2 corresponding to the two data planes, construct the functions H_{01}, H_{i1}, H_{02}, H_{i2}, and express the diffraction plane wave functions in the form (2.62); then the fact that one wave is $\exp(\pi i \lambda \Delta p^2)$ times the other [cf (2.6)] allows us to write

$$\tilde{H}_{01}(p) + \sum_i c_{i1}\tilde{H}_{i1}(p) = \tilde{H}_{02}(p)\exp(\pi i \lambda \Delta p^2)$$

$$+ \sum_i c_{i2}\tilde{H}_{i2}(p)\exp(\pi i \lambda \Delta p^2) \quad . \tag{2.63}$$

Since all functions in this equation are known, suitable sampling yields a set of linear equations for c_{i1}, c_{i2} which may be solved to determine which zeros are in fact in the upper half-plane. The process will not be fast, but should be less noise sensitive than the direct method of Sect.2.3.6 at least, and the *uniqueness* of the solution is easily explored because the equations are *linear*.

2.5.5 Two-Dimensional Extensions

The foregoing discussion is entirely one dimensional, reflecting the historical development of the material. While the extension to two dimensions is incompletely established as yet, it is already clear that the finite diffraction plane constraint becomes considerably more powerful in two dimensions than in one.

The essential point is that if the *two*-dimensional transform of $\psi_i(\underline{r})$ vanishes outside a finite region, so does the *one*-dimensional transform of any section through $\psi_i(\underline{r})$; accordingly any section of $\psi_i(\underline{r})$ may be continued off the real axis and expressed as a product over its complex zeros in the same way as the one-dimensional image wave function above. NAPIER and BATES [2.63] take central sections at different angles θ (see also [2.35,58]), while SAXTON [Ref.2.6, pp.55-57] considers sections at different "heights" y. It is reasonable to expect that the zero positions will be continuous functions of θ or y, and SAXTON shows that this is indeed the case, at least between any intersections there may be. Flipping a given zero therefore involves flipping the entire curve $z_i(\theta)$ or $z_i(y)$.

In this way, the degree of ambiguity does not rise at all when we make the transition to two dimensions: N choices remain N, instead of becoming N^2. Since intensity measurements in other planes will provide a further number of constraints that *is* proportional to N^2, we can expect overdetermination of the wave function in two dimensions, and relatively few uniqueness problems compared with the one-dimensional case.

2.6 Holography

2.6.1 The Linear Case

Holography applied to electron microscopy is reviewed fully in Chap.6, so the remarks here are limited to a number of cases of particular relevance to the strong object problem. The characteristic feature of holographic methods is complex wave determination, usually from a single image, achieved by the addition of a known "reference" wave to the image wave function, and usually carried out by operating on the image intensity in the transform plane, in which wanted and unwanted terms are spatially separated.

The principles are simply explained. If a tilted plane reference wave $g \exp(2\pi i \underline{t}_0 \cdot \underline{r})$ is added to the image wave function $\psi_i(\underline{r})$, the image intensity becomes

$$I(\underline{r}) = g^2 + \psi_i(\underline{r})^2 + 2 \, \text{Re}\{\psi_i^*(\underline{r})g \, \exp(2\pi i \underline{t}_0 \cdot \underline{r})\} \quad . \tag{2.64}$$

The corresponding diffraction plane wave is

$$g\delta(\underline{t} - \underline{t}_0) + \tilde{\psi}_i(\underline{t}) \quad , \tag{2.65}$$

and the transform of the image intensity is the autocorrelation function of this, namely

$$\tilde{I}(\underline{t}) = g^2 \delta(\underline{t}) + \Psi_i(\underline{t}) + g^* \tilde{\psi}_i(\underline{t}_0 + \underline{t}) + g \tilde{\psi}_i^*(\underline{t}_0 - \underline{t}) \quad , \tag{2.66}$$

where Ψ_i is the autocorrelation function of $\tilde{\psi}_i(\underline{t})$. If $\tilde{\psi}_i$ is of limited spatial extent, the same will apply to Ψ_i, so that a sufficiently large choice of $\underline{t}_0$ will prevent any overlap between the terms of (2.66), allowing immediate recovery of $\tilde{\psi}_i(\underline{t})$.

The entire phase determination problem is thus soluble from a single picture in principle; the practical difficulties are, however, considerable. The fine fringes in the image intensity [due to the linear phase factor in (2.64)] require a large electron dose for good recording definition, so it is essential to minimize $\underline{t}_0$ to avoid specimen damage; where the reference wave is produced by deflecting onto $\psi_i(\underline{r})$ a relatively undisturbed part of the image wave (e.g., where the specimen has a hole) using a biprism [2.64,65], spatial coherence requirements are obviously high; where these are eased by splitting off a reference wave from the main beam (e.g., by diffraction at a crystal as proposed by BATES and LEWITT [2.66]), and passing it too through the objective lens, $\underline{t}_0$ is again limited by the requirement that the *focus* spread effective in practice does not destroy the coherence between the reference and ψ_i. If the no-overlap condition for the terms in (2.66) is to be preserved, reducing t_0 means restricting the width of $\tilde{\psi}_i(t)$ somehow, and in consequence reducing the overall resolution: fortunately it is not always strictly necessary, and we now consider the circumstances in which it can be relaxed.

When $|g| \gg |\psi_i(r)|$, the term $\Psi_i(t)$ can be neglected: this is the linear case. The no-overlap condition reduces to requiring that $\tilde{\psi}_i(t_0 + t)$ and $\tilde{\psi}_i^*(t_0 - t)$ do not overlap, which is much less stringent: if $\tilde{\psi}_i(t)$ is restricted to a circular region of radius a, for example, $|t_0| \geq a$ is sufficient in this case, whereas $|t_0| \geq 3a$ is necessary in general.

A particular instance of this limit is worth comment. If a bright-field image wave function ψ_i is separated into unscattered and scattered waves $\psi_i = 1 + \psi_s$ — cf (2.34) — we may regard it as the superposition of a reference wave ($g = 1$, $t_0 = 0$) on the scattered wave ψ_s; (2.66) becomes in the present limit

$$\tilde{I}(t) = \delta(t) + \tilde{\psi}_s(t) + \tilde{\psi}_s^*(-t) \quad . \tag{2.67}$$

If the imaging is performed with a half-plane aperture, so that $\tilde{\psi}_s$ fills half the t plane only, $\tilde{\psi}_s(t)$ and $\tilde{\psi}_s^*(-t)$ do not overlap, and ψ_s may be recovered directly from I. This is the holographic description of the half-plane aperture method of Sect.2.4.1; imaging with a displaced circular aperture, or with the illumination tilted to near the edge of a central aperture, can be understood in almost identical terms — indeed the contrast transfer functions for tilted illumination modes may also be explained in substantially the same way, an effective aperture due to limited coherence replacing a physical aperture at high resolution [2.67].

It is interesting to see the holographic principle applied by BALDWIN and WARNER [2.68] to a simple problem in aperture synthesis radio astronomy.

2.6.2 The General Case

Even when Ψ_i in (2.66) is not negligible, it is sometimes possible to use holography with $|t_0|$ much smaller than is required for a purely holographic analysis, dependent on spatial separation of the terms in (2.66): Greenaway [2.53] (see also [2.59]) has shown how the ideas of Sect.2.5 can be applied to the problem. The minimum requirement is simply that t_0 lies outside $\tilde{\psi}_i(t)$ in such a way that a straight line can be drawn between them.

Firstly, suppose $|g|$ to be greater than $|\psi_i(r)|$ (but not *much* greater, which is what is required for Ψ_i to be negligible). Then any section through g exp($2\pi i t_0 \cdot r$) + $\psi_i(r)$, in a direction normal to the line separating t_0 from $\tilde{\psi}_i(t)$, has a continuation off the real axis that is free of zeros in the upper half-plane, and (2.56) — or more simply (2.40) — can be used to calculate its phase at once. This provides a holographic view of imaging *strong* objects with half-plane apertures. Iterative application of the linear theory is an alternative solution, where the process converges (cf Sect.2.4.1).

Secondly, even when $|g|$ is not so large as to eliminate zeros from the upper half-plane, it may be possible to *locate* them using the technique of Sect.2.5.4 provided there is a finite gap between t_0 and $\tilde{\psi}_i(t)$. For each section now, the minimal phase function H_0 is calculated from (2.56), the zeros of the intensity are located as in Sect.2.6.1, and the functions H_i calculated from these; then (2.62) is applied, sampled at a number of points between t_0 and $\tilde{\psi}_i(t)$, where the left-hand side is zero, to yield a set of linear equations that determine the coefficients c_i and hence which zeros of H_0 are to be moved to the upper half-plane to complete the solution for the complex image wave; $\tilde{\psi}_i(r)$ is obtained simply by subtracting the reference wave.

That the solution in this second case is in fact unique (which is admittedly not to say that the equations are well conditioned) follows from a proof given by GREENAWAY [2.69] for a related problem (cf Sect.2.6.3).

2.6.3 Some Particular Cases

BATES and LEWITT [2.66] have recently suggested obtaining a holographic reference by using a crystalline film to support the specimen, and a special diffraction plane aperture which passes a circular region around the center and one high-order diffracted beam to serve as a reference beam; they have called this technique "crystoholography". A similar proposal is made by HIDAKA [2.70]. Provided a reasonably strong diffracted beam can be found (to ensure adequate signal-to-noise ratios) this should indeed allow complete determination of the complex wave function within the central part of the aperture, after which aberration correction as usual will recover the object plane wave function $\psi_0(r)$ to the resolution the aperture permits. This wave function will of course include the crystal projection as well as that of the specimen under study, but there is no difficulty here in principle since the scattering due to the crystal can be assumed known and can therefore be subtracted from $\psi_0(r)$; alternatively, it could well be a faulted region of the crystal itself which *was* the specimen.

BATES and LEWITT propose that the reference beam be sufficiently far away for all the terms of (2.66) to be separated; this does restrict the method significantly given present instrumental stabilities (cf Sect.2.9.1). An alternative way to those of the previous section in which the reference beam may be placed less far away is suggested by GREENAWAY and HUISER [2.71], who propose recording a second image *without* the reference beam. This yields $|\psi_i(r)|^2$, and hence $\Psi_i(t)$ on transformation, allowing this troublesome term to be removed completely from (2.66) by subtracting the two recorded images.

FRANK's [2.72] bright-field/dark-field method is in fact very like this — a pair of images is recorded with the unscattered beam passed as a reference in one and intercepted at a central stop in the other. This is the previous case with the reference beam centrally placed instead of to one side; the scattered wave overlaps its conjugate in the intensity transform (2.67) and cannot be uniquely established from these data. As FRANK points out, the data fix $|1 + \psi_s|$ and $|\psi_s|$, thereby allowing a twofold solution for ψ_s at all points, though in *principle* a restriction on the width of the diffraction plane, leading to analyticity in ψ_s, will allow continuity criteria to be applied to the choice at each point, eliminating all but two solutions (ψ_s and ψ_s^*).

If FRANK's central stop is moved to one side, the process can be made viable once more: LOHMANN [2.73] moves the stop towards the outside of $\tilde{\psi}_i(\underline{t})$ — to $\underline{t}_1$, say — and again requires images with and without the stop. If this stop is small, and has area A, inserting it is approximately equivalent to adding a reference wave for which $g = -A\tilde{\psi}_i(\underline{t}_1)$ and $\underline{t}_o = \underline{t}_1$. The image without the stop identifies Ψ_i as before; the remaining (linear) terms in (2.66) overlap slightly, but the bulk of $\tilde{\psi}_i(\underline{t})$ can be recovered at once, except for uncertainty as to $\tilde{\psi}_i(\underline{t}_1)$. A third image must be recorded, with the stop moved off axis in the opposite direction, to recover the remaining part, subject to a similar uncertainty, which can now be resolved by comparing the common portions of the solutions obtained and normalizing to match the total flux in $\psi_i(\underline{r})$, say. The idea of *removing* part of a wave to serve as a holographic reference beam is certainly ingenious; it is obviously because such a reference cannot of course be moved outside the wave that complete solution is not possible without a second "stopped" image. The reference produced in this way is likely to be weak however, leading to very low signal-to-noise ratios when the Ψ_i is subtracted; it is important to note therefore that considerable freedom is possible in the precise placing of the spots, allowing any strong local peaks in $\tilde{\psi}_i(\underline{t})$ to be exploited.

Lastly, GREENAWAY [2.69] has demonstrated that the existence of an asymmetric finite gap in the transform of a wave, such as is achieved by Lohmann's stops, is in fact a strong enough constraint for the wave's phase to be uniquely determined, "in general", from its intensity (at least in one dimension). The proof is based on a formulation of zero flipping like that of Sect.2.5.4, and only requires comment here because it is incorrectly set out in the reference. A formula valid only for flipping zeros initially all in the same half-plane [Ref.2.69, Eq.(6)] is applied to explore possible flipping in either direction; on correction, the summation of Eq. (9) of the reference is split between H *and* G (GREENAWAY's notation, reassuringly symmetric; the uniqueness of the solution nevertheless follows under the conditions stated, and is certainly an impressive result. The problem of *retrieving* the solution economically, however, is not yet solved.

2.6.4 Nonplanar Reference Waves

For simplicity, the discussions above have assumed a simple plane reference wave throughout, and it should be noted here that there is no such restriction on the method in principle (e.g., BATES [2.35]). If the point reference $g\delta(\underline{t} - \underline{t}_0)$ of (2.64) is replaced by $\tilde{g}(\underline{t})$, (2.66) becomes

$$\tilde{I}(\underline{t}) = G(\underline{t}) + \Psi_i(\underline{t}) + c(\underline{t}_0 + \underline{t}) + c^*(\underline{t}_0 - \underline{t}) \quad , \tag{2.68}$$

in which G is the autocorrelation function of $\tilde{g}$, and c the cross correlation of $\tilde{g}$ and $\tilde{\psi}_i$. Again it is obvious that for sufficiently large $\underline{t}_0$ the terms are spatially separated; the term $G(\underline{t})$ is known in any case and may overlap the others without loss of information.

Similar improvements in the minimum $\underline{t}_0$ acceptable can be achieved by the use of zero location technqiues. If a minimal phase solution is to be true, however, (i.e., if zeros are to be eliminated from the upper half-plane) additional conditions on the reference become necessary; its continuation must be zero-free as well as stronger than the wave being measured.

2.7 Ptychography and Related Methods

This section considers briefly some methods for relatively direct measurement of the phases of the diffracted beam from a perfect crystal. These rely on causing neighboring diffraction orders to overlap and interfere in such a way as to reveal their relative phase.

HOPPE has given the name "ptychography" to the simplest (conceptually) of these. In the original proposals [2.74,75], a very small area of the specimen (a few unit cells) is illuminated coherently by a beam with a well-defined profile $d(\underline{r})$, the transform of which $\tilde{d}(\underline{t})$ is a uniformly bright disc; the diffraction plane then contains the disc $\tilde{d}(\underline{t})$ repeated at every reciprocal lattice site, with the modulus and phase of the corresponding diffracted beam. Thus around two neighboring reciprocal lattice sites $\underline{t}_n$ and $\underline{t}_n + \underline{g}$, where the wave diffracted by an infinite lattice would be

$$c_n\delta(\underline{t} - \underline{t}_n) + c_{n+1}\delta(\underline{t} - \underline{t}_n - \underline{g}) \quad , \tag{2.69}$$

convolution with $\tilde{d}(\underline{t})$ leads to

$$c_n\tilde{d}(\underline{t} - \underline{t}_n) + c_{n+1}\tilde{d}(\underline{t} - \underline{t}_n - \underline{g}) \quad . \tag{2.70}$$

Now if the diameter of the disc $\tilde{d}(\underline{t})$ just exceeds $|\underline{g}|$, the two discs will overlap, and in the overlap region the intensity will be simply

$$|c_n + c_{n+1}|^2 \quad ; \tag{2.71}$$

since $|c_n|$ and $|c_{n+1}|$ are known from the infinite lattice diffraction pattern, this gives a twofold choice at once for the relative phase of c_n and c_{n+1}. The twofold ambiguity is resolved by recording a further diffraction pattern with $d(\underline{r})$ moved a small distance (less than a unit cell) $\underline{r}_o$ laterally relative to the specimen: this introduces a linear phase factor in $\tilde{d}(\underline{t})$ so that the overlap region has instead an intensity

$$|c_n + c_{n+1} \exp(2\pi i\underline{g} \cdot \underline{r}_o)|^2 \quad . \tag{2.72}$$

As HOPPE points out, the choice of disc profile for $\tilde{d}(\underline{t})$ is not essential, and the light-optical simulations reported in [2.74] in fact use rectangle and disc profiles for $d(\underline{r})$ instead; although strictly the extended nature of $\tilde{d}(\underline{t})$ then means that *all* diffraction orders contribute to (2.70) or (2.71), only the neighboring orders contribute significantly. The real problem with the method is the very substantial difficulty encountered in producing the assumed illumination conditions with a conventional microscope; the STEM is much better equipped in this respect, though it is correspondingly less convenient for recording the diffraction pattern, and SPENCE [2.76] has recently outlined the necessary practical arrangements. Alternatively, simple defocusing of the CTEM diffraction pattern, though more difficult to analyze, should reveal the information in much the same way.

Among variants on the method, HOPPE also considers an extension to aperiodic specimens, using the device of a "virtual lattice" of which only one cell is illuminated. This clearly *not* viable, as Sect.2.5 showed how to construct large numbers of phase distributions consistent with such data.

A very interesting variation of the ptychographic approach has been put forward by BERNDT and DOLL [2.77]: far from requiring stringent coherence conditions, these authors positively require an extended *incoherent* source to reveal the relative phases of neighboring diffracted beams. The effect they observe, moreover, depends on the quadratic phase factors present in (2.1) and (2.2), which we have been able to neglect until now. They begin with a beam-splitting crystal in the illumination system, which produces many coherent diffracted beams of which two are selected by a suitable aperture. We treat the beam splitting as an imaging process with $\psi_o(\underline{r}_o) = \delta(\underline{r}_o - \underline{r}_s)$ and with a grating $g(\underline{r}_a)$ included in the diffraction plane, in which case applying the propagation formulae (2.1-2) gives at once

$$\psi_i(\underline{r}_i) = \exp(\pi i r_i^2/\lambda d)\tilde{g}(\underline{r}_i/\lambda d + \underline{r}_s/\lambda f) \quad . \tag{2.73}$$

If we select two beams only from $\tilde{g}(\underline{t})$, namely those at $\underline{t}_n$ and at $\underline{t}_n + g$, we can use (2.73) to evaluate their phase difference: it proves to depend linearly on $\underline{r}_s$, through a term

$$2\pi d\underline{g} \cdot \underline{r}_s/f \qquad\qquad\qquad\qquad (2.74)$$

arising from the difference of the two quadratic phase factors.

If these two diffraction spots are used in their turn as the *source* in producing a diffraction pattern from the crystal to be studied, two copies of the diffracted wave are produced, laterally displaced by an amount proportional to $\lambda d\underline{g}$. With suitable adjustment of the magnification in the beam-splitting system, it can be arranged that neighboring beams diffracted by the specimen are exactly superposed, in which case they interfere. Now the resulting intensity depends of course on the relative phase of the diffracted beams, which is what we wish to determine, but it depends also on the relative phase of the two sources, and this in turn depends on $\underline{r}_s$. As $\underline{r}_s$ varies, therefore, the final superposition diffraction pattern not only moves with it, but also varies sinusoidally in intensity, with a period $f/d\underline{g}$, referred back to $\underline{r}_s$ — cf (2.74). When the final step is taken of replacing the point source $\delta(\underline{r}_0 - \underline{r}_s)$ by an extended incoherent disc source, therefore, the superposed diffraction spots spread to form discs crossed by *fringes* the lateral position of which can be used to reveal the relative phase of the spots superposed. BERNDT and DOLL have reported both light-optical and electron optical demonstrations of the method [2.77].

Note that in both of the above methods a second experiment is necessary to relate the phases of neighboring *rows* of spots; this is probably advisable in any case to limit the accumulation of errors along the rows.

MADSEN and COTTERILL [2.78] suggest superposing in the specimen plane the crystal to be studied and a known reference crystal with similar unit cell dimensions; selecting, with a small diffraction plane aperture, a pair of nearby beams, one from each crystal, a dark-field image crossed by (moiré) fringes is obtained, the lateral position of which reveals the relative phase of the two spots. The process has to be repeated for every diffracted beam, so is likely to prove very laborious.

2.8 Bright-Field/Dark-Field Subtraction

The possibility of subtracting a pair of images recorded under identical conditions, except for the interception in one case of the primary (unscattered) beam at a small diffraction plane stop, has been discussed before (e.g., [Ref.2.51, pp.88-91]), as a means of minimizing the effects of inelastic scattering (cf Sect.2.9.2), but the full implications do not appear to have been realized. There certainly *are* practical difficulties in obtaining the data assumed, but the possible benefits seem likely to justify considerable effort in that direction. Accordingly, this survey of methods for handling images outside the linear theory ends with an examination of a method that may become important in the future.

Section 2.2 has already discussed the recovery of the object plane wave from the *real* part $u(\underline{r})$ of the image plane wave; obtaining $u(\underline{r})$ from the image intensity involved removing the nonlinear terms $|u(\underline{r})|^2$ and $|v(\underline{r})|^2$ in (2.10), and Sect.2.2 considered iterative approaches to this. Now the dark-field image corresponding to (2.10) is simply

$$I_d(\underline{r}) = |u(\underline{r})|^2 + |v(\underline{r})|^2 \qquad\qquad (2.75)$$

so that, as BURGE points out, subtracting this from the bright-field image (2.10) yields $1 + 2u(\underline{r})$ at once, without any need to iterate, and the object plane wave can be easily recovered from this through the use of the linear theory (2.11), if *more than one* such image pair is recorded.

This provides therefore an extremely simple way (mathematically) of recovering the object wave for arbitrarily strong objects. Since the solution is noniterative, there is no limit imposed on the relative size of the squared terms in (2.10) to ensure convergence. Since the full power of the linear theory is available, we may use two bright-field/dark-field pairs at different defocus values, two pairs recorded with half-plane apertures, or two pairs recorded with opposite illumination tilts, giving considerable scope for compromise with other factors (cf Sect.2.9); for the same reason we may combine the information in *more* than two pairs to reduce the sensitivity of the reconstruction to recording noise, and we can in any case optimize the treatment of noise by the use of least-squares (Wiener) filters to solve (2.11). Finally, and again because the linear theory is invoked, we can allow properly for the effects of limited coherence conditions on the recorded images (cf Sect.2.2). These are extremely strong advantages, to which must be added Burge's other point, that the inelastic image is also largely removed in the bright-field/dark-field subtraction.

The practical difficulties include reproducing the same focus level with and without the beam stop, avoiding charging of the beam stop with consequent distortion of γ (2.6) in the dark-field case, and the difficulty with radiation-sensitive materials of obtaining the many images needed (at least four, of which two are dark field). These are at least not insuperable, and the simplicity of the method must recommend it in spite of them. Perhaps the best experimental arrangement will prove to be the use of tilted illumination, with a slight movement of the (circular) diffraction plane aperture to change between bright- and dark-field conditions.

2.9 Other Perspectives

2.9.1 Coherence

This section reconsiders the foregoing methods in the light of the most serious
practical difficulties, and examines some other constraints not so far considered.
For brevity, detailed references to original papers are *not* always given here, but
they can be found in other reviews, e.g., MISELL [2.3], SAXTON [2.6], if not elsewhere in this volume. The limited coherence of the imaging system is treated first.

While a reasonably concise description of the effects on the image of the coherence conditions is possible within the linear theory (cf Chap.1), the same is
not true in the general case. It remains a simple matter to write down general expressions for the image intensity, involving suitable extensions of the linear theory
envelope factors but even computing the intensity from them is a lengthy process.
However, a simple physical picture can yield a number of useful results.

We are concerned, in assessing the effects of the limited coherence, to average
the image intensity over a range of illumination directions and a range of focus
levels. Now Fig.2.6 shows sections through the wave aberration function γ of
(2.6) for three defocus values near the "Scherzer" focus. We take a particular component of the *linear* terms in the image intensity, formed by interference between
the primary (unscattered) beam and a scattered beam at $\underline{t}_0$, say, and suppose axial
illumination, so that the primary beam passes through the center of the diffraction
plane. We can see at once how focus fluctuations cause a fluctuation in the relative
phase of the two beams, so that the corresponding image fringe system moves to and
fro laterally, and disappears completely if the fluctuations are large enough; the
tolerable level of focus fluctuation obviously decreases sharply with t_0, i.e., with
the resolution sought. Varying the direction of the illumination is equivalent to
moving both beams laterally across the figure, and the phase fluctuations arise
because of the different *slopes* in γ at the origin and at t_0, so that if t_0 falls
at the minimum of the γ surface, the spread of illumination directions does not affect the image intensity at all, but a rapid fading is observed as t_0 grows beyond
this. Paradoxically, the range of directions tolerable decreases sharply with the
resolution sought, because of the increasing slope of γ, i.e., a progressively
larger coherence area is required to image progressively *smaller* details; this makes
clear the dangers of a loose treatment based simply on reference to coherence areas.

These effects currently limit the linear transfer of specimen information to the
image to a resolution of around 0.3 nm in good 100 kV microscopes. The difficulty
in grasping the effects of the illumination divergence and focus spread in the
general case arise because the squared terms in the image intensity contain contributions from all possible pairs of positions in the diffraction plane: the mutual
interference of two beams equidistant from the axis, e.g., at A,A' in Fig.2.6, is
not affected by focus fluctuations, for example, because both beams fluctuate in

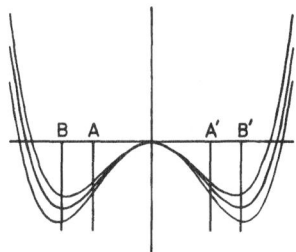

Fig.2.6. The wave aberration function γ for $\Delta = 0.9$, 1.0, and 1.1 (see text for details)

step, and beams at B,B' (on the "achromatic circle") are not even affected by variations in illumination direction, because γ has a zero slope there. Thus an image can often be obtained showing fringes twice as fine as the linear theory will permit, if the scattering is strong enough for the squared terms in the image intensity to be significant. This is particularly true if the illumination is *tilted*, because the *primary* beam can then be one of the two interfering, and the linear terms too can include contributions to much higher resolution than in the axial mode, albeit only to one side of the transform plane.

There are important consequences for at least two of the reconstruction schemes mentioned previously. Firstly, it is wasteful to use axial illumination with half-plane apertures, as suggested in Sect.2.4: if one half of the diffraction plane is being discarded anyway, it is clearly more sensible to use a normal central circular aperture and tilt the primary beam towards its edge, as proposed by HOPPE [2.52]. Secondly, it is unwise to use a well off-axis reference beam, as in BATES' and LEWITT's [2.66] crystoholography, to interfere with a central region of the diffraction plane (or the other way about, as [2.70] suggests): to place the reference beam and the area being studied symmetrically either side of the optical axis will reduce the coherence requirements enormously.

In the special case when the *object* plane is used as one data plane (cf Section 2.3.3), the image/diffraction method is little affected by the coherence problem, but otherwise the methods of Sects.2.3-5 are incapable of treating it properly, and cannot safely be used therefore at high resolution. The only methods at present capable of making satisfactory allowance for limited coherence conditions are those that rely on extracting *linear* terms from the image intensity, since the effect of the coherence on *these* is well understood and can be compensated (cf Chap.1): such methods include iteration of the linear theory solution (Sect.2.2), holography (Sect.2.6) with a reference sufficiently distant for all the terms in (2.66) to be distinct, and the bright-field/dark-field subtraction method (Sect.2.8).

The method of BERNDT and DOLL (Sect.2.7), which *depends* on a substantial illumination divergence, clearly challenges the assumption that high coherence levels are essential to high-resolution reconstruction, but remains at present an isolated case.

2.9.2 Inelastic Scattering

Electrons scattered inelastically at the specimen contribute to the image mainly as a low resolution (> 2 nm) background, since the energy losses involved are such that the inelastic image is heavily defocused under normal conditions, and often contains a wide range of energies (leading to a wide focus spread) besides. The scattering process itsèlf is relatively delocalized, compared with the elastic process, though less so in disordered materials, so even the indirect contribution made by inelastic scattering through *removal* of electrons from the elastic beam may not be a serious nuisance; this contribution has not yet been thoroughly assessed, however.

The proportion of electrons scattered inelastically exceeds that scattered elastically only for specimens of low atomic number. The inelastic image is therefore a problem relevant only to low-resolution imaging of light materials, such as unstained or lightly stained biological materials. MISELL [Ref.2.3, p.264] asserts that the problem does not arise for periodic specimen either, where the diffraction plane is sampled only at reciprocal lattice sites; it seems improbable, however, that the inelastic image of such a specimen will not share the periodicity of the specimen.

Apart from looking forward to the widespread advent of energy-filtered images, we can note some cases in which this problem is greatly reduced. It is generally true of holography, for example, that the inelastic image will matter little, because it will contribute only near the center of the image transform (because of its low-resolution character), and while it is not easily separated from the autocorrelation term Ψ_i of (2.66), it does not affect the linear terms from which the reconstruction is carried out.

On the ground that the inelastic image does not change rapidly near the normal (elastic) focus, and is substantially the same in bright- and dark-field conditions (because there is no primary wave at the same energy[2]), the subtraction of image pairs at comparable focus levels, in bright- or dark-field conditions has been proposed (e.g., [2.51]) as a means of eliminating the inelastic image: one application of this is GREENAWAY's and HUISER's [2.71] variation of crystoholography (Sect.2.6.3), in which the subtraction that eliminates the unwanted autocorrelation term also largely eliminates the inelastic image. Another is the bright-field/dark-field subtraction method of Sect.2.8, applicable to a wide range of imaging modes.

---

2 Whether this is in fact the case must be questioned in view of certain features of diffractograms from tilted illumination micrographs, which appear to require a *linear* inelastic image for their explanation (e.g., [2.79]).

2.9.3 Recording Noise and Radiation Damage

Noise sensitivity is a more important property of a reconstruction method in electron microscopy than it is in many other fields, because for some specimens, though
by no means all, a fundamental limit appears to be set on the noise level attainable at a given resolution, by the fact that the specimen is damaged by inelastic
scattering events, putting a limit on the total number of electrons available to
form images before the specimen is damaged beyond use.

It should be noted here, therefore, that methods requiring two images are more
restrictive, in terms of the specimens to which they can be applied, than those
requiring one; experience of reconstructions within the linear theory suggests that
more than two may often be necessary, which is still less promising. The only cases
in which damage-sensitive materials can yield images with "arbitrarily" low noise
levels are those in which the noisy images of many like specimens can be averaged
— which is simple of course only when the specimen is periodic.

It may be expected in general that half-plane aperture (or tilted illumination)
methods will be less sensitive to noise than defocus methods, because the linear
theory transfer functions are so much more favorable. Methods producing a solution
serially, working from point to point (e.g., the "direct" method of Sect.2.3.6,
and ptychography) will be relatively seriously affected by noise. The logarithmic
forms of Hilbert transform integral (2.40) and (2.56) are very noise sensitive
where the image intensity is locally small.

If the recording noise is attributed wholly to the electron statistics, the
signal-to-noise ratio for a given electron dose at the specimen differs by a factor of $\sqrt{2}$ only between bright- and dark-field conditions (in favor of bright).
In practice, unless the specimen is very strong, the much lower intensity in a
dark-field image necessitates a considerably higher specimen dose to produce a
detectable signal on a photographic plate, so dark-field images must be considered
more expensive than bright, in this respect. Diffraction patterns, on the other
hand, are relatively inexpensive, because the scattered electrons are concentrated
in a few well-defined directions.

Holographic methods, allowing reconstruction from a single image, appear at
first sight well suited to minimizing the specimen exposure. This is not in fact
the case, because in relying on a widely spread out image transform the methods
imply higher than usual resolution in the recorded image intensity, which in turn
requires a larger number of electrons for adequate definition of the fine detail.

2.9.4 Practical Details of Computer Processing

There are three aspects needing consideration at least, namely the alignment of
image pairs, the assessment of the wave aberration function γ of (2.6), compensation
for which has been assumed as a final step in most of the methods described in the
chapter, and the time required to carry out the whole process.

Alignment of *linear* images by correlation methods is a well-established technique by now (cf Chap.5), and capable of great accuracy. The effect of differing imaging conditions between the two images is understood, and does not normally prevent alignment; where they cause serious distortion of the correlation peak (as with half-plane aperture pairs, for example), the situation can be remedied by correcting the images, even very roughly, for the imaging conditions in question before attempting correlation. The question is how far the correlation method, dependent on matching the linear parts of two images, will succeed in the general case.

Where both images contain linear as well as squared terms (e.g., in the two-defocus method), correlation may be expected to succeed by matching the linear terms, whether the squared terms match or not. Where the squared terms only are present (e.g., in dark-field conditions), experience shows that correlation succeeds for moderate differences in imaging conditions at least, though a proper analysis of the effect is not easy. What seems likely to be troublesome is attempting to correlate one image in which the linear terms dominate with another in which the squared terms dominate (e.g., bright-field and dark-field images of a *weak* object), as the positive and negative regions in the linear terms will both tend to be positive in the squared terms, leading to a general cancellation.

For the assessment of the wave aberration function, it appears to be indispensable to have, near the specimen area to be analyzed (or under it, as a support film), a relatively thin disordered region with light atoms only; the diffraction pattern of the image intensity (diffractogram) for such an area reveals the squared modulus of the weak phase object transfer function for the imaging mode in question, e.g., as the characteristic "Thon" rings for the normal axial mode. Worse, it appears necessary to use bright-field conditions: no such distinctive effect occurs with dark-field images, so that focus assessment for this must be based either on a bright-field image recorded subsequently without refocusing, or simply an accurate stepping of the lens current from a reproducible reference focus, if one can be established. In bright-field conditions at least, it appears to be possible to assess any additional phase shifts introduced into the transform through the charging up of apertures (e.g., in the half-plane aperture methods) by superposing the diffractograms of two images [2.48,80]. The problem of assessing γ with tilted illumination, when few or no Thon rings are visible, is also quite tractable [2.81].

Processing time merits some discussion if only because the serious limitations it imposes have not always been considered in the past. In spite of the wide variations between different computers and different algorithms, a useful comparison is possible on the basis of how rapidly the number of operations to be performed grows with the number of data points in the problem, N.

The fastest procedures one could hope for would be those in which the number of operations was simply proportional to N: none of the methods in this chapter is as fast as this, the fastest being those based on Fourier transformation, possibly iterated, and requiring a number of operations proportional to $N \ln N$; this includes most of the methods. The "direct" methods of Sect.2.3.6 exemplify procedures where the number of operations is proportional to N^2 — in that case because of the direct evaluation of correlation sums; even this modest increase in the dependence on N can easily mean a hundredfold increase in computing time in a reasonable two-dimensional problem. Methods based on linear equation sets are worse still, the proportionality increasing to N^3; the impact of this on Greenaway's reformulated zero flipping methods (Sect.2.5.4) is greatly abated, however, by the fact that the number of unknowns remains proportional to the number of points per row, rather than the total number of data points, in two dimensions (Sect.2.5.5). The fact that some methods that have been proposed involve times proportional to $\exp(N)$ can only serve as a horrific warning!

2.9.5 Other Constraints

The X-ray crystallographer might reasonably feel that the electron microscopist, able to refocus his diffracted beams so that they interfere, should have little really to complain of. The fact that the electron microscopist can study crystal structure on a local basis, using extremely small areas at a time, gives additional importance to any possible solution to the phase problem in his case. Unfortunately the constraints that have proved so powerful for the crystallographer (see, for example, [2.25], or [2.82]) carry relatively little weight for the electron microscopist.

The most powerful of the constraints in solving a structure from X-ray data is the fact that the object to be recovered (the specimen electron density) consists of a number of well-separated peaks, all of which are positive. This by itself is enough to unravel a perfectly resolved Patterson function (the autocorrelation function of the object, calculated by Fourier transformation of the diffraction pattern intensity) completely and unambiguously; in practice the many peaks of the function are *not* usually all resolved even in the X-ray case. The constraint also leads to large numbers of inequalities, equalities, and "probable equalities" among the complex structure factors (diffracted beams), the power of which, *given adequate resolution*, are well demonstrated by DORSET and HAUPTMAN [2.83] who use them to solve the structure of some small unit cell crystals on the basis of electron diffraction data to about 0.1 nm. It must be accepted, however, that, given the lower resolution prevailing in electron micrographs, constraints of this type are unlikely to prove very powerful in interpreting aberrated images. On the contrary, what is more likely is that the phases determined, for the relatively low structure factors accessible to electron microscopy, using the methods of this chapter, will then provide the basis for a complete solution using X-ray diffraction (see, for example, [2.32]).

Symmetry considerations, for periodic specimens or isolated macromolecules, will on the other hand be valuable enough constraints where they are applicable: the centrosymmetric case, in which transform phases are either zero or π *besides* being centrosymmetric, is particularly powerful, and has been sufficiently important in the X-ray case to have accumulated many methods peculiar to it.

The finite size of the object shows little sign of being a strong constraint in practice, however attractive it is mathematically in producing an analytic diffraction plane wave, with the possibility of superresolution, and considerable simplification of theoretical uniqueness problems. LANNES [2.84] shows that it is sufficient to recover a complex object wave from a single defocused image, and gives an algorithm for the recovery process, of the iterative transform type: practical experience of reconstructing even *weak* objects from *pairs* of images suggests that too much is being expected of the constraint, however. BARRETT and MISELL [2.85] report limited success only with a technique for providing high-resolution phases for a structure by transforming a low-resolution structure, limited to the appropriate region.

The recent development of detectors for the STEM giving a signal proportional to the specimen phase *gradient* [2.86-88] is an important development. Interesting proposals for solutions of the phase problem based on detectors not yet in existence are given by GALE et al. [2.89], who require the *transverse* current density in the image plane, and GRÜNBAUM [2.90], who requires the diffraction intensity *fluctuation* level due to thermal motion of the atoms in the specimen.

Finally, the use of the "maximum entropy" criterion for providing phases where none exist, or simply selecting a "best" solution to fuller data in the presence of noise [e.g., 2.91], should be mentioned, since recent formulations have now made the approach computationally feasible [2.92,93]. In this a statistical model for the object is put forward that allows an a priori likelihood to be assigned to different objects, and the most likely object consistent with the data, partial or otherwise, and known noise levels is selected: this is usually, though not always, unique. ABLES [2.94] gives an entertaining discussion of the principles. The statistical models used for the object have so far been real and positive, however, so a substantial reformulation is needed if the method is to be applied to recovery of the wave function at the object plane; moreover the maximum entropy constraint appears less powerful when applied to objects not consisting of well-separated peaks, so the method may not prove to be well suited to electron microscopy, though it is certainly worth further investigation. Applied to the X-ray problem, it could be a useful alternative to the methods currently used to unravel Patterson functions: GULL and DANIELL [2.93] show a field containing about fifty well-separated stars (atoms) recovered directly from the transform intensities only.

2.10 Conclusions

Given the wide range of techniques described in this chapter, it is only possible
to generalize ruthlessly in summarizing the position as regards recovery of information on strongly scattering objects. We may say that several ways exist to recover the object plane wave function, and that a useful range of specimen thick-
nesses exists for which this is interpretable; that the recovery methods that are
simplest experimentally (the image/diffraction and the two-defocus method) still
lack a sufficiently fast and reliable algorithm for finding the solution, quite
apart from occasional uniqueness problems; that methods easier to analyze are experimentally less straightforward (the half-plane aperture methods; holography;
bright-field/dark-field subtraction). The first resort should in practice usually
be to iterated application of the linear theory, and probably after that to the
two-defocus method, provided that the resolution is not such that coherence is a
serious factor; if the experimental difficulties of the other methods can be overcome, however, they will well repay the effort.

References

2.1 P.W. Hawkes: *Electron Optics and Electron Microscopy* (Taylor and Francis, London 1972)
2.2 D.L. Misell: J. Phys. *D9*, 1849-1866 (1976)
2.3 D.L. Misell: In *Advances in Optical and Electron Microscopy*, ed. by V.E. Coss-lett, R.E. Barer, Vol. 7 (Academic, London 1979) pp. 185-279
2.4 D.F. Lynch, A.F. Moodie, M.A. O'Keefe: Acta Cryst. *A31*, 300-307 (1975)
2.5 P. Goodman, A.F. Moodie: Acta Cryst. *A30*, 280-290 (1974)
2.6 W.O. Saxton: *Computer Techniques for Image Processing in Electron Microscopy*, Advances in Electronics and Electron Physics Supp. Vol. 10, ed. by L. Marton (Academic, New York 1978)
2.7 D.L. Misell, R.E. Burge, A.H. Greenaway: J. Phys. *D7*, L27-30 (1974); an almost identical paper appears in Nature *247*, 401-402 (1974)
2.8 D.L. Misell, A.H. Greenaway: J. Phys. *D7*, 832-855, 1660-1669 (1974)
2.9 A. Lannes: In *Proc. 4th Int. Cong. High Voltage Electron Microscopy, Toulouse 1975*, ed. by B. Jouffrey, P. Favard (SFME, Paris 1976) pp. 155-158
2.10 A. Lannes: J. Phys. *D9*, 2533-2544 (1976)
2.11 P. van Toorn, A.M.J. Huiser, H.A. Ferwerda: Optik *51*, 309-326 (1978)
2.12 A.M. MacLeod, J.N. Chapman: J. Phys. *E10*, 37-42 (1977)
2.13 R.W. Gerchberg, W.O. Saxton: Optik *35*, 237-246 (1972)
2.14 R.W. Gerchberg, W.O. Saxton: In *Image Processing and Computer-Aided Design in Electron Optics*, ed. by P.W. Hawkes (Academic, London 1973) pp. 66-81
2.15 D.L. Misell: J. Phys. *D6*, L6-9 (1973)
2.16 D.L. Misell: J. Phys. *D6*, 2200-2216, 2217-2225 (1973)
2.17 R.W. Gerchberg, W.O. Saxton: J. Phys. *D6*, L31 (1973)
2.18 R.W. Gerchberg: Nature *240*, 404-406 (1972)
2.19 J.N. Chapman: Philos. Mag. *32*, 527-540, 541-552 (1975)
2.20 N. Toms, J.N. Chapman, R.P. Ferrier: Manchester (1972), pp. 422-423
2.21 A.M.J. Huiser, A.J.J. Drenth, H.A. Ferwerda: Optik *45*, 303-316 (1976);
     A.M.J. Huiser, H.A. Ferwerda: Optik *46*, 407-420 (1976);
     A.M.J. Huiser, P. van Toorn, H.A. Ferwerda: Optik *47*, 1-8 (1977);
     P. van Toorn, H.A. Ferwerda: Optik *47*, 123-134 (1977)

2.22   J.A. Venables (ed.): *Developments in Electron Microscopy and Analysis* (Academic, London 1976)
2.23   M. Horner: In Ref. 2.22, pp. 209-212
2.24   P. Schiske: Optik *40*, 261-275 (1974)
2.25   H. Lipson, W. Cochran: *The Determination of Crystal Structures* (Bell, London 1966)
2.26   E.N. Fowle: IEEE Trans. IT-*10*, 61-67 (1964)
2.27   P. Schiske: J. Phys. *D8*, 1372-1386 (1975)
2.28   A.J.J. Drenth, A.M.J. Huiser, H.A. Ferwerda: Opt. Acta *22*, 615-628 (1975); B.J. Hoenders, H.A. Ferwerda: Opt. Acta *23*, 445-456 (1976); P. van Toorn, H.A. Ferwerda: Opt. Acta *23*, 457-468, 469-481 (1976)
2.29   S.R. Robinson: J. Opt. Soc. Am. *68*,87-92 (1978)
2.30   E.M. Hofstetter: IEEE Trans. IT-*10*, 119-127 (1964)
2.31   H.B. Voelcker: Proc. IEEE *54*, 340-353, 735-755 (1966)
2.32   J. Gassmann: Optik *48*, 347-356 (1977)
2.33   J. Gassmann: Acta Cryst. *A33*, 474-479 (1977)
2.34   W.J. Dallas: Optik *44*, 45-49 (1975)
2.35   R.H.T. Bates: Optik *51*, 161-170, 223-234 (1978)
2.36   W.J. Dallas: Opt. Commun. *18*, 317-320 (1976)
2.37   H.A. Ferwerda: Jerusalem (1976), Vol. 1, pp. 1-3
2.38   R.A. Gonsalves: J. Opt. Soc. Am. *66*, 961-964 (1976)
2.39   J. Frank: Biophys. J. *12*,484-511 (1972)
2.40   W.O. Saxton, A. Howie, A. Mistry, A. Pitt: In *Developments in Electron Microscopy and Analysis 1977*,ed. by D.L. Misell, Institute of Physics Conference Series 36 (Institute of Physics, London 1977)
2.41   H.A. Ferwerda, B.J. Hoenders, A.M.J. Huiser, P. van Toorn: Photogr. Sci. Eng. *21*, 282-289 (1977)
2.42   H.A. Ferwerda: "The Phase Reconstruction Problem for Wave Amplitudes and Coherence Functions", in *Inverse Source Problems in Optics*, ed. by H.P. Baltes, Topics in Current Physics, Vol. 9 (Springer, Berlin, Heidelberg, New York 1978) pp. 13-39
2.43   B.J. Hoenders: "The Uniqueness of Inverse Problems" in *Inverse Source Problems in Optics*, ed. by H.P. Baltes, Topics in Current Physics, Vol. 9 (Springer, Berlin, Heidelberg, New York 1978) pp. 41-82
2.44   W.O. Saxton: J. Phys. *D7*, L63-64 (1974)
2.45   J.C.H. Spence: Opt. Acta *10*, 835-837 (1974)
2.46   W. Hoppe, R. Langer, F. Thon: Optik *30*, 538-545 (1969)
2.47   D.L. Misell: J. Phys. *D7*, L69-71 (1974)
2.48   K.H. Downing, B.M. Siegel: Optik *42*, 155-175 (1975)
2.49   W.O. Saxton: "The Theory of Image Formation and Processing" in Ref. 2.22, pp. 191-196
2.50   R.E. Burge, M.A. Fiddy, A.H. Greenaway, G. Ross: J. Phys. *D7*, L65-68 (1974)
2.51   R.E. Burge: In *Principles and Techniques of Electron Microscopy: Biological Applications*, ed. by M.A. Hayat, Vol. 6 (Van Nostrand-Reinhold, New York 1976) pp. 85-116
2.52   W. Hoppe: Z. Naturforsch. *26a*, 1155-1168 (1971)
2.53   A.H. Greenaway: "Determination of the Image Wavefunction with Particular Reference to Electron Microscopy"; Thesis, Queen Elizabeth College, Univ. London (1976)
2.54   A. Walther: Opt. Acta *10*, 41-49 (1963)
2.55   B.J. Hoenders: J. Math. Phys. *16*, 1719-1725 (1975)
2.56   R.H.T. Bates: Mon. Not. R. Astron. Soc. *142*, 413-428 (1969)
2.57   R.H.T. Bates, P.J. Napier: Mon. Not. R. Astron. Soc. *158*, 405-424 (1972)
2.58   R.H.T. Bates, P.J. Napier, A.E. McKinnon, M.J. McDonnell: Optik *44*, 183-201 (1976)
       A.E. McKinnon, M.J. McDonnell, P.J. Napier, R.H.T. Bates: Optik *44*, 253-272 (1976)
2.59   R.E. Burge, M.A. Fiddy, A.H. Greenaway, G. Ross: Proc. R. Soc. Lond. A *350*, 191-212 (1976)
2.60   J.S. Toll: Phys. Rev. *104*, 1760-1770 (1956)

2.61 G. Ross, M.A. Fiddy, M. Nieto-Vesperinas, M.W.L. Wheeler: Proc. R. Soc. Lond. A *360*, 25-45 (1978); a shorter account is given in Optik *49*, 71-80 (1977)
2.62 M.A. Fiddy: Personal communication
2.63 P.J. Napier, R.H.T. Bates: Astron. Astrophys. Suppl. *15*, 427-430 (1974)
2.64 H. Wahl: Optik *39*, 585-588 (1974)
2.65 J. Munch: Optik *43*, 79-99 (1975)
2.66 R.H.T. Bates, R.M. Lewitt: Optik *44*, 1-16 (1975)
2.67 R.H. Wade, W.K. Jenkins: Optik *50*, 1-17 (1978)
2.68 J.E. Baldwin, P.J. Warner: Mon. Not. R. Astron. Soc. *175*, 345-353 (1976)
2.69 A.H. Greenaway: Opt. Lett. *1*, 10-12 (1977)
2.70 T. Hidaka: J. Opt. Soc. Am. *66*, 147-150 (1976)
2.71 A.H. Greenaway, A.M.J. Huiser: Optik *45*, 295-300 (1976)
2.72 J. Frank: Optik *38*, 582-584 (1973)
2.73 A.W. Lohmann: Optik *41*, 1-9 (1974)
2.74 W. Hoppe: Acta Cryst. *A25*, 495-501 (1969);
     W. Hoppe, G. Strube: Acta Cryst. *A25*, 502-507 (1969);
     W. Hoppe: Acta Cryst. *A25*, 508-514 (1969)
2.75 R. Hegerl, W. Hoppe: Manchester (1972), pp. 628-629
2.76 J.C.H. Spence: Optik *49*, 117-120 (1977)
2.77 H. Berndt, R. Doll: Optik *46*, 309-332 (1976); *51*, 93-96 (1978)
2.78 J.U. Madsen, R.M.J. Cotterill: Acta Cryst. *A34*, 378-384 (1978)
2.79 W. Krakow, D.C. Ast, W. Goldfarb, B.M. Siegel: Philos. Mag. *33*, 985-1014 (1976)
2.80 K.H. Downing, B.M. Siegel: Optik *38*, 21-28 (1973)
2.81 W.K. Jenkins: "Contrast Transfer in Bright-Field Electron Microscopy of Amorphous Objects"; Thesis, Univ. Cambridge (1979)
2.82 A. McPherson, Jr.: In *Principles and Techniques of Electron Microscopy: Biological Applications*, ed. by M.A. Hayat, Vol. 6 (Van Nostrand-Reinhold, New York 1976) pp. 117-240
2.83 D.L. Dorset, H.A. Hauptman: Ultramicroscopy *1*, 195-201 (1976)
2.84 A. Lannes: Opt. Commun. *20*, 356-359 (1977)
2.85 A.N. Barrett, D.L. Misell: In Ref. 2.22, pp. 213-216
2.86 N.H. Dekkers, H. de Lang: Optik *41*, 452-456 (1974); Philips Tech. Rev. *37*, 1-9 (1977)
2.87 J.N. Chapman, P.E. Batson, E.M. Waddell, R.P. Ferrier: Ultramicroscopy *3*, 203-214 (1978)
2.88 P.W. Hawkes: J. Opt. (Paris) *9*, 235-241 (1978)
2.89 W. Gale, E. Guth, G.T. Trammell: Phys. Rev. *165*, 1434-1436 (1968)
2.90 F.A. Grünbaum: Proc. Nat. Acad. Sci. USA *72*, 1699-1761 (1975)
2.91 B.R. Frieden: "Image Enhancement and Restoration", in *Picture Processing and Digital Filtering*, ed. by T.S. Huang, Topics in Applied Physics, Vol. 6 (Springer, Berlin, Heidelberg, New York 1975) pp. 177-248
2.92 S.J. Wernecke, L.R. d'Addario: IEEE Trans. C-*26*, 351-364 (1977)
2.93 S.F. Gull, G.J. Daniell: Nature *272*, 686-690 (1978)
2.94 J.G. Ables: Astron. Astrophys. Suppl. *15*, 383-393 (1974)

3. Computer Reconstruction of Regular Biological Objects

J. E. Mellema

With 24 Figures

Direct observation of biological structure made possible by electron microscopy has always played an important role in structural research. Modern electron microscopes render biological objects directly visible on the fluorescent screen and in many cases the electron micrographs of the objects are sufficient to support the observations. For example, if the question concerns the determination of the shape of a virus particle, spherical or elongated, the answer can be obtained in a straightforward way. If the details to be derived are small compared to the size of the biological structure a much more elaborate approach will be required. In that case, many electron micrographs have to be recorded and the result will be based on the analysis of many micrographs. In the field of molecular biology, an example is the electron microscopy of nucleic acids, in which the localization of a non base-paired region within a double helical structure has to be mapped.

In the last 10 years another approach for analyzing biological structure by electron microscopy has become an important field. The methods used in this field are derived from the techniques with which the X-ray crystallographers determine crystal structure. For a successful application the biological objects must exhibit regularity and the images of the structures are analyzed by computer after being digitized. The ways in which the analysis can be performed are quite diverse and to some extent, the type of symmetry present in the object will determine the character of the structural data derived. The aim of the digital analysis is to retrieve the maximum amount of information from the electron image. The methods have been developed by KLUG and collaborators in Cambridge and have shown their usefulness in quite a number of cases in structural biology (3.1-4). Two important approaches can be distinguished: in one, two-dimensional information about the specimen is derived, whereas in the other case, a three-dimensional reconstruction of the object is carried out from its two-dimensional projections.

In this chapter only the basic theory of the image processing in the field of biological structure will be presented and the type of biological object amenable to this approach will be discussed. Before embarking on the use of image analysis methods some crucial questions concerning the state of the object in the microscope, the effects of radiation damage, and the type of information recorded on the micrograph must be answered. Finally a number of recent results will be discussed as much of the older literature has been covered in the reviews by CROWTHER and KLUG, and CROWTHER [3.5-6].

3.1 The Biological Object

3.1.1 General Remarks

In order to be suitable for electron microscopic analysis, biological material has to be processed by chemical and physical methods. Generally these combinations of treatments are known as preparation techniques. The intention of the particular preparation method is to make the object under study suitable for examination in vacuum with electrons, ideally in such a way that the minimum number of electrons is required to image the structure. This last requirement is necessary to limit the irreversible changes due to radiation damage, which ultimately effect the biologically relevant information content of the image. In the last 10 years the structural biologist has become aware of the necessity to pay attention to the radiation damage [3.7] and various methods of limiting the damage are now known. As biological material consists of relatively light elements which do not scatter electrons appreciably, procedures have to be carried out in order to enhance the scattering power of the object or parts of it. Heavy metal in solution or as vapor can be used to achieve this. However, it has recently been shown that contrasting with these high Z materials is not strictly necessary when redundancy of the object is used to improve the weak scattering signal of an individual unit. Phase contrast by defocusing can also be used to enhance image details especially in the high resolution range (spacings smaller than about 2 nm).

The vacuum conditions in the microscope require that the specimen be investigated in a dry state, although there are now a number of methods of substituting for the water molecules of the hydrated biological structure. This may prevent a total collapse of the three-dimensional structure when the specimen is introduced into the vacuum.

Before discussing the various methods of reconstruction, it must be stressed that the different preparation methods form an integral part of the structural studies by image processing as they in fact determine the meaningful resolution that can be obtained by such an analysis. Moreover, the development of these preparation methods is among the most important aspects of molecular electron microscopy as the application of image processing methods only serves to obtain an average result or an image which can be interpreted in a straightforward way.

It is generally accepted that under standard conditions the resolution of biological details is limited to about 2 nm (see for example [3.7]). However, a number of methods are available which make it possible to surpass this limit [3.8,9].

Finally it must be made clear that this presentation is not exhaustive, especially the section on recent applications, but in the opinion of the author this review covers the most important developments in computer reconstruction of regular biological specimens now and most probably in the near future.

3.1.2 Regular Biological Objects

The reconstruction and subsequent interpretation of images of biological objects has been very succesful for biological objects with some symmetry. It is instructive to consider the basic principles which underly the architecture of regular biological structure. The symmetry operations in biology consist of translations and rotations, whereas mirror symmetry is not allowed. The reason for the absence of mirror-symmetric elements in nature is that biological structures are built of molecules which are different from their mirror images. The only kind of allowed operations are therefore rotations and translations or combinations of both. The combinations of these operations can be represented by point, line, plane, and space groups. Figure 3.1 shows for every class mentioned an example represented in a schematic way. The line or helical groups are built according to a generalized screw operation (rotation plus translation parallel to a line) and this may be combined with a cyclic or dihedral point-group symmetry. There is no restriction on the number of asymmetric units that can be placed in a turn of the helix, nor need this number be integral. This configuration leads to an assembly of molecules of indefinite size, but generally a second type of component interacts in a specific way with the main component, which limits the length. Helical aggregates for example occur in striated muscle and many rodlike viruses are built according to this principle [3.10,11].

Plane groups are defined by two nonparallel translation vectors and may include also two-, three-, four-, and sixfold rotational axes. These types of assemblies occur in nature as cell walls, membranes in which the subunits form a two-dimensional interface. The extension of plane groups with another translational vector to form space groups is the basic principle of all true regular three-dimensional structures, whatever their nature. As has been remarked with the line groups, the plane and space groups lead to aggregates of indefinite length, although in any real case they are terminated by various mechanisms.

Particles which have a point-group symmetry have a finite size and their symmetry elements consist of a set of rotational axes which share a common point. This combination leads to a class of structures with a finite extension. Every type of rotational symmetry is allowed. The best-known class of particles studied by electron microscopy which have a point-group is the spherical viruses. The symmetry axes of these are five-, three-, and twofold axes, which define the icosahedral surface lattice. Many organized biological structures are formed by making use of these principles. The stability of these complexes is governed by weak noncovalent inter-

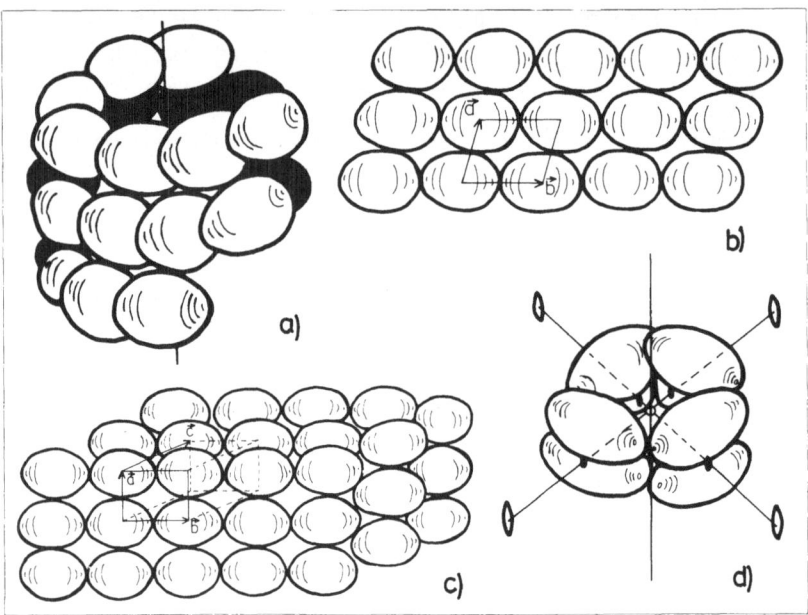

Fig.3.1a-d. Examples of regular biological assemblies, arranged according to line, plane, space, and point groups of symmetry. The asymmetric units are schematically represented by ellipsoids. (a) An assembly generated by a translation and a rotation along a line, which results in a helical aggregate; (b) a two-dimensional array of units defined by two nonparallel translations a and b; (c) a three-dimensional array of elements generated by the operation of three nonparallel translations a, b, and c; (d) a finite aggregate consisting of eight units arranged according to the point group 422

actions between the constituent parts. Also, ordered aggregates can be formed in vitro under special conditions. These aggregates have been prepared deliberately in order to use image analysis methods to elucidate their structure (see for example [3.12,13]).

3.1.3 Chemical and Physical Processing of the Object

The aim of the preparation procedure is to enhance the scattering power of the object and simultaneously substitute the volatile components attached to the object, in order to preserve its three-dimensional structure in the microscope. In practice there are a number of different methods for biopolymers which can be successfully used. These methods are schematically presented in Fig.3.2. In one method, the negative staining technique, the stain forms an amorphous envelope of heavy metal salt around the object. Moreover, this negative stain preserves in some way the three-dimensional structure of the biological object. As has been remarked previously (Sect.3.1), regular biological objects are stabilized by weak interactions and therefore are vulnerable to small changes in solvent conditions. As the

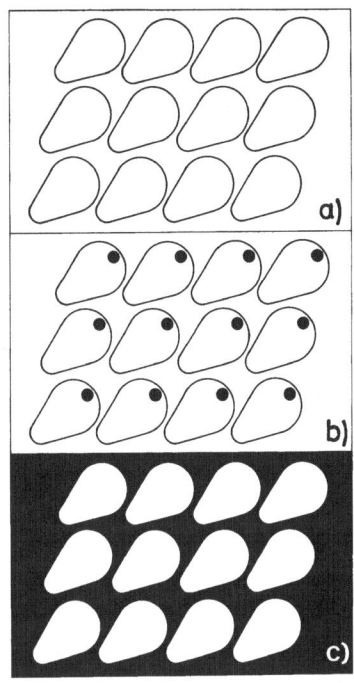

Fig.3.2a-c. A schematic presentation of the ways of preparing regular biological objects for electron microscopy. The object is a two-dimensional lattice of "pear-shaped" units which may represent protein molecules. (a) The two-dimensional specimen; the space between the units may be occupied by for example a sustaining agent like glucose. (b) The two-dimensional specimen in which the units are labelled in a specific way by, for example, a heavy atom marker or an antibody. (c) The two-dimensional specimen embedded in an amorphous layer of heavy metal salt (negative stain)

negative staining technique uses solutions of heavy metal salts like uranyl or tungstate ions, the ionic conditions of these solutions have to be controlled because they may be the reason for a dissociation of the biological object during preparation. In another approach, a heavy metal label or a larger marker such as an antibody may be used to react to a specific part of the biological object. This marker may be located and its position will provide information on a particular site of the structure. Recently a low density, nonvolatile material such as glucose has been used to fill the space between the units in the biological structure. This approach, in which unstained biological material has been investigated, will only be possible if the weak contrast of the regular object can be enhanced by image processing methods. There have been also two other attempts to retain the three-dimensionality of biological specimens. In one of these methods, the water in the biological material is replaced by a liquid such as CO_2 under pressure, which has its critical point at $31^\circ C$. When the temperature is raised above this value, where two fluid phases cannot coexist, the ambient gas is allowed to escape. In this way the specimen will not be wrecked as no surface of a phase boundary is formed. Another procedure involves rapidly freezing the specimen and pumping off the water vapor from the solid specimen. The phase boundary between ice and water vapor will pass through the specimen, which retains much of its three-dimensional structure, although perhaps not so well as in the critical point method [3.14,15].

3.1.4 Contrast in Bright-Field Images

The contrast in the electron microscopic bright-field image of a thin biological object can be attributed to the combined presence of phase and amplitude contrast components. In the absence of phase effects the image density will depend on the differential scattering of electrons from regions of the specimen with differing mass thickness. A clear account of this type of image contrast has been presented by HEIDENREICH [3.16]. The principle has been used to determine the mass of biological specimens such as erythrocytes and viruses at low resolution [3.17,18]. In general the assumption that only so-called amplitude effects are operating is not strictly valid. In order to assess the effects of phase and amplitude contrast of a thin object it will be very instructive to use the results derived by COWLEY [3.19].

For a thin object the effect of the electron beam on the specimen may be expressed by the transmission function of the object

$$q(x,y) = \exp[-i\sigma\phi(x,y) - \mu(x,y)] \quad . \tag{3.1}$$

The beam is coherent and has an amplitude of unity. In (3.1) $\sigma = \pi/(\lambda E)$ in which E is the accelerating voltage of the beam, $\phi(x,y)$ is the projection of the potential distribution of the object in the direction of the beam, and $\mu(x,y)$ is the projection of an absorption function. This function describes all the processes in which the electrons are prevented from taking part in image formation.

It can be shown that in the case of thin objects (thickness about 20 nm) the amplitude distribution in the back focal plane of the imaging system of the microscope is

$$\Psi(X,Y) = \delta(0,0) - M(X,Y)\cos_\chi + \sigma\Phi(X,Y)\sin_\chi \quad . \tag{3.2}$$

In this expression M and Φ are the Fourier transforms of the absorption function μ and the potential distribution φ, respectively. The unscattered part of the electrons in the diffraction pattern is represented by $\delta(0,0)$. The values of $\sin_\chi$ and $\cos_\chi$ depend on the defocusing and spherical aberration constant of the objective lens and determine the relative weights of the various contrast components. Graphs of $\sin_\chi$ and $\cos_\chi$ as a function of the reciprocal spacing show, especially at the higher spacings, a modulated appearance. The aim of the defocusing is to obtain an image intensity distribution as close as possible to that of an amplitude object. In general $\mu(x,y)$ will be much smaller than $\sigma\phi(x,y)$; the contrast produced by lens aberrations and defocus on $\sigma\phi(x,y)$ is usually greater than that for the so-called absorption effect.

In the ideal case the intensity distribution in the image would have the form

$$I(x,y) = 1 \pm 2\sigma\phi(x,y) \quad , \tag{3.3}$$

neglecting absorption effects. This will occur if the value of $\sin\chi$ is made close to +1 or -1 over the largest possible range in reciprocal space. For low values of (X,Y) where there are large deviations from unity, compensating influences may be expected from $M(X,Y)\cos\chi$, as $\mu(x,y)$ and $\phi(x,y)$ are likely to have the same form. Therefore to a first approximation an optimum defocused image will be similar to the amplitude object and the image interpretation of biological structure is based on this rationale. A more complete treatment of the effects of phase contrast of thin specimens has been given by ERICKSON [3.20]. See also Chap.1.

3.1.5 Radiation Damage

It has been generally accepted that the effects of the electron beam on the biological object will cause a loss of information. A number of techniques have been used to monitor the damage to the specimen such as studying the fading of reflections in the electron diffraction patterns and measuring the mass loss of the specimen. Moreover, in the case of regular biological objects, optical diffraction can be used to assess the damage. By analyzing the optical transform of images of a helical virus, UNWIN has put forward the model that the negative stain is able to migrate over the specimen due to the irradiation [3.21].

Up to now, no standard method is known for eliminating the irreversible effects of irradiation, although not much experimental work has been performed to overcome them. Among the possible approaches may be the use of "energy" scavengers, beam-resistant stains, and spatial averaging in order to reduce the electron dose during recording. A number of instrumental developments such as the use of more sensitive recording devices [3.22] and scanning transmission microscopy [3.9] are also promising in this respect.

3.2 Fourier Processing of Electron Micrographs

One of the most powerful methods of assessing the information content of electron images of regular objects is to use the Fourier transform of the recorded image. Most of the applications of image analysis are based upon this approach, which also has a natural counterpart in image formation. Moreover, the use of Fourier series to solve structural problems is quite common in X-ray crystallography, from which many of the methods in this field are derived. For this reason only the Fourier transformation will be discussed, although a number of other orthogonal transformations have been proposed [3.23]. The theory of Fourier transformations has been well covered by a number of authors [3.24-26], so that in this chapter only the basic principles will be described. There exist fast algorithms to carry out digital Fourier transformations, which are designated in the literature as FFT [3.27]. Even on minicomputers, subroutines using this principle may be implemented and the application of digital image analysis by means of Fourier transforms is no longer limited to large computer systems.

3.2.1 Quantization and Preprocessing

The first step in the digital analysis of an electron micrograph is the conversion of the image into a set of numbers, equivalent to the image intensity. The continuous signal is represented by a limited set of intervals and the number and the distance between the intervals are based on a number of considerations. The conversion of the image into this form takes place in a densitometer, an instrument which is commercially available and which outputs the result on a medium compatible with a digital computer system.

The normal and most widely used form of quantization is uniform sampling in which the intensity intervals of the image are on a linear scale and are represented by the value at the middle of the interval. Other schemes of digitalization are known, such as those which represent the most frequently encounterd intensity values with the highest accuracy [3.28].

The next step in the analysis is to mask the image and separate the area to be transformed from the background [3.29]. This step is usually followed by a so-called floating procedure. The aim of the procedure is to minimize the effects of the mask in the Fourier transform of the object, a phenomenon which is quite dominant for example in optical transformation [3.30]. In the floating procedure, the average intensity value around the image is determined, and then subtracted from the area to be transformed.

3.2.2 The Whittaker-Shannon Sampling Theorem [3.31]

The process of Fourier transformation of a two-dimensional function can be described as follows. The image $f(x,y)$ will be sampled by the densitometer by means of the following function

$$s(x,y) = \sum_{m_1=-\infty}^{\infty} \sum_{m_2=-\infty}^{\infty} \delta(x - m_1\Delta x, y - m_2\Delta y) \quad . \tag{3.4}$$

This sampling function $s(x,y)$ is a two-dimensional array of delta functions, which are separated by Δx and Δy in the image. In Fig.3.3 this function is shown. The Fourier transform of $s(x,y)$ is $S(X,Y)$ and will have the form

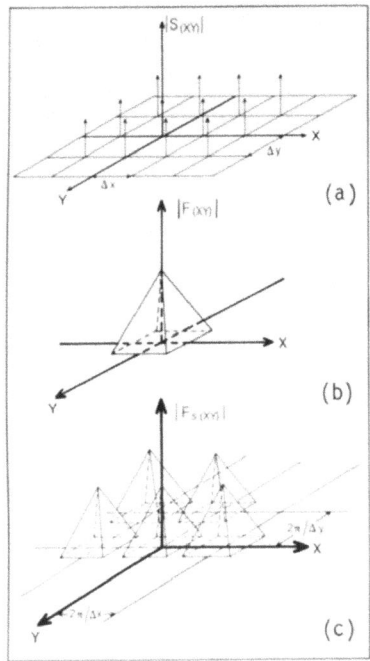

(a)

(b)

(c)

Fig.3.3a-c. (a) The sampling function s(x,y)
consists of a two-dimensional array of delta
functions spaced at values Δx in x and Δy in y.
(b) The modulus |F| of a two-dimensional Fourier
transform presented in a schematic way. (c) The
effect of the convolution of S(X,Y) and |F(X,Y)|
presented in a schematic way. The individual
transforms are repeated in Fourier space at distances 2π/Δx along X and 2π/Δy along Y

$$S(X,Y) = \frac{4\pi^2}{\Delta x \cdot \Delta y} \sum_{m_1=-\infty}^{\infty} \sum_{m_2=-\infty}^{\infty} \delta\left(X - \frac{2\pi m_1}{\Delta x}, Y - \frac{2\pi m_2}{\Delta y}\right) . \tag{3.5}$$

The sampled image can be written as

$$f_s(x,y) = f(x,y) \cdot s(x,y) , \tag{3.6}$$

and the Fourier transform of $f_s(x,y)$ will have the following form

$$F_s(X,Y) = \frac{4\pi^2}{\Delta x \cdot \Delta y} \sum_{m_1=-\infty}^{\infty} \sum_{m_2=-\infty}^{\infty} F\left(X - \frac{2\pi m_1}{\Delta x}, Y - \frac{2\pi m_2}{\Delta y}\right) . \tag{3.7}$$

Expression (3.7) shows that the Fourier transform of the sampled image is a repeating series of F(X,Y), separated by 2π/Δx and 2π/Δy. In Fig.3.3 the effect of the
convolution of F(X,Y) with the function S(X,Y) is schematically represented. If the
transform of f(x,y) is band limited, then no overlap of the periodic signal will
result. This can be expressed by the two conditions

$$\frac{2\pi}{\Delta x} \geq X_c \quad \text{and} \quad \frac{2\pi}{\Delta y} \geq Y_c . \tag{3.8}$$

In these expressions X_c and Y_c represent the spatial frequencies for which $F(X,Y)$ is actually zero. This means that $\Delta x \leq \pi/X_c$ and $\Delta y \leq \pi/Y_c$, which requires that the sampling must be equal to or smaller than half of the periodicity of the fastest intensity fluctuation in the image.

The original image can be reconstructed from the sampled image by convolution with the impulse response, $r(x,y)$, of the reconstruction filter. The expression is

$$f_r(x,y) = f_s(x,y) * r(x,y)$$

in which $*$ represents the convolution operation. In the Fourier domain this expression has the following form

$$F_R(X,Y) = F_S(X,Y) \cdot R(X,Y) = \frac{4\pi^2}{\Delta x \cdot \Delta y} R(X,Y) \sum_{m_1=-\infty}^{\infty} \sum_{m_2=-\infty}^{\infty} F\left(X - \frac{2\pi m_1}{\Delta x}, Y - \frac{2\pi m_2}{\Delta y}\right) .$$

$$(3.9)$$

The Fourier transform of $r(x,y)$ has been represented by $R(X,Y)$.

The Fourier transform of the reconstructed image can be made equal to that of the original image if there is no overlap and $R(X,Y)$ has such a form that it will transmit only the spectrum specified by $m_1 = m_2 = 0$. For example, if $R(X,Y) = 1$ for the values $|X| \leq X_1$ and $|Y| \leq Y_2$, whereas $R(X,Y) = 0$ everywhere else, then an exact reconstruction will be possible if

$$\frac{\pi}{\Delta x} \geq X_1 \geq X_c \quad \text{and} \quad \frac{\pi}{\Delta y} \geq Y_2 \geq Y_c . \tag{3.10}$$

The reconstruction will take place with the aid of sinc functions in the (x,y) domain and the impulse response of the filter will have the form

$$r(x,y) = \frac{X_1 Y_1}{\pi^2} \cdot \frac{\sin(xX_1)}{(xX_1)} \cdot \frac{\sin(yY_1)}{(yY_1)} . \tag{3.11}$$

Exact recovery of a band-limited function can be achieved from an appropriately spaced rectangular array of its sampled values. This result is known as the Whittaker-Shannon sampling theorem.

3.2.3 Fourier Transforms of Regular Objects

Before proceeding to the description of the theoretical background of a number of image processing methods, it is useful to summarize a number of properties of Fourier transforms. This summary is presented in Table 3.1. The relationships given in the table are completely reciprocal in that they will also hold if the headings real space and reciprocal space are interchanged.

Table 3.1.

Operation in real space	Effect in reciprocal space
Rotation about axis	Rotation about parallel axis with same frequency
Change of scale in one direction	Reciprocal change of scale in same direction
Translation	Modulus unchanged, phase change of Fourier component

By application of the convolution theorem the transform of a regular object can be easily derived. It will be the product of the unit cell transform with the transform of the lattice function. This means that the product will be zero except at the reciprocal lattice points and at these positions in the Fourier domain it will be the product of the corresponding value of the Fourier transform of the unit cell and the amplitude of the reciprocal lattice function. This last function will be determined by the magnitude of the translational vectors of the regular object. For example, in the case of a two-dimensional array (see Fig.3.1), a and b specify this reciprocal relationship.

In all cases regular biological objects are built from three-dimensional units such as protein or nucleic acid molecules. During recording in the microscope projections of these three-dimensional objects are obtained, as the size of the biological specimens is smaller than the depth of the focus of the electron microscope [3.16]. In order to arrive at a three-dimensional structure determination from these projections, a number of methods have been proposed, of which those making use of the Fourier transform have been applied with much success [3.1].

The proposed method makes use of the principle that a two-dimensional projection of a three-dimensional object corresponds to a central section of the three-dimensional Fourier transform of that object [3.32]. The three-dimensional structure of the object will be known if sufficient projections can be obtained, which fill Fourier space in a homogeneous way. This means that in general the views of the object have to be collected by tilting in the electron microscope, but in the case of regularly built objects, the symmetry will drastically limit the number of projections required. Two cases can be distinguished: projections of structures with helical symmetry, in which the three-dimensional reconstruction by means of the Fourier transform will follow naturally from the transformed data, and reconstructions of objects with point-group and translational symmetry. In the first case a few projections are required which may be collected from a field of particles in which all kinds of different orientations are present. If the objects have a larger spatial extension, such a plane layers, tilted images will be required, as these objects have a preferential orientation on the specimen support film. The tilt angles should span 180°, which is a physical impossibility due to the construction of the microscope. In some instances the number of tilts may be reduced, especially in the case of two-dimensional regular objects, where the Fourier transform perpendicular to the plane has a continuous form, so that it is not necessary to obtain central sections through a number of well-defined reciprocal lattice points.

The helical and icosahedral reconstruction schemes will be discussed in Sects. 3.2.6 and 3.2.7. Projections of regular two-dimensional arrays and those derived from particles exhibiting point-group symmetry can be investigated by two-dimensional filtering methods with much profit, without arriving at a three-dimensional model.

3.2.4 Processing of Two-Dimensional Structures with Translational Symmetry

The preparation methods for electron microscopy, e.g., the negative staining technique (Fig.3.2), cause variations in the image details recorded. Sometimes this variability arises from perturbation and distortion of the biological object and sometimes from variations in the penetration of the contrasting material into the regular objects. In some cases the recorded images do not represent clear patterns of one layer, but are superposition patterns of two or more layers. If a helical particle is imaged in a direction perpendicular to the cylindrical axis, superposition patterns will also be present in these images, because the two sides of the particle will project on top of each other [3.33].

By Fourier transforming the image and separating the regular features from the Fourier transform, followed by a back transformation of these regular data, a kind of spatial averaging will be achieved, from which the image details can be derived with much more confidence than from the unprocessed image. Also the overlapping layers in a projected structure may be recovered in this way.

Figure 3.4 represents the outline of the method for the case of two overlapping lattices, which are related by a rotation. At a number of reciprocal lattice points the contributions of the individual lattices may coincide (see Fig.3.4), so that the individual sets of Fourier components may not be recoverable. The steps taken in the procedure are that a new Fourier transform is generated, with values of zero everywhere except at or around the reciprocal lattice points. An inverse Fourier transformation of this data set will yield the reconstructed image. Carried out in this way, the process is formally equivalent to optical filtering as first described by KLUG and DeROSIER [3.33]. The disadvantages of the method are that low-frequency components of noise are not removed and that lack of phase coherence across each diffraction maximum will remain. The low-frequency terms are sometimes large and are due to stain variations on the sample. As the Fourier components are known in the procedure, the method can be refined by averaging the amplitudes over a maximum and corresponding phases can be calculated as weighted averages across the peaks. These averaged Fourier components can be substituted at refined reciprocal lattice posi-

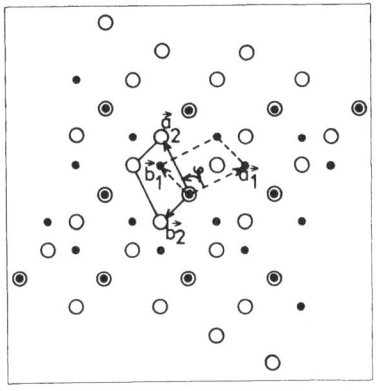

Fig.3.4. A schematic representation of the modulus of the Fourier transform of two superimposed, regular, two-dimensional arrays, which do not have the same orientation. The rotation angle of the two layers is specified by ϕ. Two reciprocal lattices can be drawn. The vectors a_1, b_1 determine the reciprocal lattice points with the filled circles. In a number of cases dots and circles coincide, so that at these places the individual Fourier components cannot be recovered

tions. As a quantitative measure of the quality of each reconstruction, the weighted average phase for each reflection may be compared to each of the individual phases used in calculating the average [3.36]. The weighted mean phase angle discrepancy and the weighted mean cosine of this angle (figure of merit) become very important measures of the validity of each reconstruction. The weighted mean phase angle discrepancy will be calculated as

$$\Delta\phi_{hk} = \frac{\sum\limits_{i,j} |F_{ij}|^2 [\phi_{ij} - \bar{\phi}]}{\sum |F_{ij}|^2} \tag{3.12}$$

in which $|F_{ij}|$ and ϕ_{ij} represent the amplitude and the phase angle, respectively, of the Fourier component (i,j); $\bar{\phi}$ is the average phase angle over (i,j). The index (h,k) represents the particular reciprocal lattice point centered at the set (i,j). The individual figure of merit (m_{hk}) for the lattice point (h,k) follows from

$$m_{hk} = \frac{\sum\limits_{i,j} |F_{ij}|^2 \cos(\phi_{ij} - \bar{\phi})}{\sum\limits_{ij} |F_{ij}|^2} \tag{3.13}$$

and the overall figure of merit for the reconstruction can be calculated according to

$$m = \frac{\sum\limits_{h,k} |F_{h,k}|^2 m_{hk}}{\sum\limits_{h,k} |F_{hk}|^2} \quad . \tag{3.14}$$

In a study of acetylcholine receptors, CREPEAU et al. [3.36] determined values of m ranging from 0.95 to 0.99 with no individual reflection below 0.80. In comparison the overall figure of merit for a reconstruction of an unordered area ranged from 0.2 to 0.81.

Examples of two-dimensional filtering have been carried out on a variety of regular objects, e.g., [3.34-37] and a number of them will be discussed later on.

Another type of filtering, applied to biological structures with helical symmetry, is schematically presented in Fig.3.5. The Fourier transform of a cylindrical structure will exhibit intensity modulations sampled on a set of equidistant
lines [3.38,39]. If the structure can be approximated by a mass distribution
present at one radius, two reciprocal nets will emerge from the Fourier transform
(see Fig.3.5). These nets are derived from the so-called near and far sides of the
structure with respect to the supporting film [3.30]. In the same way as described
for the two-dimensional arrays, the contributing parts to the projection may be
separated and inversely transformed, resulting in a better definition of the details present in the projected surface lattices. In practice subsidiary spots will
be present in the diffraction pattern, which correspond to secondary maxima of
Bessel functions and which make the analysis more complicated. Also the "one-radius"
model is a very simple approximation, which will probably never occur in practice.
This also complicates the interpretation of the Fourier transform, because contributions due to mass modulations at different radii will be present on the layer
lines. Nevertheless, in some instances, digital analysis of the type of structure
may be helpful in deriving surface lattices which originate at different radii.

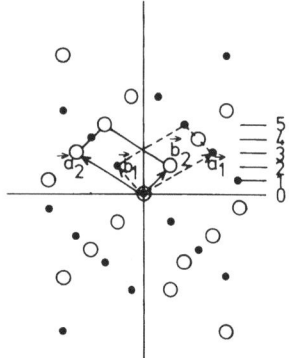

Fig.3.5. A schematic representation of the modulus
of the Fourier transform of a projection of a thin
cylindrical tube, which is constructed according to
helical symmetry. The pattern exhibits mirror symmetry along the meridional (vertical) line, which
runs parallel to the cylindrical axis of the particle. The circles and dots specify the surface
lattice of the two sides of the particle. The unit
cells of the two sides are indicated by a_1, b_1, and
a_2, b_2. The numbers in the figure indicate set
of horizontal lines, layer lines, on which the
moduli are sampled and which specify the axial
repeat distance of the particle

3.2.5 Rotational Filtering

A particularly elegant application of two-dimensional filtering is the analysis of
projections of assemblies with a point-group symmetry. Projections of these aggregates down an axis of symmetry will exhibit symmetry relationships and in a number
of cases the analysis of these images by digital filtering has been very rewarding
[3.40]. The analysis takes the following form. The image intensity I in polar coordinates (r,ϕ) is expanded according to

$$I(r,\phi) = \sum_{-\infty}^{+\infty} g_n(r)\exp(in\phi) \quad . \tag{3.15}$$

In this expression, $g_n(r)$ represents the weight of the n-fold component at a radius r of the structure. The relative weights of the various n-fold components present in the image can be conveniently expressed as

$$P_N = \varepsilon_N \int_0^a |g_n(r)|^2 r \; dr \quad , \tag{3.16}$$

in which $\varepsilon_n = 1$ for $n = 0$ and $\varepsilon_n = 2$ for $n \geq 0$. A graph of P_n as a function of n will be referred to as the power spectrum and ideally this graph ought to show maxima at values of n which are integer multiples of the symmetry On the basis of this symmetry a filtered image can be calculated by means of a Fourier transform of the image intensity, in which the transform is expanded as follows:

$$F(R,\Phi) = \sum_{-n}^{+n} G_n(R)\exp\left[in(\Phi + \frac{\pi}{2})\right] \quad ; \tag{3.17}$$

$g_n(r)$ and $G_n(R)$ are related by a Fourier-Bessel transform. For an explanation of the variables in (3.17) see the text accompanying (3.18). In the reconstruction, only values of n which are integer multiples of the symmetry are used. Thus, allowed values for g_n in the case of m-fold symmetry are g_0, g_m, g_{2m}, g_{3m}, etc. There will be a value of n beyond which the rotational components may be neglected. because they are effectively zero in the power spectrum. An essential step in the analysis is that the origin of the coordinate system should lie on the symmetry axis. For this reason a search procedure is carried out in which a residual is calculated on the basis of m-fold symmetry. This residual measures the departure from the assumed symmetry in the transform by summing the differences in magnitude of the Fourier components at a particular radius on the basis of this symmetry. By shifting the origin and recalculating this residual the best origin on the basis of the assumed symmetry will be found. The process may be repeated for every type of symmetry so that a well-determined choice of symmetry in the structure will be possible.

3.2.6 Three-Dimensional Reconstruction of Objects with Helical Symmetry

If a structure contains symmetrically arranged subunits, such as assemblies with helical symmetry, one single electron image will effectively contain many different views of that structure. The symmetry of this regular biological structure can be introduced, allowing the three-dimensional structure to be reconstructed from a single view, or a small number of them.

This method was proposed by DeROSIER and KLUG [3.1] and applied to the contractile protein assembly of bacteriophage T4. Since then, the technique has been applied to a variety of biological structures which all possess helical symmetry [3.12,13, 41-46]. A refinement of the method has been described by AMOS and KLUG [3.47], in which the statistical weight of the reconstruction was improved by averaging the Fourier transform of a number of different structures after the orientation parameters had been determined.

In the mathematical formulation of the reconstruction, it is advantageous to use cylindrical polar coordinates. The density of the object is denoted by $\rho(r,\phi,z)$ and its three-dimensional Fourier transform by $F(R,\Phi,Z)$. The z axis runs parallel to the helical symmetry axis. The theory used in the reconstruction has been worked out in [3.29]. In short, the reconstruction can be formulated as follows. A structure with helical symmetry will have a Fourier transform sampled on layer lines $Z = 1/c$, in which 1 is the layer line number and c the axial repeat of the structure. Then,

$$F(R,\Phi,\frac{1}{c}) = \sum_{n} G_{n,1}(R)\exp\left[in(\phi + \frac{\pi}{2})\right] . \qquad (3.18)$$

The selection rule

$$1 = tn + um \qquad (3.19)$$

will link the layer line number 1 with n, the order of $G_{n,1}$. In (3.19), t represents the number of turns of the basic helix per axial repeat and u the number of subunits per repeat. The allowed values of m are any integers that satisfy (3.19). The reconstruction involves the determination of $G_{n,1}(R)$ from $F(R,\Phi,1/c)$ followed by a calculation of $\rho(r,\phi,z)$ through a Fourier-Bessel transformation

$$\rho(r,\phi,z) = \sum_{1} \sum_{n} g_{n,1}(r)\exp(i\phi)\exp(-2\pi ilz/c) \quad , \qquad (3.20)$$

where

$$g_{n,1}(r) = \int G_{n,1}(R)J_n(2\pi Rz)2\pi R \; dR \quad . \qquad (3.21)$$

If, to the working resolution (which will be in most cases about 2.5 nm), only one value of n exists for which $G_{n,1}$ is nonzero, then a three-dimensional structure can be derived from one view. In this case (3.18) will reduce to the form

$$F(R,\Phi,\frac{1}{c}) = G_{n,1}(R)\exp\left[in(\Phi + \frac{\pi}{2})\right] . \qquad (3.22)$$

In some structures, in which two $G_{n,1}(R)$ terms are present per layer line and within the resolution range, a single view may again yield the individual values of $G_{n,1}$

provided one order is odd and the other is even. Otherwise, more than one view will be required [3.48].

The Fourier transform of the digitized image will yield two estimates of $G_{n,1}$ if (3.22) holds. Thus, from one particle two reconstructions will be feasible, one corresponding to the "near" and the other to the "far" side of the particle [3.30].

In order to improve the statistical weight of the reconstruction, an averaging procedure can be carried out in which averaged values of the required $G_{n,1}$ terms are determined from data of different particles. For the procedure the relative orientations of the projections are required, which will be derived by computation [3.47]. With one set of layer line data as a reference, the layer line data of the other particles are modified to account for a change in angular orientation, a phase change in the axial direction, and a change of the radial scale of the Fourier components. Each of these three parameters will be varied in order to find a set of values which minimize a phase residual R

$$ R = \left(\frac{\sum_i |F_i| \cdot [\Delta\phi_i]^2}{\sum_i |F_i|} \right)^{\frac{1}{2}} . \tag{3.23} $$

The summation involves all corresponding points (i) of the two transforms to be compared. In (3.23) $|F_i|$ denotes the average amplitude of the two points in the reciprocal space and $\Delta\phi_i$ the phase difference of the Fourier components. After this fitting procedure, the transforms will be modified so that they all are in the same orientation and on the same scale. The averaged set of layer lines obtained in this way will be used in the final reconstruction.

In a similar way, an internal correlation of the near and far sides of the layer lines of a given projection is possible, so that the quality of the transformed data can be expressed in a quantitative way. For example, in a helical structure with an even cylindrical symmetry, phase differences between meridional symmetric maxima on a given layer line must be ideally 0 or 2π as follows from (3.22). These phase differences will be π if the symmetry of the structure is odd. Deviations from these ideal values are likely to be caused by a lack of preservation of the structure during preparation but may also arise by a tilt of the particle axis out of the plane perpendicular to the electron beam [3.28].

3.2.7 Three-Dimensional Reconstruction of Particles with Icosahedral Symmetry

In order to obtain three-dimensional information about objects other than those with helical symmetry, several independent views must be combined in general. This leads to mathematical difficulties, as the Fourier components must be collected on a regular sampling lattice and the projections do not necessarily provide the information on such a sampling lattice. Therefore, an interpolation scheme has to be designed, which yields the required values of the Fourier components from a set of projections [3.48,49]. The number of projections necessary to solve the interpolation problem will depend on the size and symmetry of the object to be re-constructed.

Particles with icosahedral symmetry belong to the point group 532, which means that the symmetry axes are five-, three- and twofold ones. One single view of the particle will give a number of central sections of the three-dimensional transform. In the general case one projection yields 59 other symmetry related views and the three-dimensional transform can be filled in accordingly. If the direction of the view is along a rotation axis, the number of views will be reduced.

Electron micrographs of negatively stained virus particles which have the point group 532 generally contain projections in all kinds of orientations. To determine the relative orientation of a number of particles, use is made of the symmetry pro-perties of the Fourier transform. The transform of the projected density will always contain a set of pairs of lines, along which the transform should have identical values. By searching computationally for this set of lines the orientation relative to the symmetry axes of any view of the virus can be determined. Moreover, in this way the degree of specimen preservation can also be checked by measuring the degree of icosahedral correlation of the data along these lines.

The three-dimensional transform of the virus will be sampled on a set of planes perpendicular to the fivefold axis, whereas within these planes Fourier components are determined on annuli of constant R. As in the case of a helical reconstruction, cylindrical polar coordinates are used in such a way that Z coincides with the five-fold axis.

The Fourier transform $F(R,\Phi,Z)$ is then expressed in the form given by (3.18). The form of this equation implies that around an annulus of fixed R and Z the transform is a known function of Φ. In general, some or all of the central sections of the transform computed from the given views of the particle will intersect any particular annulus at points Φ_i, whose position are determined by the projection. On each an-nulus F_i may be written as

$$F_i = \sum_n (B_{in} G_n) \ , \quad B_{in} = \exp\left[in(\Phi_i + \frac{\pi}{2}) \right] \tag{3.24}$$

and one set of linear equations will exist for each annulus. From these equations, G_n are obtained as there are as many measured transform values F_a as unknowns G_n. The three-dimensional density distribution follows from equations of the type (3.20) and (3.21). In a number of cases, the values Φ_i are not well spaced over the annulus and consequently G_n will be poorly determined. The determination can be expressed as an eigenvalue problem and criteria may be derived which relate the number of particular views required to solve G_n with a known accuracy.

The method described has been applied to particles with 532 symmetry, but in principle the method may be extended to systems with lower point-group symmetry. More projections must then be combined to achieve the same degree of resolution.

3.3 Recent Applications to Image Processing of Regular Biological Structure

In order to discuss a number of recent applications of image processing in biological structural research, it is necessary to outline the biological background of
the particular problem. Therefore this section will be devoted to a description of
the results of each analysis in combination with its biological relevance.

3.3.1 One-Dimensional Filtering: Tropomyosin Paracrystal Structure [3.50]

Tropomyosin is a rod-shaped molecule with a molecular weight of about 65 000;
the molecule has a length of about 40 nm. Together with two other proteins, actin
and troponin, the tropomyosin molecule plays an important role in regulating the
contraction of vertebrate skeletal muscle. The tropomyosin molecule exists in two
forms, which have nearly identical structures. Crystals of tropomyosin can be
easily obtained and studied by X-ray diffraction analysis and electron microscopy.
The ordered aggregates formed by the action of magnesium ions are particularly suited
for electron microscopy and have a bipolar structure. The relative orientation of
the tropomyosin molecules in the lattice is known. Examples of a negatively stained
image of a paracrystal can be seen in Fig.3.6. The image exhibits a typical band-
ing pattern with a repeat of about 40 nm. Selective staining of the crystal was
carried out with a mercury-containing marker, which was known to locate the binding
site of the thiol groups in the tropomyosin. Reacted and unreacted crystals were
negatively stained and the electron images were analyzed by optical diffraction.
It appeared that, to a resolution of about 2.5 nm (see Fig.3.6), the two types of
lattices are isomorphous and that a comparison can therefore be made of the mass
distribution of the two types of cells. In order to carry out the location studies
of the heavy metal atom marker, low dose electron micrographs were taken (1-5 C/m^{-2})
and the images digitized. The mass distribution projected on a line parallel to the
orientation of the rodlike molecules was used for further digital processing. For
these images no negative staining was used. as the uranyl atoms would have obscured
the presence of the mercury atoms. Filtered projections were calculated from the
line projections of the treated and untreated crystals. From the results, it became
clear that in the mercury images an increased density pattern is present, which is
localized about 40 nm apart. The various filtered one-dimensional images and the
results of the subtraction of the two images are presented in Fig.3.7. The results
made it possible to derive the orientation of the rodlike tropomyosin molecules in
the paracrystal and in this way it was concluded that the troponin binding site
was situated approximately 14 nm from the carboxyl terminus of the molecule.

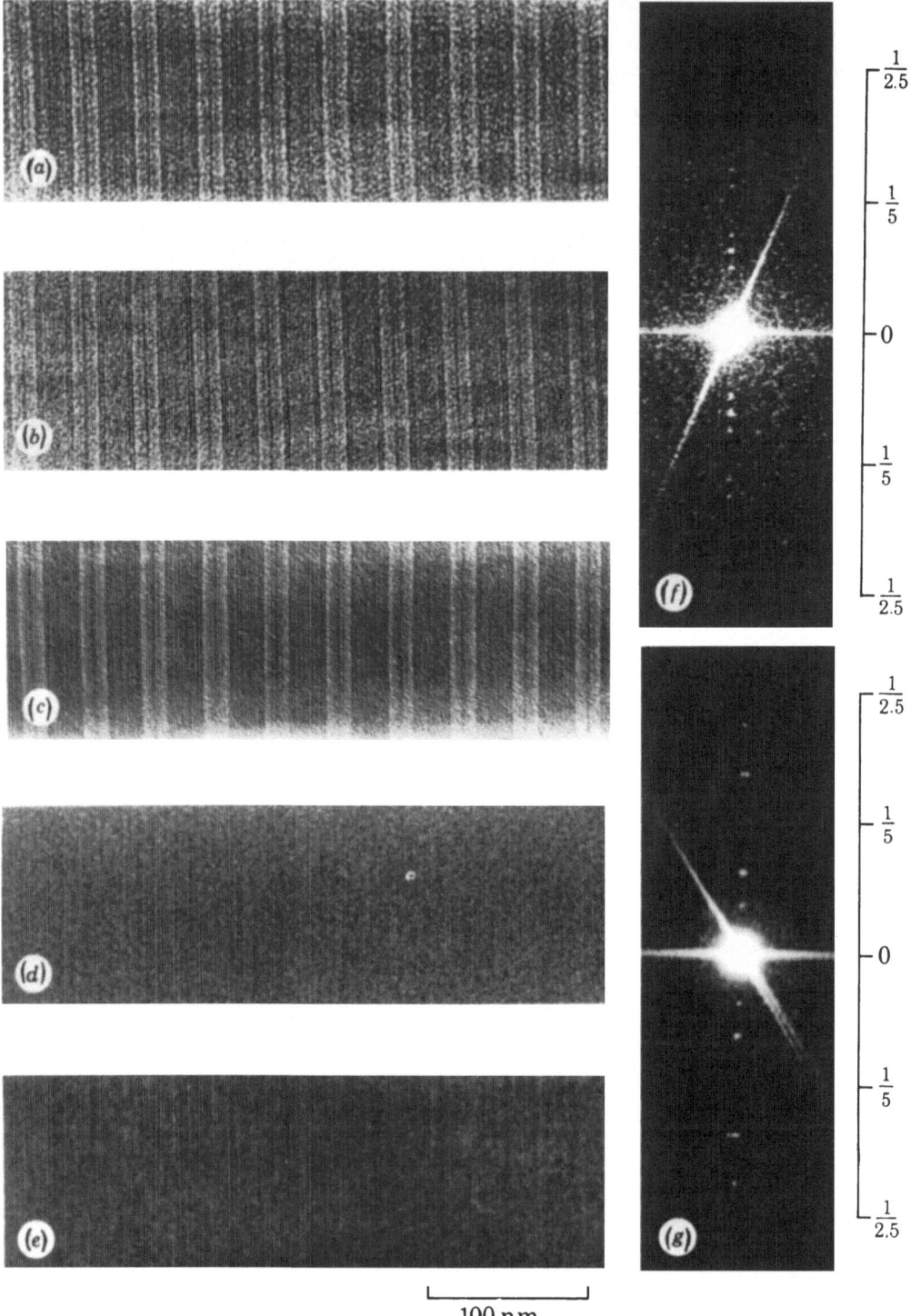

100 nm

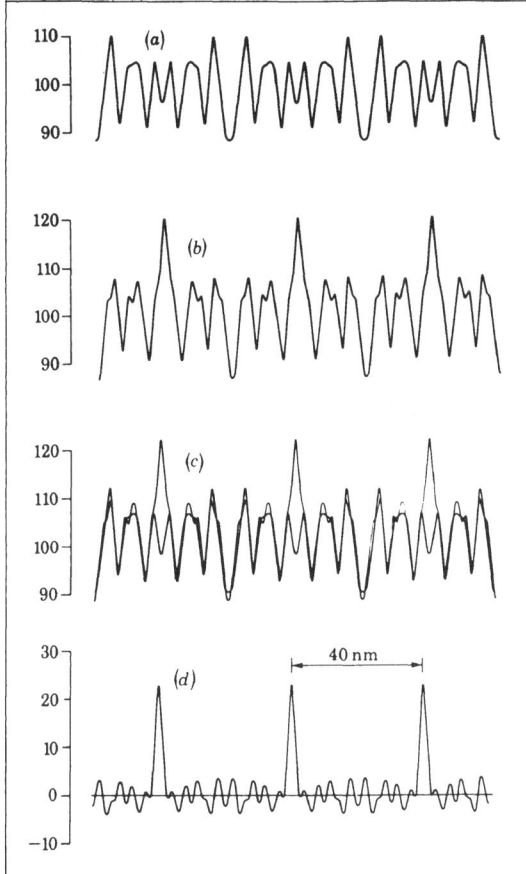

Fig.3.7a-d. The results of the
digital analysis of the line projections of the images shown in
Fig.3.6d and e. In (a) a control
image is shown in which no heavy
atom marker is present; (b) shows
the projection in which the
marker is present and (c) shows
the images (a) and (b) superimposed. The lower density variation
(d) resulted from subtracting (a)
and (b) and clearly shows the
increased density at 40 nm intervals. [3.50] (By courtesy of the
Royal Society)

Fig.3.6a-g. Crystals of tropomyosin and optical diffraction patterns derived from
the images. (a), (b), and (c) are crystals positively stained with uranyl acetate.
In (a) an untreated crystal is presented, whereas in (b) and (c) the crystals were
treated with the labelling agent under different conditions. In (c) the crystal
was treated with a thiol blocking agent prior to exposure with the labelling reagent,
so that no heavy metal is expected to bind the SH groups. In (d) and (e) unstained
images are presented of specimens; (d) was processed similarly to (c) while (e)
corresponds to the crystals in (b). The optical transform (f) was derived from
(d) and the transform (g) was obtained by analyzing (e). Reciprocal spacings in
nm^{-1} are shown at the right sides of both transforms. [3.50](By courtesy of the
Royal Society)

3.3.2 The Structure of Polyheads [3.51]

Polyheads are aberrant, hollow cylindrical structures, which are related to the cap-sids of T-even phages. These structures are convenient model systems for studying the morphogenesis of the bacteriophage capsid, which finally results in an icosa-hedral protein shell in which the genetic material (DNA) is packaged. The assembly of these structures is a series of chemical reactions in which cleavage and as-sociation of proteins play an important role. The major capsid protein has a mole-cular weight of 55 000 and is assembled in a hexagonal surface lattice in the mu-tant polyhead. The sequence of events in the morphogenesis of the polyheads can be extrapolated to the case of the icosahedral capsids. The lattice changes its dimensions (about 12%) during the morphogenesis and the unit cell content has a variable appearance in the pathway due to proteolysis and specific association of the 55 000 daltons molecular weight unit with other proteins. In Fig.3.8, a sche-matic pathway is presented for the various steps in the maturation of the polyheads. The results were mainly obtained by making use of electron microscopy combined with image analysis. At least four steps can be distinguished, based on the results of digital filtering of electron images. The steps in this scheme must be more com-plicated as only two-dimensional information was derived from the electron images.

Figure 3.9 shows examples of polyheads derived from classes II and III (see Fig.3.8); the optical transforms of the two types of structures are also shown. The transforms are clearly different, because in the upper one very strong second orders of the hexagonal lattice are present, which are not present in the lower one. On this basis, so-called narrow coarse-surfaced and expanded smooth-surfaced structures can be distinguished. As the protein walls of the cylindrical struc-tures are relatively thin, the images can be considered as two overlapping lattices which can be reconstructed as outlined in Fig.3.4. The right part of the figure shows near and far sides of both types of structure and a clear difference in mor-phology can be derived, which was thought to substantiate the reaction II → III in the scheme.

In this study a refinement procedure was used for the position of the lattice points in the digital Fourier transform. The search procedure used was based on maximizing the sum of intensities of all grid points in the hexagonal lattice.

Fig.3.8. A scheme for the structural transition of the surface lattice during the maturation of the capsid as derived from electron microscopy and image analysis. The surface lattices are all hexagonal; the class II to class III transition has been analyzed in the following figure. In going from II to III the lattice has ex-panded and the distribution of protein in the cell is different. It is believed that a new capsid protein is needed to stabilized the structure derived in IV [3.51]. (By courtesy of MIT Press)

111

I
uncleaved
("second
order")

II
cleaved
but
anchored
("second
order")

III
cleaved
and
transformed
("third
order")

IV

"fourth
order"

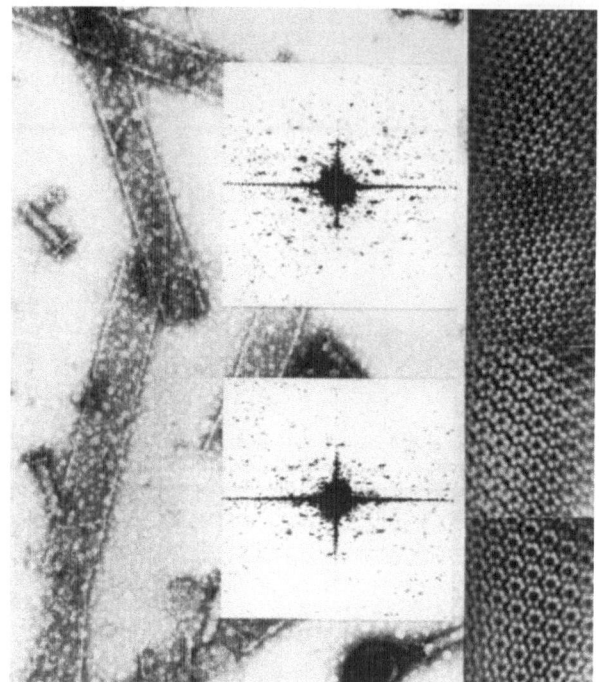

Fig.3.9. Negatively stained polyheads. The optical transforms also shown in the figure are representative of two types of surface lattices. The upper transform shows relatively strong second orders of a hexagonal lattice, whereas these features are absent from the lower transform. Filtered images of both sides of the structures are presented in the right part of the figure. The near and far side filtered images of both types of structure are clearly different. Not only is the unit cell size different but also the content of the cells of the two upper images differs from that of the two lower ones. Protein is white. The size of the hexagonal cells is approximately 11 nm and 12.5 nm for the upper and lower set of filtered images, respectively [3.51]. (By courtesy of MIT Press)

3.3.3 The Structure of Ribosomes [3.52,53]

Ribosomes are complex particles consisting of protein and nucleic acid molecules which are present in every type of cell. The function of these nucleoprotein particles is to direct cellular protein biosynthesis, a complex series of reactions in which a number of other types of biopolymers are also engaged. In recent years much biochemical knowledge about the composition of the ribosomes has been obtained. It is generally accepted that every ribosomal particle consists of about 50 different protein molecules and three types of ribonucleic acid molecules.

Thus far, electron microscopic observations have been carried out with single isolated ribosomal particles and although a number of useful results have been obtained in this way —for example about the relative localization of a number of proteins in the complex —quite a lot remains to be learned about the structure.

Recently, small ordered arrays of eukaryotic ribosomal particles have been obtained and by using image analysis methods the electron images of these two-dimensional crystals have been investigated. It is hoped that the size as well as the ordering of the crystals can be improved so that it may be possible to extend the analysis to a higher resolution than obtained thus far. The regular information in the microcrystals is present to a resolution of about 5 nm, as can be derived from the Fourier transforms of digitized images. The crystalline sheets are composed of two layers and in each of these the ribosomes are organized as tetramers on a P4 type of lattice. The two layers face in opposite directions and tend to be related to one another crystallographically, generating a family of P422 crystals of different unit cell dimensions. In Fig.3.10 a sheet of negatively stained ribosomes is presented. The image appears to be composed of a number of different superposition patches, due to the moiré effects of the two overlapping layers. Dotted lines in the electron image have been drawn to mark boundaries between different crystalline patches. Also arrays of dots have been drawn in the figure to indicate that similar lattices can be placed over the patches. It has been possible to determine the projected structure of one layer from the negatively stained images by separating the contributions from each layer in the digital Fourier transforms.

The ribosome appears to be constructed from a small and a large unit as indicated in Fig.3.11. The large subunits (L) make contacts within the tetramer. From the result it was concluded that the large-subunit small-subunit axis lies approximately parallel to the plane of the sheet.

3.3.4 The Structure of the Purple Membrane [3.54,55]

It has been generally accepted that radiation damage during imaging is one of the main reasons for specimen destruction. A particularly useful method of studying unstained biological structure has been put forward by HENDERSON and UNWIN [3.54]. Their method consists of embedding the specimen in amorphous glucose and then imaging the object with a very low electron dose (about 8 C/m^2). Another condition for the successful application of the method is that the specimen has to be regular. By using a regular object consisting of n repeating elements it can be shown that the signal-to-noise ratio of the Fourier components of the repeating element will be improved by a factor of $(n)^{\frac{1}{2}}$. To determine the Fourier components of the elementary cell a large number of these cells have to be present in the image, about 10^4 in the cited studies. Large periodic arrays of the membrane derived from *Halobacterium* were used for such an approach. Electron diffraction patterns and defocused bright field images were recorded under low dose conditions. The micrographs recorded appear featureless by eye, as the phase contrast effects due to defocusing are small compared to the statistical fluctuations of the number of electrons striking the photographic emulsion. The information from each individual unit cell has to be combined by densitometry and Fourier processing. A three-dimensional picture of the unstained

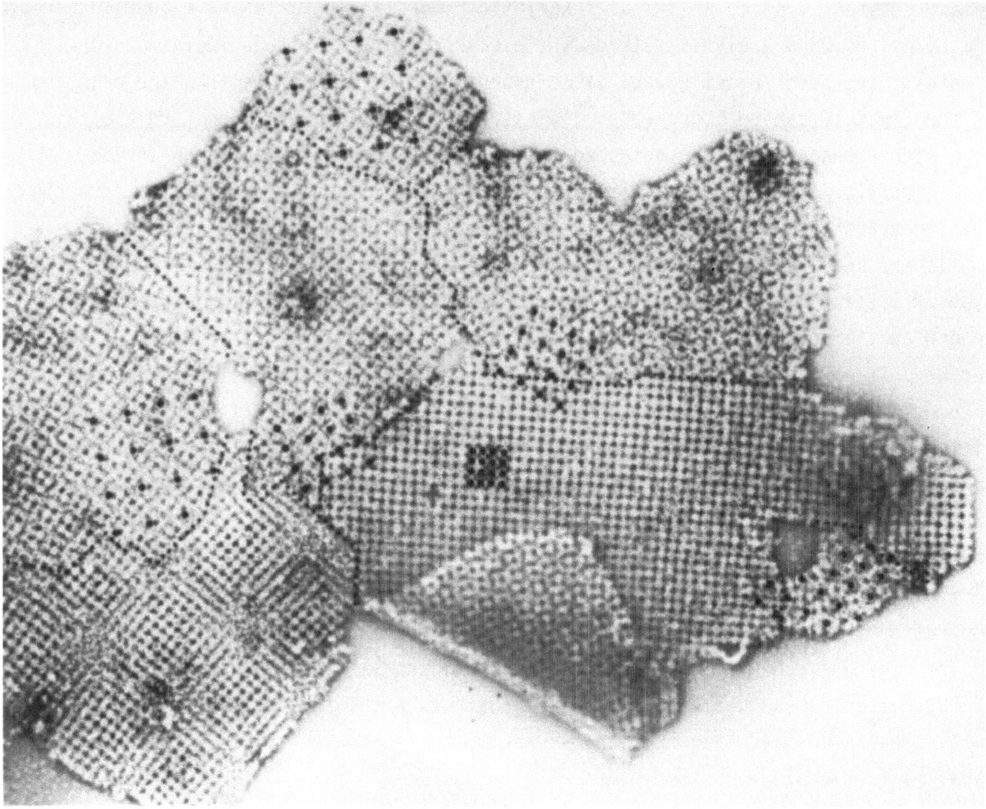

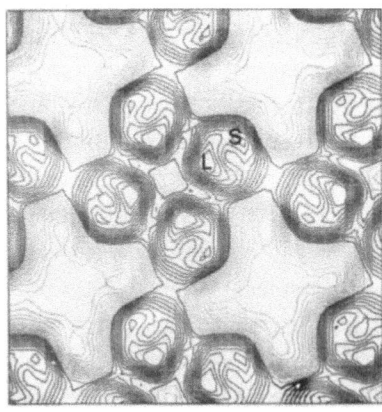

Fig.3.10. A sheet of negatively stained ribosomes which show a number of different superposition patterns due to two overlapping layers. Dotted lines have been drawn to show the boundaries between different crystalline patches. The arrays of dots in the micrograph have been drawn to show that similar lattices can be placed over the patches [3.52]. (By courtesy of Academic Press)

3.11 (10 a)

Fig.3.11. The projected layer of crystalline ribosomes obtained after a two-dimensional Fourier synthesis. The unit cell dimension from the center of one tetramer to the next is about 60 nm. Each ribosome is divided into a large (L) and small (S) subunit as indicated in the figure. The stain-excluding regions are indicated by the thicker lines [3.52]. (By courtesy of Acadmic Press)

material can likewise be derived from a series of tilted images obtained in a similar way. The principle for this approach has been outlined by DeROSIER and KLUG [3.1] as discussed in Sect.3.3. A number of images of the membrane are presented in Fig.3.12. It can be observed that the density pattern shows no indication of regularity. However, after Fourier processing and correcting the phases for the effects of defocusing, a potential distribution of the unit cell as shown in Figure 3.13 can be derived. This two-dimensional map shows the location of the protein and lipid components and also gives some information about the arrangement of the polypeptide chain within each protein molecule. Although the resolution has been limited to about 0.7 nm, the regular information in the diffraction pattern was present to spacings of about 0.3 nm. In order to derive a three-dimensional map, a series of tilts of the membrane was collected over a range from 0 to 57° with respect to the electron beam. The tilted images gave information about the continuous variation of the Fourier transform of the unit cell in three-dimensional Fourier space. By sampling the continuous curves at appropriate intervals, it was possible to calculate the three-dimensional transform as if for a three-dimensional repeating structure with a repeat distance along the C axis (perpendicular to the membrane plane) of arbitrary length but greater than the thickness of the membrane layer. Owing to physical limitations, tilt angles beyond 57° were impossible; this meant that a cone of reciprocal space remained unmeasured. Although to a resolution of 0.7 nm this volume is about 37% of the total sphere, the effect of omitting these data from the three-dimensional Fourier synthesis is negligible as the amplitudes in this region are very small. In this way, a 0.7 nm resolution map of the three-dimensional membrane structure was derived. This showed that the membrane protein consisted of seven closely packed α-helical segments which extend approximately perpendicular to the plane of the membrane for most of its width. The space between the protein molecules is filled by lipid bilayer regions.

This study showed that electron microscopy of unstained biological material is feasible and that meaningful biological information can be derived by using well-controlled irradiation conditions combined with image processing methods.

3.3.5 A Correction for Distorted Images [3.56]

The surfaces of bacterial cells are covered with regular arrays of protein subunits and the biological function of these layers is not well understood. The regular arrays can be prepared relatively easily and studied by electron microscopy using for example the negative staining technique. A number of problems have to be overcome in order to derive a structural model of these layers, which is necessary to understand their function. These problems arise because the images are very noisy and moreover overlapping parts of different layers obscure the details in many cases. This last obstacle may be overcome by reconstructing the individual layers of the overlapping components with the method outlined in Sect.3.2. Also, when per-

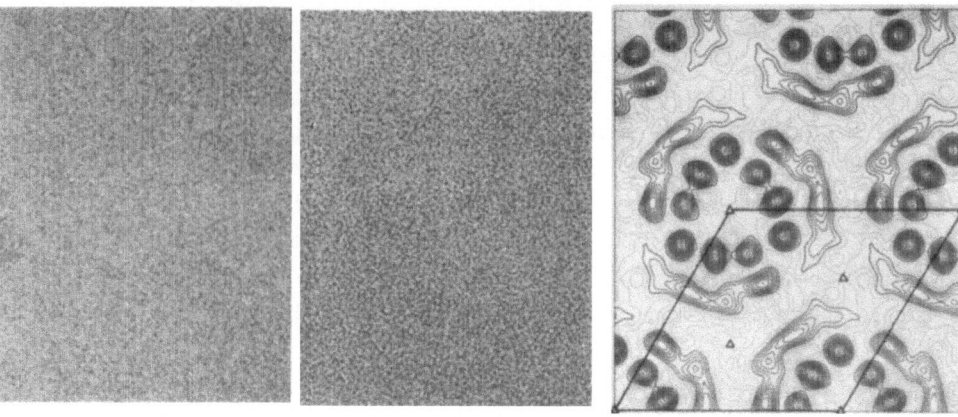

Fig.3.12.   Fig.3.13.

Fig.3.12. Bright-field electron images of glucose-embedded purple membrane. Left:
a low dose image (about 8 C/m^2) and right the image obtained at a higher irradiation
level (about 4000 C/m^2). The magnification is about 300 000 × [3.55]. (By courtesy
of Academic Press)

Fig.3.13. The potential distribution of the projected membrane structure at a resolution of 0.7 nm. Positive contours are indicated by thicker lines and are due
to the presence of high concentrations of scattering material such as protein. The
unit cell dimensions are 6.2 × 6.2 nm^2 [3.55]. (By courtesy of Academic Press)

forming this analysis a spatial averaging is achieved, which results in a better
signal-to-noise ratio. There is another image defect, inherent in the use of these
types of specimens, which arises through their tendency to curve or buckle over a
few unit cells. It is this type of distortion that can be corrected for by computer.

Figure 3.14 shows two negatively stained images of tetragonal layers from *C.
thermosaccharolyticum* and the corresponding optical diffraction pattern. The procedure proposed by CROWTHER and SLEYTR first constructs a low-resolution lattice of
the array by using large circular apertures in the filter mask and this lattice
enables defects to be mapped out. A search for the centers of gravity of each unit
cell is the next step. A mean undistorted lattice is constructed with unit cell
parameters equal to the average of the unit cell vectors found in the piece of distorted lattice. The mean lattice is placed over the distorted lattice in such a way
that the centers of gravity coincide. The root mean square displacement of the distorted lattice from the average undistorted lattice will give a measure of the distortion. There are two possible ways of constructing an image of the average unit
cell from these data. In the first method, a bilinear interpolation of density sets
of corresponding points within each distorted cell followed by an averaging of the
resulting contributions will yield the average information. This approach is outlined in Fig.3.15. The second method consists of reinterpolating the entire image
to remove the distortions, followed by Fourier transformation and Fourier synthesis
using the data at and around the exact reciprocal lattice points. A further rotational averaging may also be performed as the layers appear to possess hexagonal

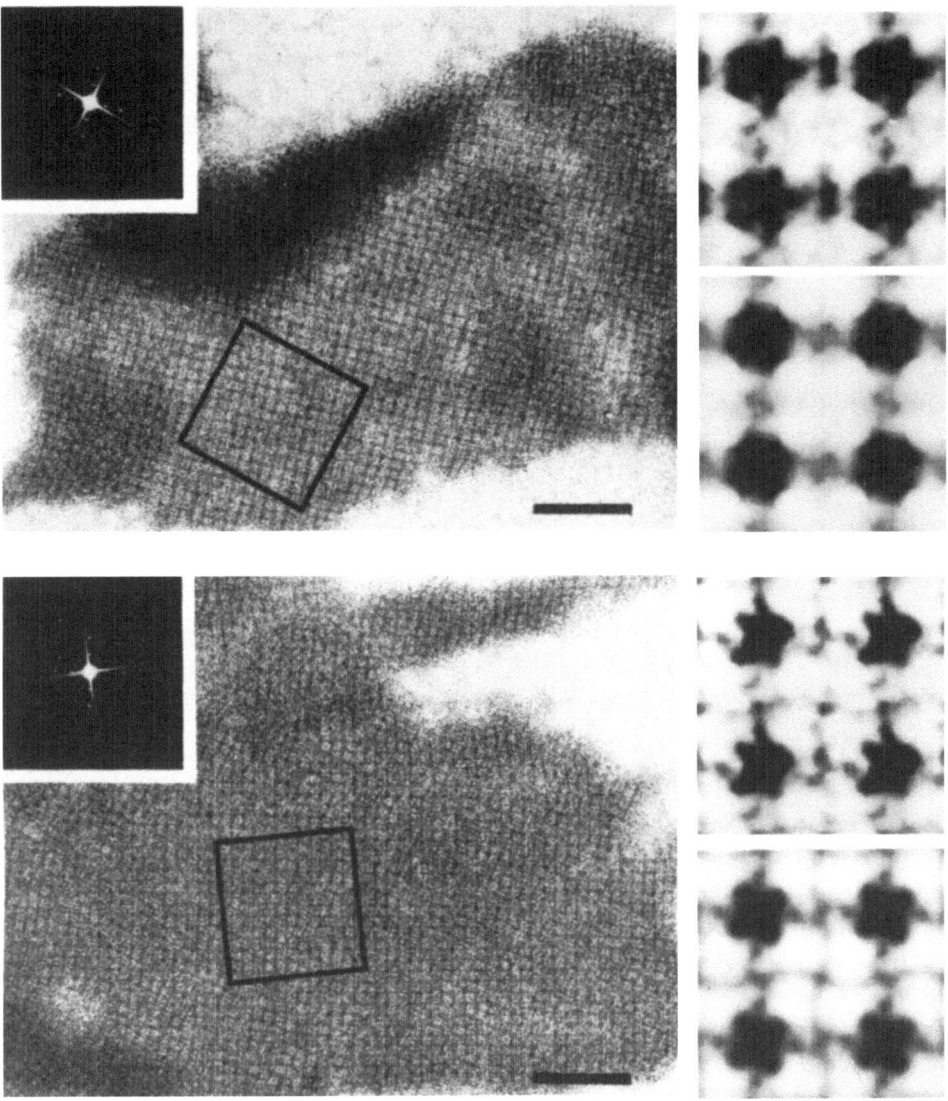

Fig.3.14. Negatively stained images of surface layers of *C. thermosaccharolyticum*. The mean tetragonal unit cell size is about 11.5 nm. The insets are optical diffraction patterns of the corresponding boxed areas. The upper part of the figure shows images derived from a single layer. The lower part is a superposition of two layers. At the right side of the figure, translationally and rotationally averaged versions of the corrected images are presented. The bar in the micrographs represents 100 nm [3.56]. (By courtesy of Academic Press)

or tetragonal symmetry. Figure 3.14 also presents the results of this procedure applied to the negatively stained images. The upper part of the figure consists of a single array, whereas the lower image is a superposition pattern derived from two layers, with a difference in orientation of about 25°. This can be clearly seen in

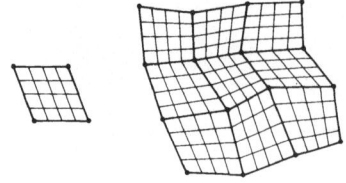

Fig.3.15. A scheme which indicates how a distorted lattice (right) and an undistorted average unit cell (left) are correlated. The dots represent the centers of the repeating motif as determined from the low-resolution image (lattice image), whereas the heavy lines represent the edges of the unit cells. Each cell edge is divided into a number of equal parts. By joining corresponding pairs of points on opposite cell edges a distorted sampling grid is generated. The density values at these grid points can be interpolated from the array of densitometer samples and used to determine the average unit cell [3.56]. (By courtesy of Academic Press)

the optical transform. The right part of the figure shows the corrected translational average and also the rotationally averaged version. In the processed images clear tetragonal morphological units are visible which are joined by fine bridges.

3.3.6 Rotational Filtering of Base Plates [3.57,58]

A large number of biological structures have been assembled according to a point-group symmetry (see Fig.3.1). This means that the resulting structure has a finite extent and that the symmetry operations which describe the structure only consist of rotational symmetry axes. Electron micrographs of these types of protein aggregates will exhibit rotational symmetries, provided a view of the protein assembly is obtained along one of these axes. A method of analyzing these images has been proposed by CROWTHER and AMOS and involves the decomposition of the image into circular harmonics. The relative weights of the circularly symmetric components in a given image up to a certain resolution can be used to draw conclusions about the symmetry. Filtered images can be synthesized by combining those components which are consistent with the symmetry.

The method is based on the following decomposition (see also Sect.3.2):

$$OD(r,\phi) = \sum_{-\infty}^{+\infty} g_n(r)\exp(in\phi) \tag{3.25}$$

in which OD denotes the optical density of an image element (r,ϕ). The weight or power of a particular component with symmetry n can be expressed as

$$P_n = \int_0^a |g_n(r)|^2 r\,dr \quad . \tag{3.26}$$

In this expression, a represents the maximum radius of the projection to be analyzed. The base plate, which has a hexagonal shape as can be observed from the micrograph, is a structure that contains about 14 different types of proteins. This structure together with other structural components from the bacteriophage plays a role in the attachment of the virus to a bacterial cell which is required for injection of the genetic material into the host cell. The base plate undergoes a transition from a hexagonal to a star-shaped configuration during the infection cycle.

In Fig.3.16, a negatively stained specimen of base plates is presented. A number of well-defined hexagonal particles can be observed. The result of the analysis described above is shown in Fig.3.17. This graph represents a power spectrum of the different rotational frequencies present in the image. The maxima in this graph are clearly situated at n = 0 and integer multiples of 6, which reflects the sixfold symmetry of the protein aggregate. Filtered images of two of the base plates are also presented in Fig.3.16; these images were obtained by using only the sixfold symmetrical components. In this way the basic morphology of the aggregates can be derived and by combining this method with genetic analysis it has been possible to map a number of the structural proteins. Also the rearrangement of the hexagonal into the starlike aggregates in terms of the constituent protein molecules can be explored. The analysis was carried out with a number of mutant base plates in which specific structural proteins are lacking.

The result of the analysis is summarized in Fig.3.18, in which both configurations are presented, together with a number of descriptive names for the parts. It has been possible to allocate three types of protein molecules to the features presented in the figure. For example one (gene product 9) occupies a peripheral position in both the hexagons and stars.

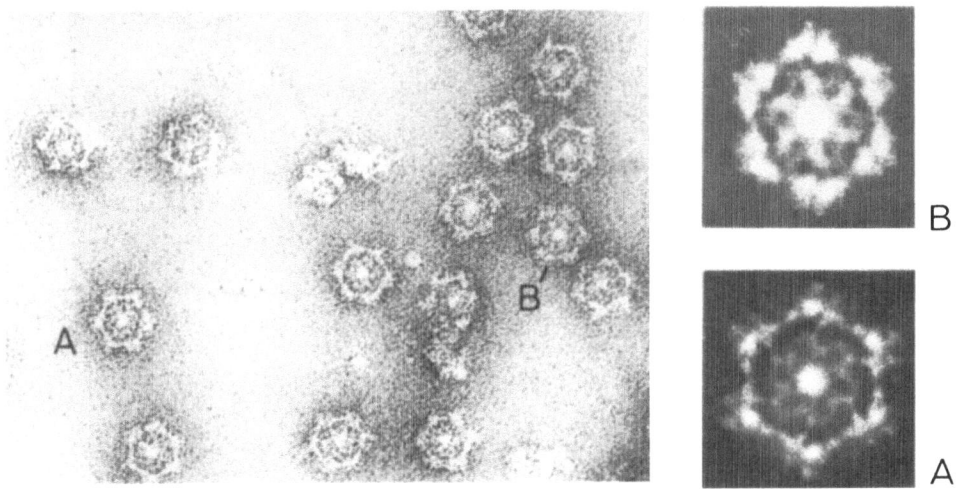

Fig.3.16. A negatively stained preparation of base plates from bacteriophage T4. The diameter of the hexagonal particles is about 40 nm. On the right, two rotationally filtered images are shown which were derived from the micrograph. The density levels of plotting are different in the filtered images; this explains the differences in appearance, although the main features are similar [3.57]. (By courtesy of Academic Press)

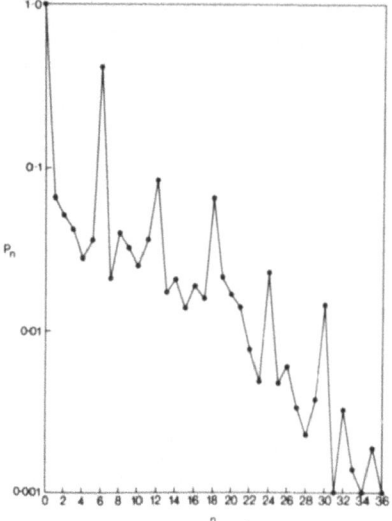

Fig.3.17. A plot of the power spectrum of a bacteriophage T4 base plate (particle A in the previous figure), in which the sixfold character is clearly present. The curve is normalized with $P_0 = 1$ and the power associated with $n = 37$ and higher is less than 0.001 [3.57]. (By courtesy of Academic Press)

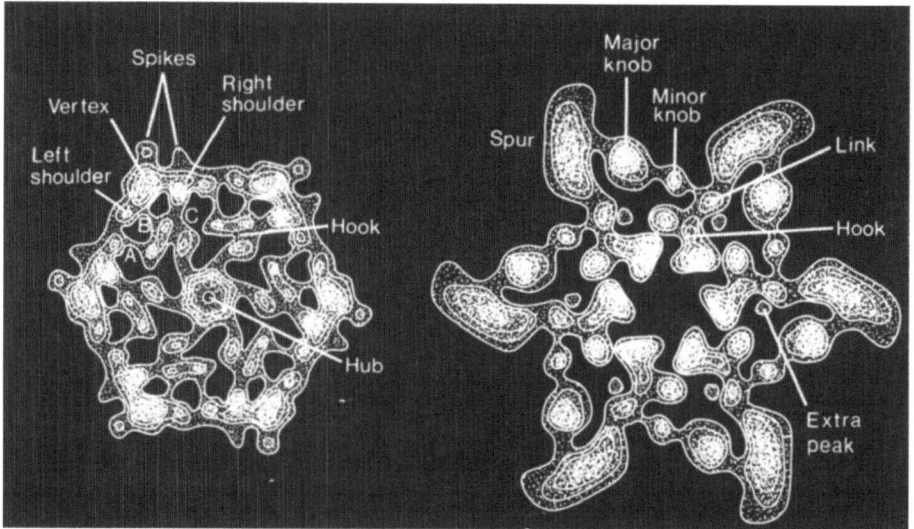

Fig.3.18. A schematic diagram of the gross morphology of the hexagon (left) and star (right) forms of the base plates. The features seen in the images were derived by digital filtering of the two types of structures. The scale is such that 1 cm corresponds to approximately 9 nm. Protein is white [3.50]. (By courtesy of Academic Press)

3.3.7 The Structure of the Contractile Sheath from Bacteriophage Mu [3.59,60]

Many bacteriophages, such as the T-even family and Mu, possess a contractile tail which plays a role in the injection process of the virus DNA in the host cell. The tail can be considered as a kind of primitive muscle and the structural transition during contraction can be studied from negatively stained images. By using the

three-dimensional reconstruction technique a detailed picture can be derived from
the two states of the contractile structure. In Fig.3.19, contracted and extended
tails of bacteriophage Mu are presented. The tail consist of a tail tube and a contractile sheath, which consists of 192 identical protein subunits with a molecular
weight of about 50 000 daltons.The structural transition does change the quaternary
structure, which can easily be determined by optical diffraction of sheath images.

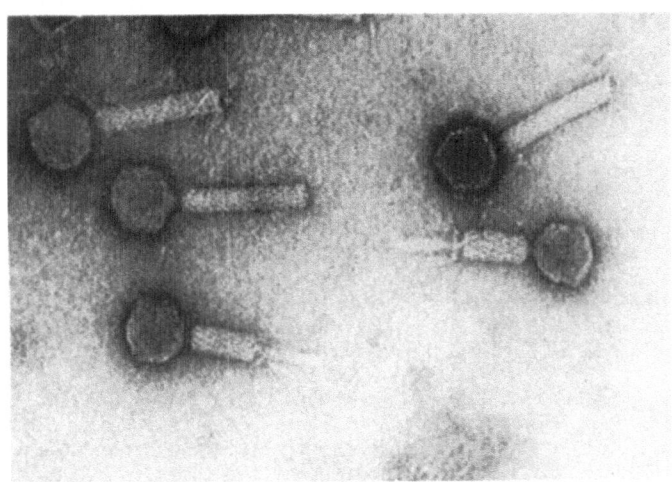

Fig.3.19. A negatively stained specimen of an extended and contracted tail of bacteriophage Mu. The phage particles consist of a head and a tail of variable length;
the short tails are contracted and another phage component, the tube, becomes visible in these cases

The results of this type of analysis as well as the prospects for a three-dimensional analysis are determined to a great extent by the length of the structure
to be analyzed. This is illustrated in Fig.3.20, where a single contracted sheath
and a polymer of these sheath particles are shown, together with the optical transforms. The surface lattice can be derived with much more confidence in the case
of the polymer.

The results of a comparison of the shape of the protein subunits in the extended
and contracted sheath of bacteriophage Mu are shown schematically in Fig.3.21. In
the extended sheath, the subunits are more curved and the long axis of the molecule
runs roughly normal to the helical axis. Six subunits are arranged in an annular
ring and rings of subunits are stacked on top of each other with a fixed azimuthal
disposition from ring to ring. Similarly, the contracted sheath structure can also
be described by a stack of sixfold rings. However, in this form the protein subunits appear to have a much more elongated form, whereas their long axis points
more in a radial direction and is slightly tilted out of a horizontal plane.

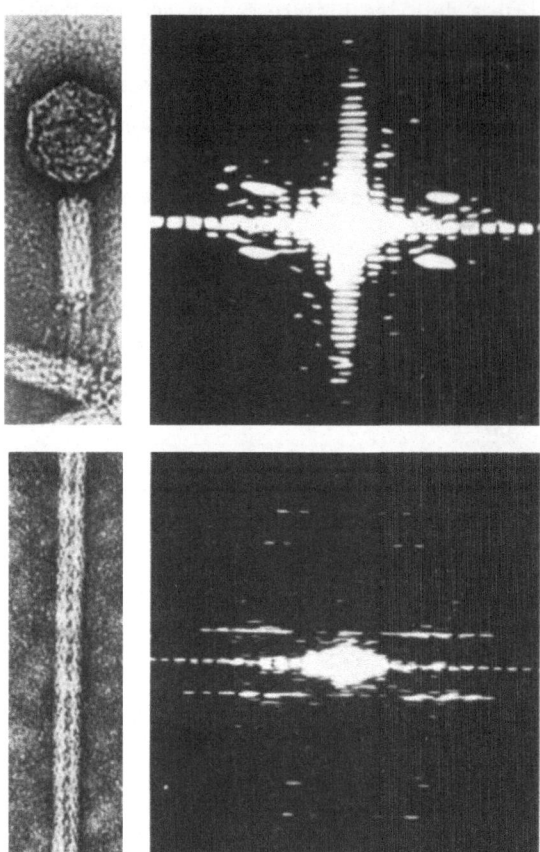

Fig.3.20. Single and polymerized contracted sheath structures derived from bacteriophage Mu as imaged by the negative staining technique. At the right side of each image the corresponding optical transform is shown. The diameter of the tubes is about 25 nm. The effect of the polymerization on the layer line pattern in the optical transform can easily be seen

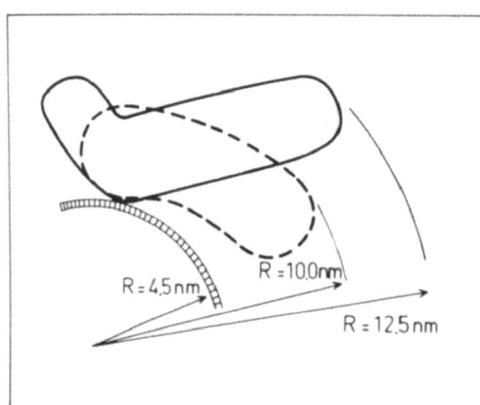

Fig.3.21. A scheme of the tentative shapes of the sheath subunit in the extended and contracted configuration as derived from the results of the three-dimensional reconstruction of the sheath structure. The approximate diameters of the two types of assemblies are indicated in the figure. The shape of the protein subunit in the extended state is drawn by a dashed line, the contracted configuration is shown by a solid line

It is likely that the subunits change shape when the transition from extended to contracted sheath takes place. This conclusion emerges from the three-dimensional reconstruction studies of negatively stained images; evidence from other biophysical techniques has supported this conclusion.

3.3.8 The Three-Dimensional Structure of an Icosahedral Virus Particle [3.61,62]

A large number of plant and animal viruses have icosahedral symmety, which means that the coat is constructed from a number of proteins which are related by point-group symmetry 532. In all cases the number of identical coat protein molecules is a multiple of 60T in which T can take only integer values, of which the smallest are 1, 3, 4, and 7. The coat protein molecules are clustered over the spherical lattice and in principle three extreme situations of the clustering pattern can be distinguished. Dimer, trimer of hexamer-pentamer clustering patterns are possible and these patterns may be observed in negatively stained specimens. Figure 3.22 shows a field of spherical particles which are derived from *Nudaurelia capensis* β virus (NβV). In most of the projected images a superposition pattern is visible which makes it almost impossible to derive the surface topology of the virus. However, in some cases the particles show a symmetrical pattern because a projection down a symmetry axis is observed.

These symmetrical projections are a good starting point for further analysis, which can be performed by analyzing the digital Fourier transforms of the particles. The "common lines" technique yields the relative orientation of a number of particles and generally a few (three or four) different projections are sufficient to determine the three-dimensional Fourier transform of the virus particle. Back transformation of these data will yield a three-dimensional density map, which provides information about the surface structure of the spherical virus. It has not been

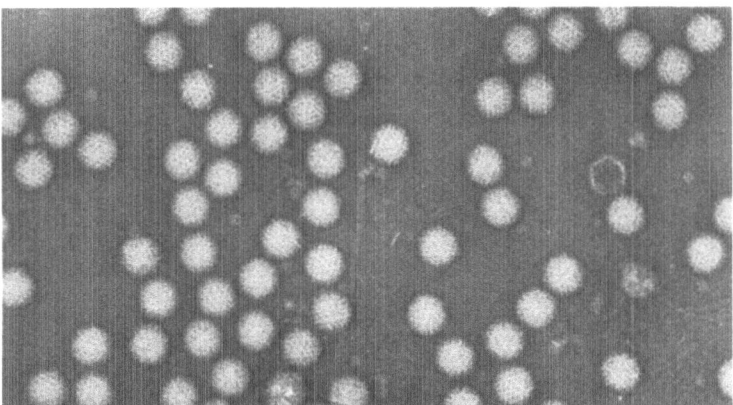

Fig.3.22. A negatively stained preparation of Nβ virus. The approximate diameter of the virus is 30 nm [3.61]. (By courtesy of Cambridge University Press)

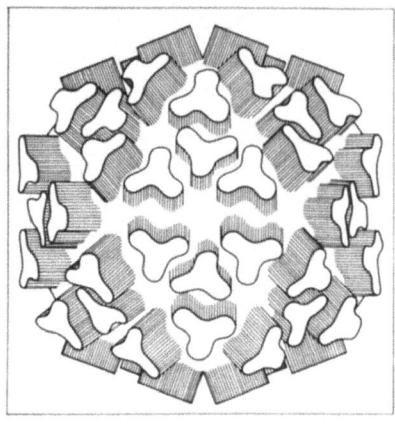

Fig.3.23. A schematic drawing,
which summarizes the result
of the surface morphology of
Nβ virus. [3.61] (By courtesy
of Cambridge University Press)

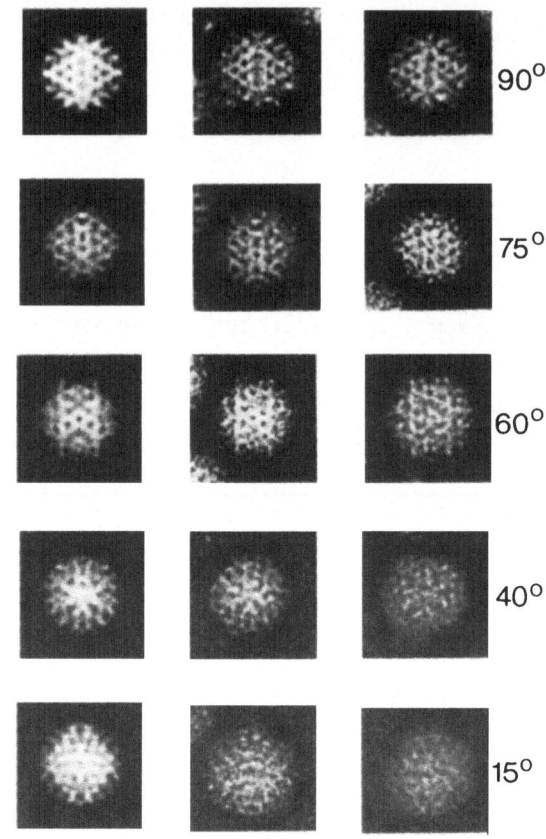

Fig.3.24. A comparison of the projections of the reconstructed Nβ virus and a number
of electron images of the virus. The left vertical row consists of a number of different projections, specified by values of θ = 90 and variable values of φ as given
in the right part of the figure. The convention of KLUG and FINCH [3.62] was used.
The two columns (right) are selected particles from the electron micrographs, which
show a striking similarity to the results obtained from the model [3.61]. (By courtesy
of Cambridge University Press)

possible to derive data at very small radii, where the nucleic acid is present. It
is known that the nucleic acid is not packed in a regular way, so that on averaging
the data from the projections no specific information will result on this part of
the structure. The three-dimensional map of the virus structure makes it possible
to derive the clustering pattern of the protein and a large number of viruses have
been analyzed in this way. Dimer and hexamer-pentamer clustering patterns have been
frequently encountered.

The three-dimensional data also enable one to simulate projections and the simulated patterns can be compared with the projections present in the micrographs.
Figure 3.23 shows the result of a three-dimensional reconstruction of the Nβ virus
in a schematic way. The outer part of the surface has a typical appearance in the
form of Y-shaped columns. These protein columns are confined within the triangular

faces of the circumscribing icosahedron and are centered around the threefold axis. Therefore coat structure can be described by a trimer clustering pattern divided over a T = 4 icosadeltahedron. There are 80 of these Y-shaped (trimer) columns, representing the morphological features of the 240 identical protein subunits which are quasi-equivalently related in the surface lattice. In Fig.3.24 a comparison of the projections of the three-dimensional model and a number of selected particles from the electron micrographs are presented. There is a striking correspondence between the projections down the twofold axis (ϕ = 90°); however, the number of striking resemblances to the simulated patterns was very small, apparently because of the fairly even distribution of the surface detail.

3.4 Outlook

The rationale behind the digital image processing techniques is based on the theory of the Fourier transform. This transformation is particular useful in the case of regular objects. It can be expected that in the coming years much more biological-ly relevant information to a resolution of about 2 nm will be gathered from elec-tron micrographs by using the methods described in this chapter, or more refined versions of them. No doubt, these methods will also be required in the study of un-stained material, in order to surpass the 2 nm resolution level; such an advance will be of fundamental importance for the study of biological structure by electron microscopy.

References

3.1   D.J. DeRosier, A. Klug: Nature *217*, 130-134 (1968)
3.2   R.A. Crowther, D.J. DeRosier, A. Klug: Proc. R. Soc. London A*317*, 319-340 (1970)
3.3   A. Klug, J.E. Berger: J. Mol. Biol. *10*, 565-569 (1964)
3.4   L.A. Amos, A. Klug: Manchester (1972) pp.580-581
3.5   R.A. Crowther, A. Klug: Ann. Rev. Biochem. *44*, 161-182 (1975)
3.6   R.A. Crowther: Proc. Third John Innes Symp. (1976) pp.15-25
3.7   M. Beer, J. Frank, K.J. Hanszen, E. Kellenberger, R.C. Willams: Quart. Rev. Biophys. *7*, 211-238 (1975)
3.8   P.N.T. Unwin, R. Henderson: J. Mol. Biol. *94*, 425-440 (1975)
3.9   A.V. Crewe: Quart. Rev. Biophys. *3*, 137-175 (1970)
3.10 J.D. Bernal: Quart. Rev. Biophys. *1*, 81-87 (1968)
3.11 A. Klug: "Symmetry and Function of Biological Systems at the Macromolecular Level", in *Nobel Symposium 11*, ed. by A. Engström, B. Strandberg (Almqvist and Wiksell, Stockholm 1968) pp.425-436
3.12 P.B. Moore, H.E. Huxley, D.J. DeRosier: J. Mol. Biol. *50*, 279-295 (1970)
3.13 J.E. Mellema, A. Klug: Nature *239*, 146-150 (1972)
3.14 T. Anderson: Trans. N.Y. Acad. Sci. *16*, 242-249 (1954)
3.15 J. Kistler, U. Aebi, E. Kellenberger: J. Ultrastruct. Res. *59*, 76-85 (1977)
3.16 R. Heidenreich: *Fundamentals of Transmission Electron Microscopy* (Inter-science, New York 1964) pp.16-47
3.17 E. Zeitler, G.F. Bahr: J. Appl. Phys. *30*, 940-949 (1959)
3.18 G.F. Bahr, E. Zeitler: Lab. Invest. *14*, 955-977 (1965)
3.19 J.M. Cowley: Acta. Cryst. A*29*, 529-536 (1973)

3.20 H.P. Erickson: Adv. Opt. Electron Microsc. *5*, 163-199 (1973)
3.21 P.N.T. Unwin: J. Mol. Biol. *87*, 657-670 (1974)
3.22 I.A.M. Kuo, R. Glaeser: Ultramicroscopy *1*, 53-66 (1975)
3.23 H.C. Andrews: *Computer Techniques in Image Processing* (Academic Press, New York 1970)
3.24 J.W. Goodman: *Introduction to Fourier Optics* (McGraw-Hill, New York 1968)
3.25 R.N. Bracewell: *The Fourier Transform and Its Applications* (McGraw-Hill, New York 1965)
3.26 A. Papoulis: *The Fourier Integral and Its Applications* (McGraw-Hill, New York 1962)
3.27 J.W. Cooley, J.W. Tukey: Math. Comput. *19*, 297-301 (1965)
3.28 J. Max: IRE Trans. *IT-6*, 7-12 (1960)
3.29 D.J. DeRosier, P.B. Moore, J. Mol. Biol. *52*, 355-369 (1970)
3.30 A. Klug, J.E. Berger: J. Mol. Biol. *10*, 565-569 (1964)
3.31 C.E. Shannon: Proc. I.R.E. *37*, 10-21 (1949)
3.32 H. Lipson, W. Cochran: *The Determination of Crystal Structures* (Bell, London 1966)
3.33 A. Klug, D.J. DeRosier: Nature *212*, 29-32 (1966)
3.34 M.J. Ross, M.W. Klymkowsky, D.A. Agard, R.M. Stroud: J. Mol. Biol. *116*, 635-659 (1977)
3.35 H.P. Erickson: J. Cell. Biol. *60*, 153-167 (1974)
3.36 R.H. Crepeau, B. McEwen, G. Dykes, S.J. Edelstein: J. Mol. Biol. *116*, 301-315 (1977)
3.37 U. Aebi, P.R. Smith, J. Dubochet, C. Henry, E. Kellenberger: J. Supramol. Struct. *1*, 498-522 (1973)
3.38 W. Cochran, F.H.C. Crick, V. Vand: Acta Cryst. *5*, 581-586 (1952)
3.39 A. Klug, F.H.C. Crick, H.W. Wyckoff: Acta Cryst. *11*, 199-213 (1958)
3.40 R.A. Crowther, L.A. Amos: J. Mol. Biol. *60*, 123-130 (1971)
3.41 P.N.T. Unwin, A. Klug: J. Mol. Biol. *87*, 641-656 (1974)
3.42 J.A. Lake, H.S. Slayter: J. Mol. Biol. *66*, 271-282 (1972)
3.43 T. Wakabayashi, H.E. Huxley, L.A. Amos, A. Klug: J. Mol. Biol. *93*, 477-497 (1975)
3.44 L.A. Amos, A. Klug: J. Cell. Sci. *14*, 523-549 (1974)
3.45 J.A. Spudich, H.E. Huxley, J.T. Finch: J. Mol. Biol. *72*, 619-632 (1972)
3.46 R. Sperling, L.A. Amos, A. Klug: J. Mol. Biol. *92*, 541-558 (1975)
3.47 L. Amos, A. Klug: J. Mol. Biol. *99*, 51-73 (1975)
3.48 R.A. Crowther, D.J. DeRosier, A. Klug: Proc. R. Soc. London A*317*, 319-340 (1970)
3.49 R.A. Crowther: Phil. Trans. R. Soc. London B*261*, 221-230 (1971)
3.50 M. Stewart: Proc. R. Soc. London B*190*, 257-266 (1975)
3.51 U.K. Laemmli, L.A. Amos, A. Klug: Cell *7*, 191-203 (1976)
3.52 P.N.T. Unwin: Nature *269*, 118-122 (1977)
3.53 P.N.T. Unwin, C. Taddei: J. Mol. Biol. *114*, 491-506 (1977)
3.54 R. Henderson, P.N.T. Unwin: Nature *257*, 28-32 (1975)
3.55 P.N.T. Unwin, R. Henderson: J. Mol. Biol. *94*, 425-440 (1975)
3.56 R.A. Crowther, U.B. Sleytr: J. Ultrastruct. Res. *58*, 41-49 (1977)
3.57 R.A. Crowther, L.A. Amos: J. Mol. Biol. *60*, 123-130 (1971)
3.58 R.A. Crowther, E.V. Lenk, Y. Kikuchi, J. King: J. Mol. Biol. *116*, 489-523 (1977)
3.59 G. Admiraal, J.E. Mellema: J. Ultrastruct. Res. *56*, 48-64 (1976)
3.60 A.F.M. Cremers, A.M.H. Schepman, M.P. Visser, J.E. Mellema: Eur. J. Biochem. *80*, 393-400 (1977)
3.61 J.T. Finch, R.A. Crowther, D.A. Hendry, J.K. Struthers: J. Gen. Virol. *24*, 191-200 (1974)
3.62 A. Klug, J.T. Finch: J. Mol. Biol. *31*, 1-12 (1968)

4. Three-Dimensional Structure Determination by Electron Microscopy (Nonperiodic Specimens)

W. Hoppe and R. Hegerl

With 23 Figures

The three-dimensional reconstruction of individual molecules from a set of electron micrographs is comparable to the "reconstruction" of structures in X-ray crystallography. Basically the electron microscope is equivalent to a diffractometer, since the specimen is illuminated by a parallel primary beam with a well-defined wavelength and orientation. The Fourier inversion of an electron micrograph delivers the distorted structure factors in a well-defined plane or weakly curved surface of the three-dimensional reciprocal space. Extraction of the undistorted structure factors and reconstruction is possible if a sufficient number of micrographs with different orientations of the object is available. Structure factors within a certain resolution in reciprocal space must be known to give a well defined reconstruction. Unlike X-ray crystallography, no "phase problem" exists since the phases of the structure factors can be deduced from the micrographs.

In the case of electron microscopy with non-periodic specimens the crystal degenerates to a single unit cell filled with the finite object. Although no crystallization is necessary, the same Fourier transformation on a number of discrete structure factors leads to the reconstructed object. There is, however, the problem of interpolation, since the reciprocal space cannot be scanned completely due to experimental restrictions. Several reconstruction methods are already well known and some new methods will be proposed in this chapter. These methods depend on axial or conical tilting.

As in conventional electron microscopy, the significance of the reconstructed object depends on the electron noise caused by the applied electron dose. An additional source of distortion is the so-called clutter, consisting of elongated image points and "noise-like" variations in the background. The elongation is due to the restricted tilting range while the background variations are created by large tilting increments.

128

4.1 History and General Discussion of the Subject[1]

One can state today that three-dimensional electron microscopy is the natural extension of conventional two-dimensional microscopy —especially if high-resolution work near or at atomic resolution is the ultimate aim. The reason is simply that the specimens are too thick compared with the possible resolution; therefore, serious overlaps must obscure the conclusions drawn from conventional micrographs. For two reasons the situation is much less favorable than in the related field of light microscopy: Firstly the high numerical apertures of light optical microscopes (near the limit of resolution) allow a selective inspection of different "focusing planes" in a "thick" object (through-focusing), especially if by incoherent imaging the overlapping images of the adjacent planes are nearly completely destroyed and contribute only to the general bright-field background. Secondly it is obviously possible to prepare specimens thin with respect to any light optical resolution and to study a thicker specimen in successive microtome sections. A conventional electron microscope has an aperture of only few degrees and images under nearly ideal coherent conditions. Thus the images of adjacent sections do not disappear —in a first approximation, we get a parallel projection of a three-dimensional specimen. It will, however, be demonstrated later that this coherent "overlapping" imaging is a fundamental advantage rather than a disadvantage (as perhaps suspected at first sight). Furthermore, obviously even the best ultramicrotome techniques cannot provide, for example, the necessary 0.5 nm thick sections of a macromolecule for a systematic three-dimensional high-resolution study since the cutting procedure would destroy the molecular structure. On the other hand, it has been recognized [4.2,3] that the extra information gained by optical spreading of the image along the third spatial coordinate (parallel to the optical axis of the microscope) is not associated with a corresponding increase of the necessary illumination dose (as in the light microscope). It is hardly possible to overestimate the importance of this finding for the concept of three-dimensional microscopy. Two-dimensional microscopy is in fact a waste of information. Electron microscopy —especially high-resolution microscopy —is a continuous battle against radiation damage. One of the most obvious prerequisites in this fight is to make full use of the information available for a given degree of radiation damage. We shall see later that this information leads quite naturally to another extension of conventional microscopy —to a further spread of the information at a given illumination dose along the time (dose) coordinate (trace structure analysis).

---

1 This contribution is restricted to the evaluation aspect in three-dimensional electron microscopy. For a more general survey of these topics (including a discussion of experimental procedures and of results) see a recent review [4.1].

Nevertheless, the advantage of three-dimensional extra information has its
price. A conventional electron microscope is a closed optical instrument for straightforward imaging which collects an overwhelming amount of image information on
photographic plates without further processing. The resolution of a photographic
plate is of the order $1\overset{\circ}{0}$ μm. Each micrograph (size, e.g., 6×9 cm^2) contains therefore $\sim 5 \times 10^7$ image points. Three-dimensional imaging, however, needs computer processing just as X-ray crystallography does. The "image" will be synthesized from
data collected by the "electron microscope", which acts as a measuring device rather
than an imaging device. This "computerized imaging", so well known in crystal structure analysis, needs fast and big computers, since we are confronted with multi-
dimensional problems characterized by a large number of resolution elements. So far,
the most information-rich method in structure analysis has been protein crystallography. It is easy to show that measuring devices for three-dimensional microscopy can provide in very short times much more information than X-ray diffracto-
meters. They deliver more information than the biggest computer available today
could handle. One of the reasons for this computer limit is that the single-channel
manipulation of data in existing computers is not well adapted to the multichannel
problems of structure research. However, the situation might change in the future.
The possible mass production of multifunctional chips could lead to powerful multichannel devices specially developed for the correction and computation of images.
For the moment, we process the data in commercially available computers using program systems organized like those used in crystal structure analysis. Obviously,
only restricted cutouts of the specimen can be studied — it is our philosophy to
provide sufficient experimental "raw material" in a number of micrographs and to
select a posteriori an interesting region for a three-dimensional study. Another
difficulty is that the human visual "pattern recognition computer" is not very well
adapted to intuitive recognition, comparison, etc., in three dimensions. These difficulties are well known in comparative protein crystal structure research. Comparisons of two-dimensional patterns are easy — they can be drawn to the same scale
(e.g., contour diagrams) and simply superimposed on one another. In this context,
it is an advantage that the three-dimensional image is available in digital form.
Displays already exist which allow us to superimpose and to rotate three-dimensional
patterns [4.4]. Furthermore, comparisons by three-dimensional correlation functions
or other pattern recognition functions are obviously only possible with digitized
data.

The general idea of the "reconstruction" of a three-dimensional image from diffraction data is very old. It started with X-ray crystal structure analysis in 1912,
although the fundamental optical principles — especially the important relations
between structure factors and Fourier coefficients of the electron density function —
had to be clarified during many years of initial work in structural research. In
fact, one can understand three-dimensional electron microscopy as crystallography

of crystals which contain only one unit cell (corresponding to the specimen cutout). This "unit cell" might have a substantial size (in the case of biological macro-molecules, of the order of several hundred Å). It has been shown [4.5,6] that an electron microscope can be understood as a diffractometer containing —with one ex-ception— all the elements of an X-ray diffractometer. This exception concerns the "Eulerian cradle" which allows us to change the orientation of the specimen with respect to the primary beam. In conventional microscopes the orientation of the specimen is not changed, although for special purposes "tilting stages" are in use which may be understood as inaccurate microgoniometers. The kinematical scatter-ing theory —used in every X-ray structural analysis— shows immediately that an ex-periment on a fixed object delivers only diffraction data on a surface in the reci-procal space.[2] In order to get the full three-dimensional reciprocal data (which correspond to the three-dimensional specimen), the reciprocal space must be scanned with this surface; this requires some means of changing the orientation of the ob-ject in the diffractometer. On the other hand, the electron microscope diffracto-meter delivers not only the amplitudes (like an X-ray diffractometer) but also the phases of the structural factors of the "single unit cell crystal" since the scat-tered beams and the primary beam will interfere by virtue of the direction-changing action of the objective lens.

A minor difference between X-ray and electron diffraction concerns the fact that in the latter, the scanning Ewald sphere calotte is quite flat even at atomic re-solution (accelerating voltages of the order of 100 kV). At moderate resolutions it can even be replaced by a plane. This leads to a simple translation of the scanned reciprocal space (Fourier space) into the direct space language: according to a general Fourier transform principle, the equivalent of a reciprocal plane in direct space is a three-dimensional body consisting of parallel straight lines perpendicular to the reciprocal plane with weights proportional to the projected densities of the three-dimensional body along these lines. It is of interest that the special case of three-dimensional reconstruction from parallel projections has been "reinvented" in the tomographical methods of radiology [4.7,8] without re-cognition of the intimate connection with the older and more general crystallo-graphic principles. Only in recent years has the dual representation in direct and reciprocal space (so characteristic for crystallography) been utilized. Analogous methods have been devised for radio astronomy [4.9] and other fields [4.10].

The concept of three-dimensional electron microscopical reconstruction was des-cribed simultaneously in three independent papers [4.11-13] which were quite dif-ferent in detail. It was later shown that the proposal of the "polytropic montage" in [4.11] corresponds to conventional (circular) tomography with optical back-

---

2 The kinematical theory is in general adequate since the "crystal" contains only one unit cell (weak scattering in spite of strong interaction).

projection [4.8]. Recently the polytropic montage has been modified according to the principles of computerized tomography (adopting iterative direct space reconstruction algorithms of the algebraic reconstruction technique (ART) type [4.14,15]). The approach adopted in the other two papers [4.12,13] was inspired by structure analysis. In a straightforward way, the equivalence with structure analysis can be recognized in [4.13]. The latter describes three-dimensional protein crystal structure analysis in the electron microscope by taking a set of crystal projection images in different orientations at medium resolution (approximation of the "plane" Ewald sphere) with Fourier inversion of the images and arranging the structure factors in the three-dimensional reciprocal lattice of the crystal. A conventional Fourier synthesis of these data "reconstructs" the three-dimensional lattice in exactly the same way as in X-ray crystallography. The aim in [4.12] was the determination of the distribution of the (radiation-resistant) heavy atom stain in the cavities of biological macromolecules. In restricting the analysis to highly symmetrical specimens, which show many identical projections (except for translations) in different orientations (like helical structure), it proved possible to calculate — in the extreme case — a three-dimensional reconstruction from one conventional (nontilted) micrograph only. The necessary set of projections can be generated by symmetry. The relative simplicity of the experiment (which can be done in conventional microscopes at low resolution) stimulated the use of these methods in a number of structure determinations in several laboratories (see a recent review in [4.16]).

The work of our laboratory has been focused in the last years mainly on the three-dimensional electron microscopy of individual specimens. In preparation for these studies we investigated two-dimensional reconstruction methods [4.17-20] (which succeeded our earlier work on filter methods, see [4.21-23]). Indeed, without the development and the use of these methods, individual three-dimensional microscopy would become extremely difficult even at lower but especially at higher resolution and at the same time rather inaccurate.

One might ask why we did not proceed along the lines of [4.13] (see, however, [4.24,25]), although — as shown in [4.17] — the averaging in periodic specimens allows radiation damage to be eliminated in the study of biological specimens. The temptation was indeed rather strong — and not only because our laboratory has been involved for many years in the related field of X-ray crystallography. Structure research of periodic specimens — especially single crystal analysis — is a well-advanced field, which has contributed much to our understanding of molecular architecture.[3]

---

3 There has, however, been a renewed interest in recent years in the study of crystals of native biological molecules by electron microscopy [4.26,27]. Thus, the three-dimensional structure of the purple membrane has now been determined [4.28].

In contrast with this, structure research of aperiodic specimens is in a very bad state. The well-known diffraction methods used for gases, liquids, and solutions furnish poor structural information due to averaging over all orientations of the specimen. Three-dimensional individual electron microscopy is in principle also a diffraction method — with the additional possibility of the phase measurement between primary and diffracted beams. It is even a "single-crystal" method — although of "crystals" containing only one unit cell. It has therefore — as already mentioned — the overwhelming information capacity of the single-crystal method, which is by orders of magnitude greater than the information capacity of the orientation-averaging diffraction methods mentioned above.

There is certainly a whole world of genuine aperiodic structures (which cannot be crystallized like organic and protein molecules), which needs to be explored. One might therefore understand that it was tempting to focus our attention on the aperiodic case rather than to develop a new variant of crystallography. It soon became clear, however, that many methodological problems have to be solved first, before straightforward structure analysis becomes possible. To some extent we are in a position similar to that of the early days of crystallography: the potential of the tool is evident but its use is only possible in a restricted way. This does not mean that only methodological work should be done in such a period — the crystallographers did not wait until automatic equipment and evaluation with direct methods became available. Indeed, even at this stage it has been possible to deduce interesting information by three-dimensional microscopy concerning the structure of a negatively stained molecular complex (yeast fatty acid synthetase) at resolutions of the order of 2 nm [4.29-32] and to do the first preliminary work on inorganic amorphous aggregates (carbon) at high-resolution [4.33].

What are the main difficulties in this research program? The archetypal and ever-present problem is the radiation-induced change of the specimen. It may be tolerable — as in the study of so-called stable specimens, where the main part of the radiation energy transferred by inelastic processes can be dissipated in the specimen without bond breakage. Conversely, the radiation-induced change may be so pronounced — as in the case of native biological macromolecules — that the number of scattered electrons that can be collected is only sufficient for very low resolutions. In the case of isolated single molecules, one can try to average over many incompletely determined molecular structures [4.34-36]. Another possibility, used in a different way in crystallography, would be the incorporation of structural constraints for the replacement of the missing information. These ideas, developed in a consistent way, lead to the notion of incorporating structural constraints not only in space but also in space and time. It has so far only been possible to outline the basic features of the corresponding method of "trace structure analysis" [4.37,38]. There is little doubt that the additional information gained by the spread of the structure along the time axis will lead to new structural insight. Whether the constraints will be sufficient for the recognition of the initial native structure remains to be seen.

Compared with these fundamental problems, the technical questions are less difficult to clarify. It can already be concluded from the results so far achieved that no real obstacle exists to make three-dimensional electron microscopy at atomic resolution impossible [4.33]. Two-dimensional image reconstruction provides the necessary basis for quantitative work, at least in the kinematical approximation, which is sufficient for most amorphous specimens. At the same time it allows us to overcome the resolution limit of conventional round lenses (lens systems with lens aberration correctors are also in development [4.39,40]), provided that a sufficiently monochromatic radiation source is used, for example, a field emission gun perhaps with an additional monochromator [4.41]. The difficult setting and alignment procedures in the necessary multiple series of exposures will certainly be facilitated by the use of the on-line cross-correlation computers developed in our laboratory [4.42]. In the evaluation procedures, the development of image reconstruction methods at restricted tilting angles is of importance, since in general the specimens are deposited on a planar support or are planar. The analysis already mentioned at the theoretical lower limit of the primary dose needs reconstruction of projections with low statistical significance. The correlation of slightly inclined projections for the determination of the origin [4.43] can also be used in the case of extremely noisy images.

Paradoxically the analysis will be easier at atomic resolution. The "atomic constraint" then becomes applicable, allowing the refinement of the origin, partial removal of the influence of clutter and — last but not least — checks of the acceptibility of the results by stereochemical rules.

4.2 The Fundamental Theoretical Background

4.2.1 The Use of a CTEM as a Diffractometer

Figure 4.1 shows the fundamental experimental installation schematically for the CTEM (conventional transmission electron microscope) and the corresponding representation in reciprocal space. The primary beam P in Fig.4.1a hits the object O, usually consisting of a thin film with a specimen deposited on its surface. The diffracted rays will be caused to interfere with the primary beam P by the lens L, thus creating the image I. For weakly scattering objects a linear approximation is adequate, which allows the amplitudes and phases of the scattered rays to be evaluated by Fourier transformation. Without the lens, the microscope looks like an X-ray diffractometer where the diffracted rays would be measured directly, however, and their phase information lost.

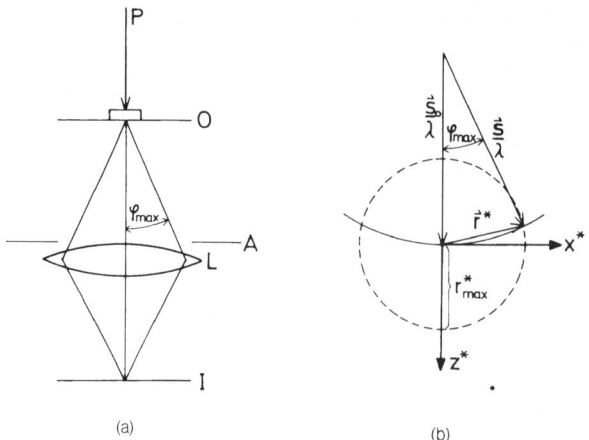

(a)   (b)

Fig.4.1. (a) Imaging with the CTEM. P primary beam, O object, L lens with aperture A, I image, φ_{max} maximum scattering angle defined by A or the envelope of the transfer function. (b) Representation in reciprocal space. Full line: Ewald calotte, dashed line: resolution sphere; for the definition of $\underline{s}_0$, $\underline{s}$, $\underline{r}^*$, x^*, and z^*, see text

In reciprocal space the geometry of the scattering experiment can be described by the Ewald sphere construction as shown in Fig.4.1b. Since elastic scattering has been assumed the electrons scattered in the direction of the unit vector $\underline{s}$ have the same wavelength λ as the electrons of the primary beam with direction $\underline{s}_0$. Thus only the points on the surface of the Ewald sphere, radius $1/\lambda$, can be observed. If we introduce for the points in reciprocal space the vector $\underline{r}^* = (x^*, y^*, z^*)$ the condition can be formulated as

$$\underline{r}^* = \frac{1}{\lambda} (\underline{s} - \underline{s}_0) \ . \tag{4.1}$$

This relation is identical with the Laue condition in the case where the object is a crystal, i.e., where $\underline{r}^*$ coincides with a reciprocal lattice point. If φ is the scattering angle, we can also write

$$r^* = \frac{2}{\lambda} \sin \frac{\varphi}{2} \ . \tag{4.2}$$

Replacing r^* by n/d, where n denotes an integer and d the lattice plane spacing, we get the well-known Bragg condition.

Due to the short wavelength of electron waves, the radius of the Ewald sphere is much larger than in the case of X-ray crystallography. Furthermore the aperture A of the lens in Fig.4.1a — or the envelope of the transfer function — limits the scattering angle to the order of $\varphi_{max} \approx 10^{-2}$. Thus only a small and flat calotte of the sphere can be measured and (4.2) may be replaced by

$$r^* = \frac{\varphi}{\lambda} \; . \tag{4.3}$$

In order to get three-dimensional information the primary beam must illuminate
the object from different directions. The corresponding calotte in reciprocal space
rotates around the origin thus scanning a sphere with the radius $r^*_{max} = \varphi_{max}/\lambda$. This
sphere, symbolized in Fig.4.1b by the dashed circle, will be called the "resolution
sphere". By analogy with X-ray crystallography, we use $1/r^*_{max}$ as a measure of the
resolution. To be exact we should call $1/r^*_{max}$ the crystallographic resolution,
which is different from other definitions usual in electron microscopy. For instance the nominal resolution defined by SCHERZER is $0.6/r^*_{max}$.

It is of importance for the phase measurement that the primary beam P should be
sufficiently monochromatic and of good spatial coherence. The divergence of the
primary beam (spatial coherence) is controlled by the illumination aperture α_{ill},
the monochromaticity (temporal coherence) by the stability of the supply for the
accelerating voltage V, by the half width of the thermal spread of electrons
emitted by the cathode ΔV_{ill}, and by the voltage itself. For thermal cathodes α_{ill}
is in the range from 10^{-4} to 10^{-3} rad and ΔV_{ill} between 0.5 and 1 V. The thermal
spread and the chromatic error of the lens in present-day CTEMs limit the envelope of
the transfer function and thus the resolution to 2-3 Å at 100 kV. The use of field
emission guns should make it possible to work with α_{ill} between 10^{-5} and 10^{-4} rad.
A reduction of ΔV_{ill} by a factor of about two is urgently needed as this would allow
us to achieve atomic resolution since the extension of the transfer function envelope is inversely proportional to ΔV_{ill}. Since this extension is also inversely
proportional to V it should be possible to achieve atomic resolution for $V \geq 200$ kV
even with thermionic cathodes. A further reduction of ΔV_{ill} should in principle be
possible by the use of monochromators [4.41].

4.2.2 The Description of Structures in Three-Dimensional Electron Microscopy

In X-ray crystallography the object is a crystal. Thus, the reciprocal space is
occupied only at the reciprocal lattice points. The amorphous nonperiodic specimen
in the electron microscope, however, fills the reciprocal space continuously.
Nevertheless, it is possible to describe the object as a fictitious crystal. Since
only limited cutouts of the micrographs will be used, we can incorporate the corresponding part of the specimen into the unit cell of a lattice and reproduce it
periodically. The lattice constants are defined by the arbitrarily chosen extensions
a and b of the cutouts and by the thickness c of the object. The advantage of this
procedure is that a discrete and (restricted resolution assumed) finite number of
structure factors is sufficient for the description of the specimen enclosed within
the unit cell. Furthermore, all Fourier integrations can be done by simple summations. For the definition of the structure factors we use, as in X-ray crystal-

lography, the concept of atomic scattering, although three-dimensional microscopy has not yet achieved atomic resolution.

Atomic Scattering Factor. Atomic scattering factors for X-rays are tabulated in "electron equivalents". For small scattering angles they are equal to the number of electrons Z. They can be converted into the absolute scale (dimension length) by multiplication with the classical electron radius r_0 (e = charge of the electron, m_0 = rest mass of the electron, c = velocity of light),

$$r_0 = e^2/4\pi\varepsilon_0 m_0 c^2 = 2.819 \cdot 10^{-6} \text{ nm} \quad . \tag{4.4}$$

Absolute atomic scattering factors for electrons are considerably larger. Their range is between 0.05 nm (hydrogen) and 1.4 nm (uranium) ($\sim 0.2 - 0.25$ nm for light atoms). They are much less dependent on Z than X-ray scattering factors (for small scattering angles they are approximately proportional to $Z^{1/3}$). Thus heavy atoms are considerably less "heavy" than in the X-ray case, which is of substantial experimental importance (e.g., better recognition of light atoms in stained specimens).

In accordance with the conventions in crystallography, the real part and the imaginary part of the atomic scattering factor correspond to the phase and the amplitude contrast, respectively. Light atoms are to a good approximation phase objects for X-rays and electrons. For heavy atoms, however, the amplitude contrast contribution (anomalous scattering) is not negligible for electrons (but small for X-rays).

Structure Factor. The structure factor F corresponds completely to the X-ray structure factor except for the replacement of the X-ray atomic scattering factor by the scattering factor for electrons. Assuming an object with atoms at $\underline{r}_j$, we write

$$F = \sum_j f_j \exp(2\pi i \underline{r}^* \cdot \underline{r}_j) \quad , \tag{4.5}$$

where $\underline{r}^*$ is a vector in the reciprocal space defined by (4.1-3).

The Potential Function. As usual in the theory of electron scattering, the object may be described by a potential function $\phi(\underline{r})$ which corresponds to the electron density function $\rho(\underline{r})$ in X-ray crystallography. Assuming a three-dimensional crystal, a Fourier expansion of the potential is possible,

$$\phi(\underline{r}) = \sum_{\underline{H}} \phi_{\underline{H}} \exp(-2\pi i \underline{H} \cdot \underline{r}) \quad . \tag{4.6}$$

The vector $\underline{H}$ corresponds to a reciprocal lattice point with the indices (h_1, h_2, h_3). Denoting by $\underline{a}^*$, $\underline{b}^*$, $\underline{c}^*$ the set of basis vectors of the reciprocal lattice, we can write

$$\underline{H} = h_1\underline{a}^* + h_2\underline{b}^* + h_3\underline{c}^* \quad . \tag{4.7}$$

If the atomic scattering factors in (4.5) are calculated in Born's approximation, the relation

$$\phi_{\underline{H}} = \frac{C}{V} \cdot F_{\underline{H}} \tag{4.8}$$

is valid with V as the volume of the unit cell in direct space. The structure factor $F_{\underline{H}}$ is defined by (4.5) for the case $\underline{r}^* = \underline{H}$. If we express $\phi_{\underline{H}}$ in volts and the volume V in nm^3 the constant $C = h^2/2\,m_0e$ with h as Planck's constant has the value

$$C = 0.48 \text{ V nm}^2 \quad . \tag{4.9}$$

The introduction of a finite body with volume V is fundamental. Reconstruction schemes tend to produce clutter and noise even in regions around the specimens where ideally the signal should be zero. It is therefore of advantage to increase V in order to avoid noise amplification in the reconstruction space by superposition of clutter and noise from adjoining "lattice" cells.[4] The choice of a parallel-epiped as "finite body" is only necessary if the reconstruction is to be done by Fourier transformation. In certain image reconstruction schemes other orthogonal function expansions are used, which lead to cylinders or spheres as "finite bodies".

4.2.3 Two-Dimensional Reconstruction (Image Filtering)

Assuming kinematical approximation and linear transfer in the microscope as well as negligible thickness of the object and curvature of the Ewald calotte, we can write for the scattered amplitude ψ_s in the two-dimensional image plane with co-ordinates $\underline{x} = (x,y)$,

$$\psi_s = \frac{i\lambda}{A} \sum_{\underline{H}}' F_{\underline{H}} \cdot T_{\underline{H}} \exp(2\pi i\underline{H} \cdot \underline{x}) \quad . \tag{4.10}$$

The prime beside the summation symbol indicates that the summation over $\underline{H}$ is confined to those lattice points within the resolution sphere that are situated in the

---

4 This effect is well known from crystallography: If a molecular pattern occupies a greater volume than the lattice cell of the point lattice, which repeats this pattern, overlaps must obviously occur.

reciprocal plane approximating the Ewald calotte. T_H describes the transfer function of the lens, corrected for chromatic aberration and partial coherence if necessary. A denotes the area of the image, i.e., the cutout which defines the fictitious unit cell mentioned above. As usual the approximation for the intensity I of a bright field image yields

$$I = 1 + 2 \, Re\{\psi_s\}$$

$$= 1 + \frac{2}{A} \sum_{H}' \, [(F_H + F_{-H}^*) \, Im\{T_H\} + i(F_H - F_{-H}^*) Re\{T_H\}] \exp(2\pi i \underline{H} \cdot \underline{x}) \quad . \tag{4.11}$$

The image contrast is a mixture of phase and amplitude contrast, which will be transferred by $Im\{T_H\}$ and $Re\{T_H\}$, respectively. If Born's approximation is valid, only phase contrast takes place since $F_H = F_{-H}^*$. In this case, all the structure factors of the sum in (4.11) can be evaluated by Fourier transformation of the image intensity and by filtering for $Im\{T_H\}$.

In general $F_H = F_{-H}^*$ is violated because the atomic scattering factors are complex or, high resolution assumed, because of the thickness of the object and the curvature of the Ewald calotte. Thus it is impossible to evaluate the structure factor completely from one micrograph, even if an ideal transfer function could be achieved. Usually, the amplitude contrast is neglected. Nevertheless, there exist several procedures which allow the calculation of all F_H [4.5,6,4.44-46]. However, these methods need the exposure of at least two micrographs.

Bright-field micrographs can conveniently be transformed into dimensionless contrast functions $(I-\bar{I})/\bar{I}$, where $\bar{I}$ is the mean of the intensity I. Thus the three-dimensional density function $\rho_c(\underline{r})$ reconstructed from these micrographs has the dimension nm^{-1}, because each image is a projection of $\rho_c(\underline{r})$. If the assumptions of (4.11) together with Born's approximation are valid and if it is possible to correct the phase transfer function $Im\{T_H\}$ to unity for all $\underline{H}$, $\rho_c(\underline{r})$ can be converted into the potential $\phi(\underline{r})$ by the use of (4.8),

$$\phi(\underline{r}) = \frac{C}{2\lambda} \rho_c(\underline{r}) \quad . \tag{4.12}$$

For $\lambda = 0.0037$ nm we obtain $C/2\lambda = 64.7$ V nm. In practice, however, this conversion can only be applied in a crude approximation, since complete correction of the transfer function would lead to noise amplification in regions with weak transfer. Furthermore, an ideal reconstruction without "blind" regions in reciprocal space would be necessary.

Summarizing we can say that the experimental structure factors (determined by Fourier transform of the micrographs and by filtering) deviate from the real structure factors. The main reasons besides the errors of measurement are the incomplete filtering and the approximation of the linear theory.

4.2.4 The Projection Theorem

A powerful tool in three-dimensional image reconstruction is the projection theorem. Under the assumption of negligible curvature of the Ewald calotte, it can be used for theoretical considerations as well as for computations. The theorem states a relation between the Fourier transform of a function and the Fourier transform of its parallel projection: The (n-1)-dimensional Fourier transform of the projection is equal to the n-dimensional Fourier transform of the projected function taken in a central plane section through the origin of the reciprocal space. Since the Fourier transformation and its inversion are symmetric operations, the theorem is valid in direct and in reciprocal space.

The theorem can easily be proved in the three-dimensional space using the integral relation between a density function $\rho(\underline{r})$ and its Fourier transform $F(\underline{r})$

$$F(\underline{r}^*) = \int \rho(\underline{r})\exp(-2\pi i\underline{r}\cdot\underline{r}^*)d^3\underline{r}^* \quad . \tag{4.13}$$

Assuming that the direction of projection is parallel to the z axis, the projection $P(x,y)$ is then given by

$$P(x,y) = \int \rho(\underline{r})\,dz \quad . \tag{4.14}$$

For the Fourier transform of $P(x,y)$ it follows successively from (4.14) and (4.13)

$$\iint P(x,y)\exp[-2\pi i(xx^* + yy^*)]dxdy$$

$$= \int \rho(\underline{r})\exp[-2\pi i(xx^* + yy^*)]d^3\underline{r} \tag{4.15}$$

$$= F(x^*,y^*,0) \quad .$$

$F(x^*,y^*,0)$ describes the Fourier transform of $\rho(\underline{r})$ in the x^*,y^* plane which is perpendicular to the z axis. The result is valid for arbitrary directions of projections since the coordinate system can always be chosen appropriately.

4.3 The Problem of Reconstruction

Let us recall that we are regarding the electron microscope as analogous to an X-ray diffractometer, with the additional facility of measuring not only the amplitudes but also the phases of the structure factors of a "crystal" containing only a single "unit cell". The obvious first step is the recording of micrographs imaging the object from different directions and the conversion of these data to the structure factors on the corresponding Ewald calottes by Fourier transformations.

140

The arrangement of the calottes depends on the experimental conditions. In Fig.4.2 a special case —rotation of the specimen around an axis (y^* axis, perpendicular to the plane of drawing) — is shown. For experimental reasons the number of electron micrographs is restricted. Let us assume for simplicity that the calottes in Fig.4.2 are taken at a constant angular interval $\Delta\alpha$. The angular range should be $-\alpha_{max} \leq \alpha \leq \alpha_{max}$ with $\alpha_{max} = 90°$ or somewhat larger (depending on the curvature of the Ewald sphere). However, in practice rotational angles of $\pm 90°$ are in most cases unattainable, since the specimen is in general planar (e.g., molecules deposited on a planar supporting foil). We find, therefore, "blind" regions in reciprocal space,[5] which are not sampled by the calottes.

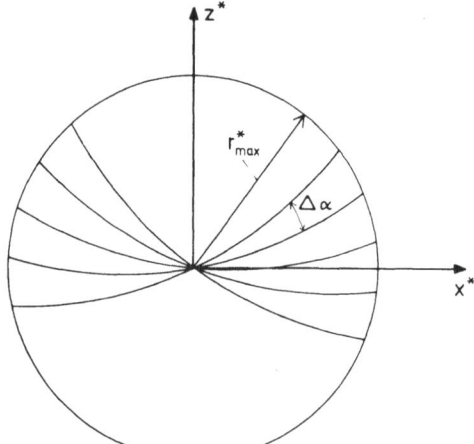

Fig.4.2. Ewald calottes sampling the reciprocal space (coordinates x^*, y^*, z^* with y^* perpendicular to the plane of the drawing) by rotation around the y axis in equiangular intervals $\Delta\alpha$

As already mentioned in the preceding section, the specimen enclosed in a unit cell can be reconstructed from a discrete and finite set of coefficients F_H by Fourier inversion. The coefficients are situated at the reciprocal lattice points corresponding to the unit cell. The reconstruction problem can, therefore, be reduced to the problem of determining this structure factor set from the continuous structure factor function on a discrete set of Ewald calottes.

4.3.1 The Whittaker-Shannon-Type Interpolation

One solution of this problem has been studied in [4.47]. It is explained for the special case of infinite depth of focus in Fig.4.3. This figure corresponds to Fig.4.2 with the exception that the Ewald calottes are replaced by planes (negligible curvature of the calottes assumed) and that the corresponding reciprocal lattice has been marked in Fig.4.3 by points. If we denote the structure factors

---

5 Since $\Delta\alpha$ can in principle be made arbitrarily small the regions between the calottes are not blind in the sense of our definition.

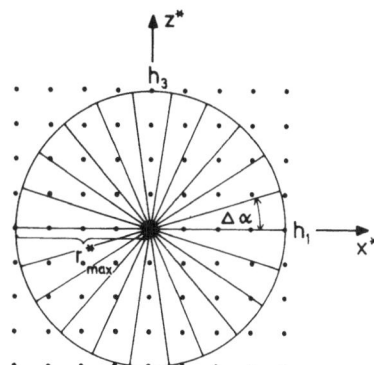

Fig.4.3. Reciprocal lattice sampled by planes.
The structure factors at the lattice points can
be calculated from the sampling planes by
Whittaker-Shannon interpolation

on the reciprocal planes by $F_{x_j^* y_j^* z_j^*}$ (the reciprocal planes will be distinguished
by number j), the structure factors $F_{h_1 h_2 h_3}$ of the reciprocal lattice can be calculated from a set of linear equations

$$F_{x_j^* y_j^* z_j^*} = \sum_{h_1} \sum_{h_2} \sum_{h_3} F_{h_1 h_2 h_3} \; C_{h_1 h_2 h_3 j} \tag{4.16}$$

$$C_{h_1 h_2 h_3 j} = \frac{\sin\pi(ax_j^* - h_1)\sin\pi(by_j^* - h_2)\sin\pi(cz_j^* - h_3)}{\pi(ax_j^* - h_1)\pi(by_j^* - h_2)\pi(cz_j^* - h_3)} \tag{4.17}$$

(a, b, c: lengths of the unit cell edges). This set of equations has been derived
in [4.47] from the Whittaker-Shannon interpolation theorem [4.48]. It can be regarded as a generalization of a two-dimensional equation set derived by HARRIS
[4.49] for the calculation of Fourier coefficients outside the aperture limit in
images ("superresolution").

This system of equations is in fact very general.[6] It allows in principle the
calculation of the structure factors not only *between* the reciprocal planes but also
in the blind region and even outside the resolution sphere (three-dimensional superresolution). The reason is that the $C_{h_1 h_2 h_3 j}$ [see (4.17)] still have finite values
at great distances from the corresponding reciprocal plane. However, as pointed out
in [4.47], the matrices become ill-conditioned since the $C_{h_1 h_2 h_3 j}$ become small and
correlated. Equation (4.16) can be regarded in this case as an extrapolation formula. If, on the other hand, the $F_{h_1 h_2 h_3}$ between two adjacent reciprocal lattice

---

6 Note that these linear equations are valid for arbitrary sets of central reciprocal surfaces with arbitrary orientations. They are not restricted to a bundle
of planes around a single rotation axis as shown in Fig.4.3. In the case of
Fig.4.3 the three-dimensional problem has been separated into a set of two-dimensional problems [see (4.19) and (4.20)]

planes are to be calculated, (4.16) acts as an interpolation formula, provided that the angular increments are small with respect to the reciprocal lattice constants. This latter condition has been formulated in [4.47] as

$$\Delta\alpha \cdot r^*_{max} \sim d^* \tag{4.18}$$

(r^*_{max} = radius of the resolution sphere, d^* = "mean" reciprocal lattice constant averaged from a^* and b^*). CROWTHER et al. [4.50] have arrived at a condition similar to (4.18) by the study of a cylindrical reconstruction body with Fourier-Bessel functions. There, d^* has to be replaced by the reciprocal diameter D^* of the cylinder.

Note that for rotation around a single axis (equal to y in Fig.4.3) the three-dimensional problem can be separated into a set of two-dimensional problems (reconstruction of discs perpendicular to the rotation axis with the resolution Δy along y as thickness of the discs). Each reciprocal plane ($F_{x^*_j y^*_j z^*_j}$) has to be replaced by a set of reciprocal lines ($F_{x^*_j z^*_j k}$) (k: number of disc) which can be calculated by linear Fourier transformation from the corresponding narrow section of the micrograph (perpendicular to y, see Fig.4.4). Consequently (4.16) and (4.17) can be simplified to

$$F_{x^*_j z^*_j k} = \sum_{h_1} \sum_{h_3} F_{h_1 h_3 k}\, C_{h_1 h_3 j} \tag{4.19}$$

$$C_{h_1 h_3 j} = \frac{\sin\pi(ax^*_j - h_1)\sin\pi(cz^*_j - h_3)}{\pi(ax^*_j - h_1)\pi(cz^*_j - h_3)} \tag{4.20}$$

Therefore b^* does not appear in the determination of the averaged distance d^*.

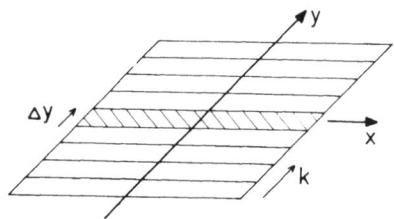

Fig.4.4. Subdivision of the micrograph into parallel strips of width Δy. For optimal information processing, Δy is half the resolution distance of the micrograph

---

7 The generalization of (4.16,17) to Ewald calottes with non-negligible curvature is obvious: The $F_{x^*_j y^*_j z^*_j}$ must simply be defined on the surface of the calottes.
Note that in this general case the separation of the three-dimensional problem into two-dimensional problems for single axis rotation is not possible!

Let us discuss (4.16) and (4.17) in more detail. It is obviously the constraint of the "finite body" which makes interpolations (and in principle also extrapolations) possible: An infinite set of reciprocal planes (or Ewald calottes[7]) need not be measured. It may seem surprising that corresponding to (4.16), the Whittaker-Shannon "interpolation" formula interpolates a certain $F_{x_j^* y_j^* z_j^*}$ from all $F_{h_1 h_2 h_3}$. Most interpolation formulae (linear, quadratic, cubic, ...) take into account only nearest neighbors. The reason is that the body is restricted by a parallelepiped with *sharp* edges. The size of the reconstructed body has a similar function in direct space to that of a restricted "resolution body" in reciprocal space. Sharp cutoffs in reciprocal space create strong diffraction ripples in direct space which in principle cover the whole direct space for each image element. In a similar way the discrete $F_{h_1 h_2 h_3}$ will be spread in reciprocal space to continuous "image points" with strong diffraction ripples. This large number of terms makes reconstruction by means of (4.16) extremely cumbersome. To our knowledge, no attempts have been made to perform real reconstructions using the complete equation system (4.16) or (4.19).

4.3.2 An Alternative Way of Incorporating the Finite Body Concept

In the preceding chapter, it has been assumed that the continuous function in the reciprocal space — and thus also the $F_{x_j^* y_j^* z_j^*}$ on the sampling surfaces — has been reconstructed from the reciprocal lattice $F_{h_1 h_2 h_3}$ of a "crystal" containing the specimen in the unit cell by Whittaker-Shannon interpolation. Consequently the $F_{h_1 h_2 h_3}$ can be found by matrix inversion from the $F_{x_j^* y_j^* z_j^*}$ [see (4.16)]. There is another way of constructing a continuous function in reciprocal space: Starting from the sampling surfaces, direct conclusions can be drawn concerning the structure factors between them, if again the specimen is restricted in size.

Let us assume for simplicity that the Ewald calottes can be approximated by planes. Each plane section through the origin of the reciprocal space is the two-dimensional Fourier transform of a projection in real space, where the direction of this projection is perpendicular to the plane. Corresponding to each projection we define a "projection body" as a three-dimensional function in real space with constant values along the direction of projection. In the planes normal to this direction, the function is equal to the projection itself. A three-dimensional Fourier transformation of this projection body again delivers the plane in reciprocal space corresponding to the projection. If, however, the projection body is restricted in size by multiplication with a shape function, the "infinitely" thin plane in reciprocal space will be broadened by convolution with the Fourier transform of the shape function. The plane shows a certain "thickness", which depends on the size of the object enclosed in the shape function. With the size known, this procedure can be applied to each sampling plane. The superposition of all broadened planes leads to a continuous function everywhere in reciprocal space, assuming that the thickness is larger than the gaps between the sampling planes.

We derive the condition for a continuous function within the resolution limit in the case of single-axis tilting, where the reconstruction can be done two dimensionally in successive discs perpendicular to the tilt axis. The measured structure factors $F_{x_j^* z_j^* k_j^*}$ of the k-th disc are arranged on straight lines. Assuming a circle with diameter D as shape function, the lines will be convoluted with the function $2J_1(\pi Dr^*)/\pi Dr^*$. Thus we obtain bars for which we can define a thickness t^* equal to the diameter of the main maximum of the function. We get roughly

$$t^* = \frac{2}{D} \quad . \tag{4.21}$$

At the resolution limit r_{max}^* of the Fourier space the gap g_{max}^* between two reciprocal lines is

$$g_{max}^* = r_{max}^* \Delta\alpha \quad . \tag{4.22}$$

If we require $g_{max}^* = t^*/2$ for complete overlap, we obtain

$$\frac{1}{D} = r_{max}^* \Delta\alpha \quad . \tag{4.23}$$

It is easy to see that 1/D corresponds approximately to d^* in (4.18). The condition (4.23) for getting a smooth function in reciprocal space is therefore the same as that for Whittaker-Shannon interpolation.

4.3.3 Back-Projection and Filtered Back-Projection

a) *Simple Back-Projection*

Nothing has been said in the preceding chapter about how to evaluate exactly the structure factors in the gaps. Superposition would be the simplest way. In this case, however, due to the linearity of Fourier transformation, the reconstruction can be done in real space as well as in Fourier space by superposition of all projection bodies. This method is well known under the name "back-projection". It is the basis of all the tomographic methods devised in the thirties and of the polytropic montage [4.11]. Once more we confine our discussion to the case of single-axis tilting with constant angular increment in two-dimensional representation. Let us assume an object consisting of a single point in the center of a circle with diameter D. The projection bodies are single lines through this point. Superimposing all projection lines we obtain a star with its center in the object point, as shown in Fig.4.5a.

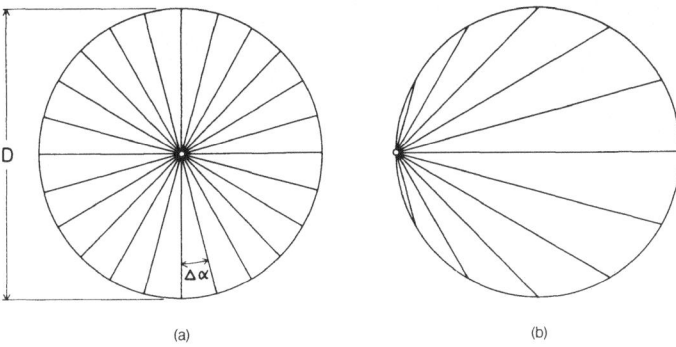

(a)   (b)

Fig.4.5a,b. Image point reconstructed by back-projection. (a) In the center of the reconstruction disc. (b) At the border of the disc. Note that in case b) also, complete overlap occurs if (4.23) is valid

Due to the finite body constraint this star is terminated by the circle with diameter D. This star is in fact the image of the object point. Stars of the same kind exist for all image points. They become, however, asymmetric if the object point is not in the center. Figure 4.5b shows the star corresponding to a border point. If there is a finite resolution (r^*_{max} in reciprocal space) the lines become bars with an approximate width $1/r^*_{max}$. The condition (4.23) means that the area of the reconstruction body is completely covered with bars, even for the asymmetric image point in Fig.4.5b. The nearer we are to the image point, the more bars overlap. Thus each image point is represented by a continuous smooth density function, which decreases approximately as $1/r$. Let us now assume that we increase D by say a factor 10 and decrease at the same time $\Delta\alpha$ to $\Delta\alpha/10$ without increasing the size of the specimen (circumscribed as before by the circle D). The only difference is obviously an enlargement of the smooth star function and a considerable reduction of the influence of a noncentral point position. All "image points" are now quite similar except for the small eccentricity already mentioned. As shown first in connection with image point problems in crystal structure analysis, the shape of image points can be changed by "modifying functions", today usually called "filter functions",[8] if all points are identical. This is approximately true in our case also.

b) *Filtered Back-Projection*

Applications of filter functions can be found today in many branches of structure analysis, e.g., correction of a bright-field image distorted by lens errors. The calculation of the filtering function is quite simple in our case. A single point

---

8 Perhaps one of the first applications of a modifying function was the "sharpening" of the Patterson maxima by filtering in reciprocal space with $1/f_0^2$ (f_0 = averaged atomic scattering factor of the structure) as early as the thirties (see textbooks of crystallography).

is represented in reciprocal space by Fourier coefficients of constant weight. Thus the reciprocal planes or — if a separation into two-dimensional x, z discs has been made — reciprocal lines sampling the reciprocal space are also uniformly weighted. Due to the overlap a smooth figure in the two-dimensional x^*, z^* space will occur by superposition. It is immediately clear that, in direct and in reciprocal space, the same shape-determining laws prevail since in both spaces uniformly weighted narrow bars overlap. The density in the radial direction decreases inversely proportional to r^* (in direct space inversely proportional to r). The filter function in reciprocal space has, therefore, to increase with r^* in order to correct for a uniform density. In direct space the distorted image points have to be convoluted with the Fourier transform of the filter function.

It is very instructive to discuss this function in some detail. Figure 4.6a shows the Fourier transform of the projection of a single point in a disc of diameter D, which is represented by a bar with length $2r^*_{max}$ and breadth $2/D$ (hatched in Fig.4.6a) occupied by the constant Fourier density F. Figure 4.6b shows the corresponding Fourier transform ρ. The main maximum determines the breadth of the back-projection ray (hatched). The product of F with the filter function gives F' (Fig.4.6c), which increases proportional to r^*. A star composed of superimposing bars at angular intervals $\Delta\alpha = 1/D \; r^*_{max}$ [see (4.23)] gives a two-dimensional function within the resolution circle, which is constant to a good approximation and which therefore corresponds to a sharp point in direct space. Let us now regard the Fourier transform ρ' (Fig.4.6d) of F' (Fig.4.6c). Compared with ρ (Fig.4.6b) we recognize two negative side minima with approximately half of the weight of the main maximum as new structural features. A back-projection of ρ' corresponding to the figures in reciprocal space (mainly consisting of a positive back-projection bar accompanied by two negative side bars, see Fig.4.6d) delivers a sharp point instead of the "star function" which decreases proportional to $1/r$. This can be easily understood intuitively, since the negative side minima are superposed on positive regions of other bars. It can even be shown semiquantitatively that the superposition function must be zero everywhere except in a small circle with the diameter $1/r^*_{max}$.

Let us consider the star figure caused by rotation of the two negative side bars only. The superposition function again decreases with $1/r$ except for a small circle (diameter $1/r^*_{max}$) in the center. Since this function is negative it cancels completely the $1/r$-function caused by the rotation of the main positive bars except in the inner circle with radius $1/r^*_{max}$. Note that the sum of the weights of both negative bars corresponds to the weight of the main bar — the scales of the positive and of the negative $1/r$ functions are thus the same.

If we have a general object, F' in Fig.4.6c has to be replaced by the Fourier transform of the corresponding projection multiplied by the r^* filter function. ρ' in Fig.4.6d has to be replaced by the Fourier transform of the filtered function. Due to the linearity of Fourier transformation, the back-projection of these modified

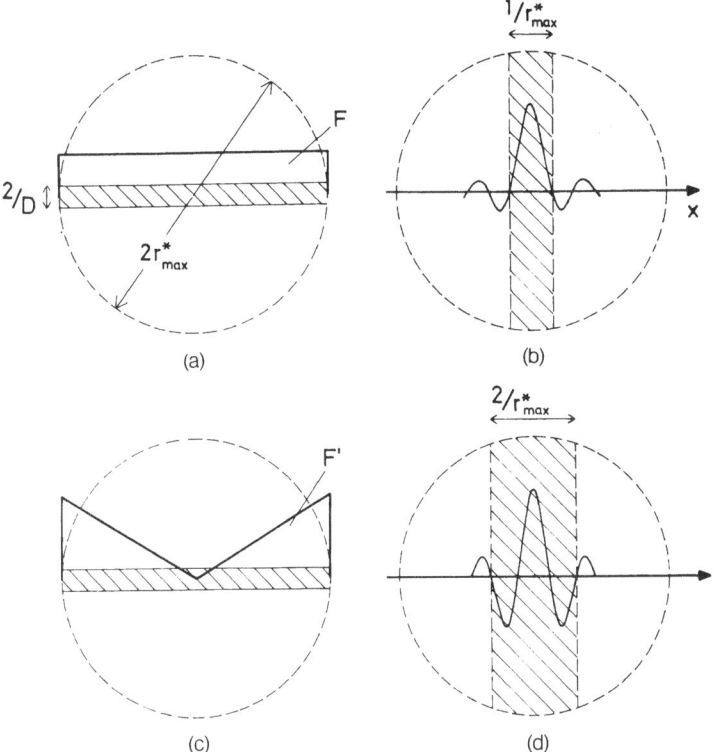

Fig.4.6a-d. Influence of r^* weighting. (a) Cutout of the Fourier density F corresponding to the projection of a single point in a disc of diameter D. (b) Projection body in direct space corresponding to (a). (c) Cutout of the r^*-weighted Fourier density F'. (d) Projection body corresponding to (c), mainly consisting of a positive bar with the width $0.75/r^*_{max}$ accompanied by two strong negative side bars (total width $2/r^*_{max}$)

projections leads to the correct image. This reconstruction method —which is a mixture of Fourier and direct space procedures — is called r^*-weighted back-projection. Since it involves only one-dimensional Fourier transformation and simple superposition (summing) operations it can also be executed in small-scale computers or in analog devices.

c) *The Influence of the Reconstruction Body*

We recall that we denote by $\Delta\alpha$ the increment between two tilting angles and by D the diameter of the reconstruction body. One may ask how the image point depends on the choice of $\Delta\alpha$ and D at a given resolution limit r^*_{max}. Obviously, the effect of increasing D for constant $\Delta\alpha$ is the same as that of increasing $\Delta\alpha$ for constant D. Thus we restrict our discussion to the case of varying D, where $\Delta\alpha$ is fixed by the number of projections. Corresponding to this $\Delta\alpha$, a diameter D_0 is defined by (4.23). Again we assume a single point in the center of the reconstruction body.

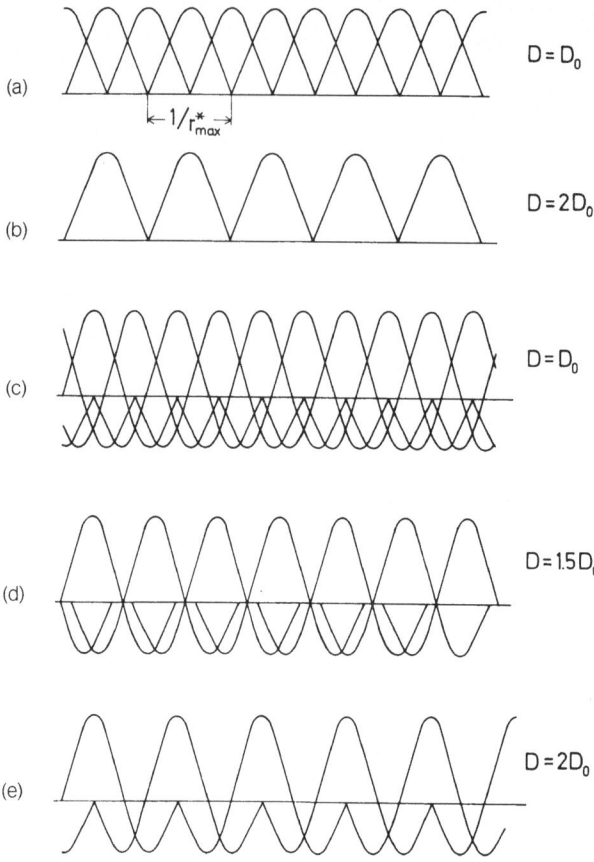

(a) $D = D_0$

$\leftarrow 1/r^*_{max} \rightarrow$

(b) $D = 2D_0$

(c) $D = D_0$

(d) $D = 1.5D_0$

(e) $D = 2D_0$

Fig.4.7a-e. Superposition of the back-projection bars of equally spaced projections of a single point along concentric circles with different diameters D. With D_0 satisfying (4.23), the following cases are shown: simple back-projection with (a) $D = D_0$, (b) $D = 2D_0$; r^*-weighted back-projection with (c) $D = D_0$, (d) $D = 1.5D_0$, (e) $D = 2D_0$

Figure 4.7 shows the reconstructed density function along the limiting circle of the reconstruction body for different diameters D.

In the case of simple back-projection, we have to superimpose functions like that of Fig.4.6b. For simplicity, this function has been truncated at its first zeros besides the main maximum. For D less than or equal to D_0, only small rippling can be seen in Fig.4.7a. The rippling increases with increasing D. From D = $2D_0$ onwards the bars are completely separated each from the other. As in Fig.4.7b, we obtain "clutter" with the maximal possible rippling.

In the case of weighted back-projection, the superposition must be done with the function of Fig.4.6d enclosing its two negative side maxima. According to Fig.4.7c the clutter is small again for D less than or equal to D_0. From D = D_0

to $D = 1.5 \, D_0$, the rippling becomes even smaller. As shown in Fig.4.7d, the density function is fairly constant for $D = 1.5 \, D_0$. On further increasing D, however, the situation becomes worse than in simple back-projection. It can be seen from Fig.4.7e that even negative values arise along circles with diameter $D = 2D_0$.

Some conclusions can be drawn from these results for filtered back-projection, if we move the image point within a reconstruction body with diameter D_0. The clutter is certainly negligible within the diameter $0.75 \, D_0$. An image point at the border of the reconstruction body, however, produces the rippling of Fig.4.7e at the opposite border. Thus the size of the object should be somewhat smaller than D_0. To judge the influence of clutter the number N of projections must be taken into account, since the maximum at an image point is N times the height of the maxima in Fig.4.7e.

There is a further argument for choosing a reconstruction body somewhat larger than the object. As already mentioned the r^*-weighting corresponds to a convolution of each projection body with the Fourier transform of r^*. This convolution also produces artefacts outside the object. If one calculates the reconstruction by the use of a computer lattice with a lattice constant equal to the size of the object, the artefacts truncated by the lattice cell will appear inside the object. The enlargement necessary to avoid these artefacts depends on the rippling of the Fourier transform of r^*. In the one-dimensional case an estimate can be derived from the function in Fig.4.6d: Neglecting all side maxima except the first, the enlargement must be at least $2/r^*_{max}$.

Let us add some ideas concerning the noise. We suppose that all projections are measured over the length D_0 with an additional white noise characterized by its variance σ^2. Since all projection bodies will be superimposed independently, each reconstructed image element within the diameter D_0 contains noise with a variance proportional to $N\sigma^2$ (N = number of projections). If the size D of the object is smaller than D_0, the noise within the object can be estimated directly from the region outside, which contains only noise. This is true for simple back-projection as well as for weighted back-projection. The weighting does not lead to noise amplification nor does it disturb the stationarity of noise within D_0. This can be explained by the fact that the noise enhancement in Fourier space with increasing r^* is compensated by a noise reduction at low space frequencies [4.3].

The "infinitely" extended supporting foil in the electron microscope prevents the measurement of projections with zero object densities at their borders. Nevertheless, estimation of the noise is possible. Let us assume that the cutouts of the projections have length $D < D_0$. If we enlarge all the projections by zeros up to D_0, the reconstruction produces noise outside the diameter D which decreases in a predictable way with increasing distance from D. Thus conclusions can be drawn from this region about the noise inside D. This must be done with care, since the noise cannot be separated from the clutter outside D.

d) *The Influence of Restricted Tilting Angle*

Scanning of the complete reciprocal space is for several reasons not possible in electron microscopy. As already mentioned in Sect.4.1, the planar supports restrict the maximal tilting angle[9] to $\alpha_{max} \approx 60°$ which leads to blind regions in reciprocal space around the optic axis (z^* axis) with a minimal opening angle of $90° - \alpha_{max}$. Reconstruction methods for experimental schemes with restricted α_{max} are, therefore, of great importance.

In the present connection, we are interested in the influence of a restricted angular range on the shape of the image point. Let us discuss some general features in Fig.4.8. Figure 4.8a shows the star in back-projection, Fig.4.8b the corresponding reciprocal space with diameters D and $2r^*_{max}$, respectively. Note that the section of the back-projected image point along x is ideal — a steep zero peak with no background — whereas the 1/r blurring takes place within the angles $90° - \alpha_{max}$ and $90° + \alpha_{max}$ (see Fig.4.8a). It can easily be predicted that convolution with the Fourier transform of r (Fig.4.6d) will produce rippling on the edge of the starlike figure. It will be especially strong near the image point along x. We expect a strong negative side maximum in this direction, since compensation by decreasing densities is not possible. Note that — due to the special shape of the Fourier transform of the filter function — the oscillation is mainly confined to the immediate neighborhood of the image point.

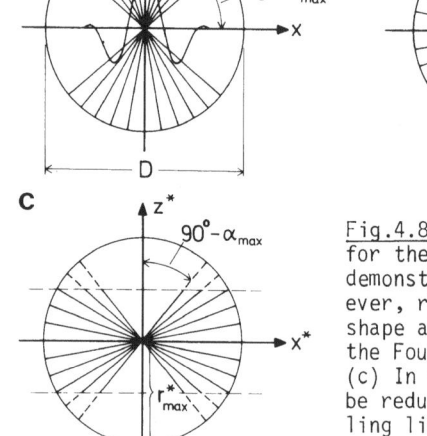
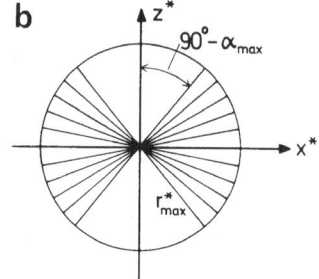

Fig.4.8a-c. Influence of restricted tilting range, shown for the case $\alpha_{max} = 50°$. (a) Back-projected image point, demonstrating that no clutter arises along x. If, however, r^*-weighting has been applied the image point shape along x is similar to the curve. (b) Sampling of the Fourier space with blind regions around the z^* axis. (c) In the case of r^*-weighting the clutter along x can be reduced by truncating the dashed parts of the sampling lines in Fourier space. The resolution along z, however, becomes worse on this case

9 We have made exploratory studies using very thin (about 20 Å) whiskers as supports which were turned around their axis [4.51].However, the difficulties turned out to be substantial.

This image point shape can also be understood from the shape of the Fourier transform in Fig.4.8b. For simplicity, we restrict our discussion to the two-dimensional case. Due to the filtering with r^*, the reciprocal space is covered by a uniform density within the angular range between $-\alpha_{max}$ and $+\alpha_{max}$. We remember now that a section in one space corresponds to the Fourier transform of the projection perpendicular to the section plane in the other space. We have already used this law to explain the general principle of back-projection (equivalence of projection and corresponding Fourier section). We have thus to project the uniform density in Fig.4.8b onto the x^* coordinate in order to get (by Fourier transformation) the density profile along x in Fig.4.8a. We see that this function increases linearly with x^* up to $r^*_{max} \sin\alpha_{max}$ and decreases to zero from $r^*_{max} \sin\alpha_{max}$ to r^*_{max}. It is again this linearly increasing function which leads to a central peak with practically only one very deep side minimum (see Fig.4.6c and 4.6d).

The dependence of the shape of the density function on these artefacts is an undesirable effect. It might be tolerable when, for example, a quaternary protein structure consists of rather spherical or ellipsoidal subunits or if an amorphous structure is composed of very small crystallites. The reconstruction will then deliver correctly the structure of the centers of gravity at least. It might even by possible to deduce the approximate shapes from a study of the distortions of bodies of typical shapes (spheres, ellipsoids at different orientations, etc.) by computer simulations. One might also contemplate using other filter functions, which would introduce a nonuniform weighting within the scanned region in Fig.4.8b, in an attempt to smooth somewhat the negative regions in the x direction without introducing too much blurring in the other directions. One possibility is shown in Fig.4.8c. If we weight the dashed regions of the Fourier lines to zero, the negative side minima become flat and broad. On the other hand, the resolution along z will obviously be reduced.

4.3.4 Conical Tilting

"Conical tilting" is a scheme for collecting tilted projections, which has been known under the name of "circular tomography" for many years in medical radiology. It was introduced into electron microscopy as "polytropic montage" by HART [4.11]. In a special version, it will be used as a "working scheme" in ring zone segment lenses for collection of three-dimensional electron microscopical data from a stationary specimen [4.52,53].

It has recently been recognized that this collection scheme—which is characterized by guiding the projection directions along the surface of a cone instead of in a plane—has an important fundamental property: with proper reconstruction methods, image points can be reconstructed that have a smooth background, like image points reconstructed from a Fourier body without blind regions; there is, however, an elongation of the image point in the z direction, which depends on the cone angle α [4.33,54]. Figure 4.9a shows back-projection lines on the surface of a cone with the cone angle α. They meet in the image point P producing a "tomo-graphical" back-projection image point (corresponding also to the image point of the polytropic montage) which has a rather unfavorable shape and which degenerates with increasing defocus Δz to circles with increasing radius. It is obvious that this type of clutter will produce dangerous artefacts especially for small Δz. However, an inspection of the corresponding Fourier body in Fig.4.9b shows that the conical shape of the blind region proves to be very favorable for the reduction of clutter.

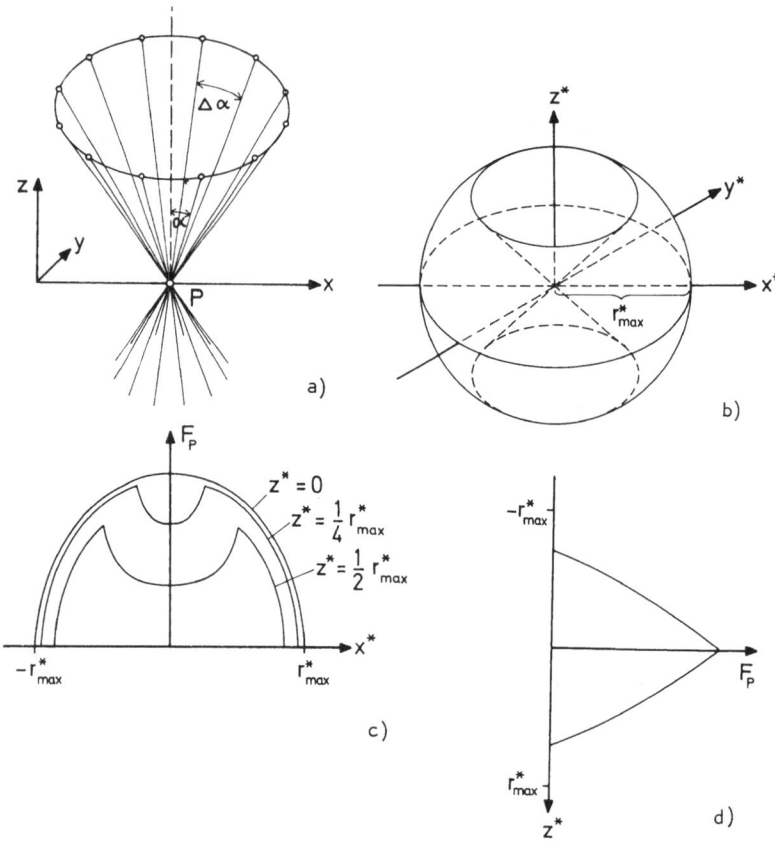

Fig.4.9a-d. Conical tilting. (a) Image point P produced by back-projection along directions on the surface of a cone with cone angle α. Equiangular sequence with the increment $\Delta\alpha$. In sections perpendicular to z, the density distribution degenerates into circles with increasing $|z|$. (b) Perspective view of the Fourier space showing the blind region forming a double cone with conical half-angle $90° - \alpha$. (c) and (d) Projection F_p of constant Fourier density onto the x^*, z^* plane. (c) Three section lines parallel to the x^* axis, (d) section line along z^*

In order to understand its influence, let us assume that the Fourier body built up by the smoothly overlapping Fourier planes (corresponding to the array of projections) has been filtered to a uniform density. According to the rotational symmetry of the cone, all plane sections through the image point containing the z axis are equal. Thus we restrict our discussion to the x, z plane which, in Fourier space, corresponds to the projection along the y^* axis. Figure 4.9c shows this projection on lines parallel to the x^* axis for $z^* = 0$ and $z^* = r^*_{max}/2$. A further section through this two-dimensional projection function along z^* can be seen in Fig.4.9d. Contrary to Fig.4.8b there is no linear increase of density starting from zero at the origin, which in turn would be responsible for deep side maxima like those of Fig.4.6d. Thus, we can already predict from Figs.4.9c and 4.9d that there will be no appreciable clutter at least along the main section lines.

Figure 4.10 shows the image point cut by the x, z plane, demonstrating that the image point is a (rotational) ellipsoid with an axis ratio that depends on α (e.g., 1:1.4 for $\alpha = 50°$). Tilting with restricted tilting angle is inevitable if the specimen resembles a platelet of "infinite" extent, as in electron microscopy. It is obviously of great importance that basically clutter-free reconstructions can be made although with a somewhat anisotropic image point.

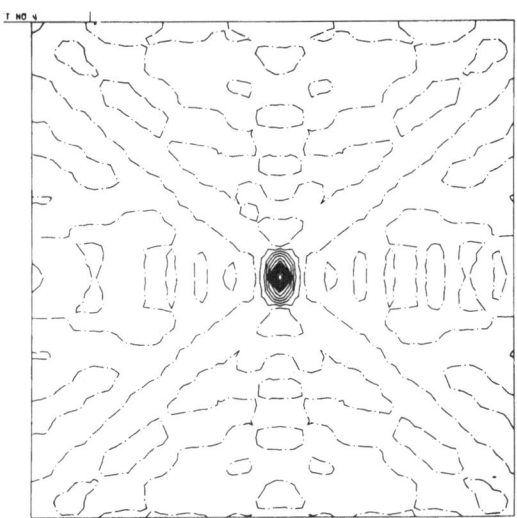

Fig.4.10. Contour plot (arbitrary units) of a section through an image point for conical tilting with α = 50°. Ideal reconstruction ($\Delta\alpha \to 0$) for a space frequency limit r^*_{max} = 0.15 nm^{-1}

It is interesting to enquire whether a reconstruction with conical tilting is a full replacement for a reconstruction based on a Fourier space without a blind region. One might argue that this must be so, since a better projection resolution in conical tilting will produce a smaller image point: If the resolution along the z axis is equal to the resolution of the spherical point, the image with elliptical points should present even more information than the spherical point reconstruction. This is in general true. However, there are exceptional cases. If all the Fourier coefficients of the specimen fall in the blind region, a reconstruction with conical tilting will not be possible no matter what resolution has been chosen. An example of such a specimen is an arrangment of parallel planes perpendicular to the z axis. In reciprocal space, there is density only along the z^* axis. From the discussion of the projections, it is already evident that no reconstruction will be possible; all projections show the same uniform density. Only projections in the direction of the planes reveal the density distribution along z. However, these cases are very rare. In general —especially at atomic resolution (if points have to be imaged) —such difficulties are very unlikely to arise.

The filter function can be determined in exactly the same way as for single-axis tilting. In the first step, one calculates the Fourier transform of the back-projected image point. The reciprocal values of this function correspond to the weighting function. For infinitely small $\Delta\alpha$ the filter function $g(\underline{r}^*)$ is given by [4.55]

$$g(\underline{r}^*) = \begin{cases} \sqrt{(x^{*2} + y^{*2})}\sin^2\alpha - z^{*2}\overrightarrow{\cos^2\alpha}/2 & \text{for} \quad \overrightarrow{\sqrt{x^{*2} + y^{*2}}} > |z^*| \cot\alpha \\ 0 & \text{elsewhere} \end{cases} \quad (4.24)$$

Figure 4.11 shows this function cut by the x^*, z^* plane as a contour line plot. It is again necessary to provide a greater reconstruction volume and to do the calculations in a lattice with enlarged unit cell in order to avoid reflections of zero artefacts into the specimen region (or to restrict the space after filtering). Note that all calculations have to be done in the three-dimensional space. Note too that for the same $\Delta\alpha$ the necessary number of projections is greater than in the single-axis tilting case. For a given $\Delta\alpha$, the number of projections that has to be measured in the single-axis case is $\pi/\Delta\alpha$, whereas conical tilting requires $2\pi\sin\alpha/\Delta\alpha$ projections (e.g., $\pi\sqrt{2}/\Delta\alpha$ for $\alpha = 45^\circ$).

It is possible to provide projection directions not only on the surface of the cone but also within the cone-proposals of this kind have been made in tomography. However, the number of necessary projections increases enormously. Thus we do not believe that such a scheme has much practical value.

4.3.5 Reconstruction by Series Expansion

Besides Fourier expansions, other orthogonal expansions have been proposed [4.50, 56-59]. One of the oldest and at the same time mathematically most elegant reconstruction methods was developed by CORMACK [4.57] and has recently been studied by SMITH et al. [4.58] and ZEITLER [4.59]. It too has found application in electron microscopy. However, for reasons to be discussed later, it is not very well suited for the reconstruction of the "infinitely" extended platelet-like electron microscopical specimens. The same is true of all other series expansion methods.

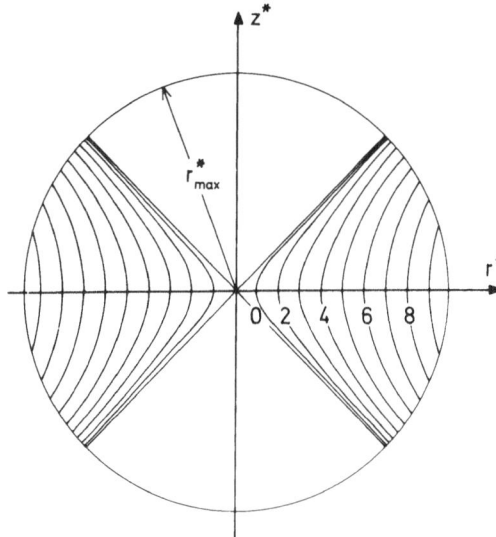

Fig.4.11. Contour plot (arbitrary units) of a section through the weighting function for filtered back-projection in the case of conical tilting with $\alpha = 45^\circ$

a) *The Cormack Method*

Let us give a brief outline of the mathematical background of the Cormack reconstruction scheme. We assume the object to be enclosed in a cylinder of radius a. As usual in the case of single-axis tilting, the reconstruction can be described two dimensionally. Thus we denote the object function by $\rho(r,\varphi)$, where r and φ are polar coordinates in the planes perpendicular to the tilt axis or cylinder axis, respectively. An azimuthal and radial expansion of $\rho(r,\varphi)$ within a circle of radius a with certain functions $R_{km}(r)\exp(ik\varphi)$ delivers a set of coefficients c_{km}.

$$\rho(r,\varphi) = \sum_{k=-\infty}^{\infty} \sum_{m=0}^{\infty} c_{km} R_{km}(r)\exp(ik\varphi) \quad . \tag{4.25}$$

The same coefficients occur in the representation of the two-dimensional Fourier-Bessel transform $F(r^*,\psi)$ of $\rho(r,\varphi)$ in the reciprocal space

$$F(r^*,\psi) = \sum_{k=0}^{\infty} \sum_{m=0}^{\infty} c_{km} A_{km}(r^*)\exp(ik\psi) \tag{4.26}$$

using the functions $A_{km}(r^*)$ with the property

$$A_{km}(r^*) = 2\pi(-i)k \int_0^{\infty} R_{km}(r) J_k(2\pi r^* r)dr \quad . \tag{4.27}$$

A similar expansion is possible, starting from the measured projections $P(\xi,\alpha)$ with $-\infty < \xi < \infty$, and $0 \leq \alpha < \pi$. The simple definition

$$\tilde{P}(r,\psi) = \begin{cases} P(r,\psi) & \text{for} \quad 0 \le \psi < \pi \\ \\ P(-r,\psi+\pi) & \text{for} \quad -\pi \le \psi < 0 \end{cases} \qquad (4.28)$$

describes all projections in a projection space with polar coordinates r and ψ, where $0 \le r < \infty$ and $-\pi \le \psi < \pi$. We assume that $\tilde{P}(r,\psi)$ can be expanded using the function $S_{km}(r)\exp(ik\psi)$ and the coefficients c'_{km}.

$$\tilde{P}(r,\psi) = \sum_{k=-\infty}^{\infty} \sum_{m=0}^{\infty} c'_{km} S_{km}(r) e^{ik\psi} \quad . \qquad (4.29)$$

According to the projection theorem a central section line through $F(r^*,\psi)$ can be obtained by Fourier transformation of $P(\xi,\alpha)$. With respect to $\tilde{P}(r,\psi)$, the corresponding relation can be formulated as

$$F(r^*,\psi) = \int_0^{\infty} [\tilde{P}(r,\psi)\exp(-2\pi ir^*r) + \tilde{P}(r,\psi\pm\pi)\exp(2\pi ir^*r)]dr \quad , \qquad (4.30)$$

where the plus sign is valid for $-\pi \le \psi < 0$, and the minus sign for $0 \le \psi < \pi$. Now we assume that the relation

$$A_{km}(r^*) = \int_0^{\infty} S_{km}(r)[\exp(-2\pi ir^*r) + (-1)^k \exp(2\pi ir^*r)]dr \qquad (4.31)$$

is valid. In this case the transformation (4.30) of (4.29) leads to the representation (4.26) of $F(r^*,\psi)$, but now with the coefficients c'_{km}. If the functions $A_{km}(r^*)\exp(ik\psi)$ form a complete and orthogonal set, the expansion (4.26) is unambiguous, which leads to $c'_{km} = c_{km}$.

The advantage of the method is that only the expansion (4.29) has to be performed explicitly. Inserting the coefficients c_{km} calculated from (4.29) into (4.25) provides the complete reconstruction, since all the Fourier transformations are carried out implicitly by an appropriate choice of the functions $S_{km}(r)$, $A_{km}(r^*)$ and $R_{km}(r)$.

One system of functions which fulfils the conditions (4.27) and (4.31) has been found by CORMACK. His proposal is well adapted to computation since $S_{km}(r)$ and $R_{km}(r)$ can be expressed in terms of Chebyshev polynomials of the second order and Jacobi polynomials, respectively. Thus these functions can be calculated very rapidly by recursion formulae.

In practice, only a discrete set of projections is available. CORMACK has therefore replaced the expansion (4.29) with respect to ψ by one-dimensional Fourier interpolations along annuli with the radii r (see Fig.4.12). Consequently only a finite set of coefficients c_{km} can be calculated unambiguously. According to [4.58] this "basic" set is determined by the condition

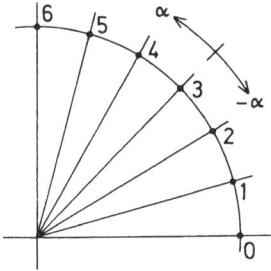 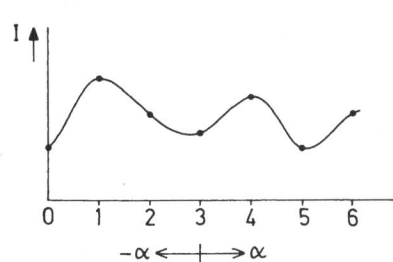

Fig.4.12. One-dimensional Fourier interpolation used by the Cormack scheme. The missing projections are "generated" by interpolation between the measured projections at constant radii

$$|k| \leq 2m \leq 2N - |k| - 1 \qquad\qquad (4.32)$$

with N as the number of projections. The name basic set is justified by the fact that all coefficients calculated for a given N remain unchanged if N is increased. Furthermore, the resolution limit is given by the condition (4.32). As pointed out by SMITH et al. [4.58], the "redundant" coefficients with $2m < |k|$ are also of interest. They should be zero since the corresponding functions $R_{km}(r)$ are zero inside the circle with radius a and the object function is zero outside. The noise of the measured projections, however, leads to deviations of these coefficients from their expected values. Thus they can be used for estimation of the noise.

Let us add some general comments about reconstruction by series expansion with special reference to the Cormack scheme, where we too have some practical experience.

In Fourier methods there is a certain freedom in choosing the shape of the reconstruction body in direct or in reciprocal space. The unit cell of a lattice is a parallelepiped with in general arbitrarily chosen axes and angles. These properties render the Fourier methods particularly well adapted to crystallography, since Nature utilizes this variability in the "construction" of crystals. In most series expansion methods, the shapes of the finite bodies are determined by the symmetry conditions of the functions — in most cases a cylindrical shape of the reconstruction body in direct space is required. Note that the symmetry condition will be destroyed if the Ewald planes are replaced by Ewald calottes: the series expansion methods cannot easily be generalized for a finite depth of focus.

It is characteristic of the Fourier methods that the image point size depends only on the projection resolution and not on the number of projections. As shown earlier, a drastic decrease of the number of projections [subject to the limit defined by (4.23)] causes a corresponding enhancement of the clutter associated with each image point. However, its shape remains independent of the position within the reconstruction body. For the explanation of this, we recall that Fourier transformation can be considered as a series expansion with trigonometric functions.

For a given resolution, the image point is formed by superposition of a certain sequence of trigonometric functions, weighted with the corresponding Fourier coefficients. If we want to reconstruct the image point with the same shape but another position within the unit cell, the same set of functions with the same amplitudes can be used. Only the phases of the Fourier coefficients have to be changed, corresponding to appropriate shifts of the trigonometric functions along the axes.

If other function systems are used for the expansion, a movement of an image point with constant shape cannot be expressed by simple shifts of the expansion functions. In the case of Cormack's system, only the dependence on the azimuthal angle φ is described by trigonometric functions whereas the functions $R_{km}(r)$ are fixed with respect to the cylinder axis. As a consequence, not only the phases but also the amplitudes of the coefficients c_{km} depend on the distance r of the image point from the center of the reconstruction disc. To keep the shape of the image point invariant, new coefficients must be calculated. This can only be done if more projections are available. Obviously the image point becomes stationary for $\Delta\alpha \rightarrow 0$.

As already mentioned, the Cormack method generates an infinite set of projections by Fourier interpolation between the measured projections. An example is given in Fig.4.12, from which we can see that this interpolation is only approximate. Figure 4.13 shows by simple geometrical arguments that the interpolation errors tend to increase with increasing distance from the rotational axis. This is a further and perhaps simpler argument for the nonstationarity of the image point. Figure 4.14 shows how the image point becomes progressively distorted with increasing r. The resolution (exactly defined only for the central point r = 0) is approximately doubled compared to the resolution defined by (4.23). It is independent of the projection resolution: Contrary to the Fourier methods (particularly the filter methods), the resolution is determined by the size of the reconstruction body and the number of projections, provided that the projection resolution is better.

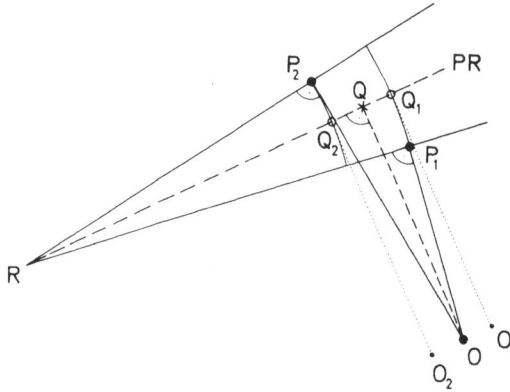

Fig.4.13. Approximation of an intermediate, unmeasured projection by angular approximation. The object point 0 occurs in two measured projections as projection points P_1 and P_2. Radial approximation leads to projection points Q_1 and Q_2 in an intermediate unmeasured projection PR. The actual projection point of object 0 would be Q. Back-projection of Q_1 and Q_2 leads to a displacement of 0 to 0_1 and 0_2. This splitting depends on the distance RQ

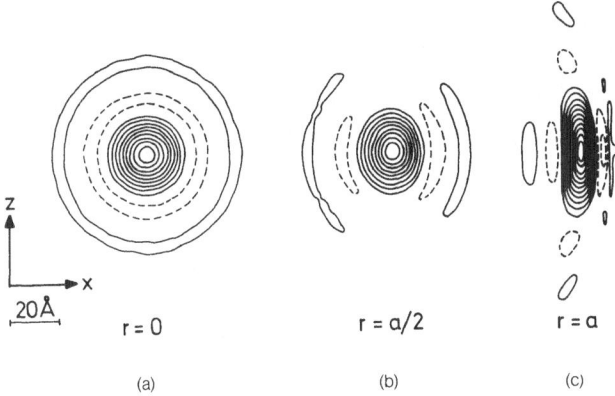

z ↑ →x

20Å

r = 0   r = a/2   r = a

(a)   (b)   (c)

Fig.4.14a-c. Contour plots (arbitrary units) of image points calculated with the Cormack scheme using the basic set of coefficients. The shape of the image point depends on its position in the reconstruction area. Complete tilting range $\alpha_{max} = 90^\circ$

b) *Aliasing*

It is well known and is in fact the basis of the numerical calculations in computers that Fourier integration of a function can be performed by calculation of a Fourier series, using as input data a set of equidistant sample values of the function (Runge method). It is further well known that for the calculation of N coefficients a division of the function into 2N sampling intervals is necessary. If one calculates more than N coefficients one simply gets a repetition of the coefficients already obtained instead of the desired high-order terms of the function (aliasing). The function in Fig.4.12 is now an equidistantly sampled periodic function and it is obvious that the number of coefficients which can be determined depends on $\Delta\alpha$. In this case we find $N = 180/\Delta\alpha$. A detailed analysis shows[10] that the two-dimensional single-tilting axis reconstruction will be controlled by two "interpolation" parameters which show certain interconnections. "Aliasing-free" reconstruction can be performed with different definitions concerning the limits of these parameters (see particularly [4.60]) which lead to different shape conditions of the image points and to different resolutions ranging from a resolution defined by (4.23) up to a resolution increased by approximately a factor of two. In the first case the image point is nearly invariant within the whole reconstruction body. An example can be given by the Cormack scheme where the interpolations along φ and r are controlled by the limits of the indices k and m, respectively. The image points in Fig.4.14 have been calculated with the aliasing-free basic set of coefficients defined by (4.32). If we want the shape of the image point to be invariant within the circle with radius a, the condition (4.32) must be replaced by

---

10 These problems have been extensively studied by CROWTHER et al. [4.56] as well as in the papers of CORMACK [4.57] and SMITH et al. [4.58].

$$|k| \leq 2m < N \quad , \tag{4.33}$$

corresponding to (4.23). As already mentioned, we have used this scheme in experimental work. A careful study of its properties has therefore been undertaken
in computer simulations [4.61].

Let us now ask what happens if we drop the postulate for aliasing-free reconstruction. One might be tempted to argue that such a reconstruction with, say,
double the number of coefficients should lead to complete nonsense, since one
half of the coefficients have not the slightest connection with the Fourier coefficients they should represent. However, let us study the case in more detail
with reference to Fig.4.12. The addition of further coefficients (which, due to
the periodicity in Fourier space, are — as already mentioned — a repetition of the
correctly determined coefficients) gradually decreases the connection lines between
the sampling points. If an "infinite" number of (periodic) coefficients were
utilized, the Fourier synthesis would furnish delta functions at the sample points
with zero gaps in between.[11] In the case already mentioned of doubling the coefficients, the interpolated value midway between two projection planes will be
"weighted down" to zero. It can easily be deduced from Fig.4.13 that near the
measured projections the interpolation is better than in the projection which bi-
sects the angle between two successive measured projections. The "weighting down"
of badly interpolated values is, however, the characteristic feature of filter
methods, where a substantial three-dimensional resolution should be achieved in
spite of a relatively small number of projections. One might therefore suspect
that a reconstruction with an enlarged set of coefficients leads in any case to an
enhanced resolution, no matter whether the $\Delta\alpha$ is sufficiently small or not. This
is obvious in the first case (non-aliasing), whereas in the second case (aliasing)
the influence of bad interpolation will be corrected by weighting. In order to
check this hypothesis one has to enlarge the set of calculated coefficients. This
can be done by introducing zero projections, which intersect the angles between the
measured projections. Thus formally the number of "measured" projection will be
increased by a factor of two.[12,13] The calculation showed, as expected, the doubled
resolution although with additional clutter at r/2 (see Fig.4.15).

---

11 For the proof of these statements see textbooks on the Fourier transform.

12 The basic set will be correspondingly enlarged since formally $\Delta\alpha$ will be halved!

13 The best procedure would be to introduce a dependence of the number of coefficients on r. Up to r/2, N coefficients are sufficient; the coefficients from N to 2N
might be formally set to zero. From r/2 to r the number of zero coefficients will
be gradually decreased, the coefficients between N and N' (N' = last nonzero coefficient) showing aliasing. This definition introduces nonzero bisecting projections.
This way of sharpening has not been tested in our laboratory.

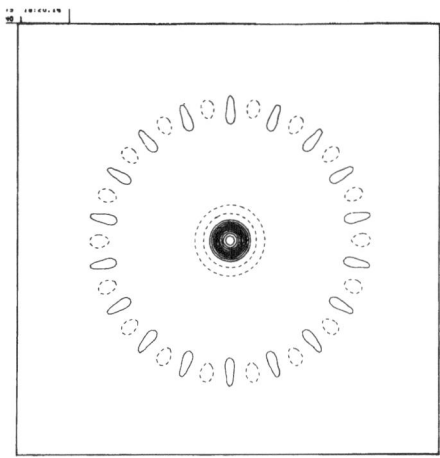

Fig.4.15. Image point reconstruction (arbitrary units) with the Cormack method from a set of projections with zero projections interlaced between the "measured" projections

This result is of fundamental importance, since it shows that even in reconstructions with series expansions, the principle of weighting down badly interpolated data (as in the Fourier methods) can be used for increasing the resolution over the "aliasing-free" resolution limit, which corresponds to twice the resolution of (4.23).

4.3.6 Algebraic Reconstruction in Direct Space

As pointed out earlier, back-projection is the simplest reconstruction scheme, corresponding to a superposition of the Fourier transforms of the Ewald planes. Back-projection can be performed by Fourier transformation of the body of the reciprocal space, generated by the bundle of reciprocal planes.[14] Since no filter function is applied, back-projection can be done more simply in direct space — starting immediately from the projections — by fast summation procedures. Figure 4.16 shows the general arrangement for single-axis tilting. The first problem is to digitize the image and to introduce a finite resolution. This is most simply done by covering the disc to be reconstructed with a square lattice with a lattice constant Δu. A linear resolution lattice (Δp) should also be introduced into the projections. Each back-projection line will thus be converted into a back-projection bar ("ray") $R_{n,j}$ which crosses the reconstruction lattice. The next definition concerns the question of how to add the projection densities spread along the rays into the reconstruction resolution elements (called "pixels"). One could use weights proportional to the fractions of the pixel areas which are covered by the ray. However,

---

14 Note in this connection that "back-projection" from Ewald calottes could be defined as the superposition of the corresponding Fourier transforms (which are not projections) in direct space. It is, however, not probable that such a generalized back-projection (without filtering) has any practical interest.

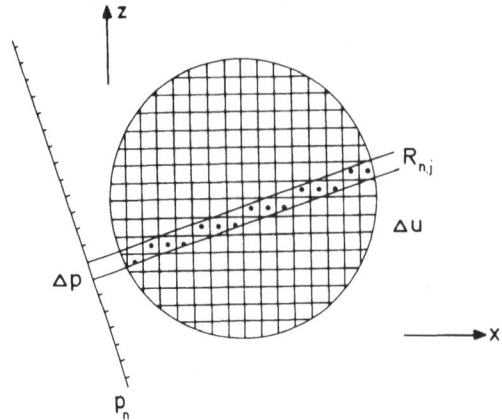

Fig.4.16. Arrangement for direct space reconstruction (single-axis tilting, negligible curvature of the Ewald calottes): The reconstruction area is divided by a grid with lattice constant Δu into pixels. The projection P_n is sampled by resolution elements Δp, which can be different from Δu. $R_{n,j}$ is the back-projection bar of the j-th element of the n-th projection

in order to simplify the calculations, a much simpler procedure is mostly used: A density is only added to the pixel if the center of a pixel lies within the ray (see Fig.4.16; these pixels are marked by points). Note an obvious variant of back-projection called r^*-weighted back-projection. Let us assume that a filter function has been applied and that each reciprocal plane has been transformed again into the direct space, leading to filtered projections (modified by convolution with the Fourier transform of the filter function). If we now perform with the filtered projections the normal back-projection scheme, then due to the linearity of Fourier transformation the correct image function will be reconstructed.

Another interesting filtering procedure has been worked out by GORDON et al. (Algebraic Reconstruction Technique ART) [4.62]. The back-projected image is obviously an approximation of the true image. If one could find a procedure for correcting it progressively in an iterative way, one could expect that after a certain number of cycles the procedure would converge to the correct solution. One of the first prerequisites for such a scheme is a criterion to tell us how wrong a solution is. For the next step one has to look for ways of modifying the solution in such a way that the criterion registers a smaller error. In the ideal case, the procedure converges after a number of cycles to the correct solution. The obvious error criteria are the projections of the reconstructed body in each cycle. If they are equal to the measured projections, the reconstruction can be accepted as correct. For the correction of the images, the ART method considers each projection separately (e.g., P_n in Fig.4.16). For these projections the error signals are the differences between the measured rays and the rays calculated from the reconstruction. In the next step, densities in the pixels along a ray are changed in such a way that the corresponding error signal becomes zero. Since in the most general case the distribution of the corrections along a ray $R_{n,j}$ is obviously unknown, ART distributes the correction uniformly over all pixels in $R_{n,j}$. This is certainly incorrect, since in the correct image the differences between the reconstructed pixels will differ along $R_{n,j}$. Only the sum of these differences will correspond to the error signal. However, experience shows that this approximation is sufficient for the convergence of the process if the data are not too noisy. The subsequent projections will be treated in turn and after a certain number of cycles minimal error increments will be reached.

For the mathematical formulation let ρ_{pq} denote the object density of the pixel with indices p and q. ρ_{pq} is related to the density P_{nj} of the j-th image element of the n-th projection by

$$P_{nj} = \sum_p \sum_q w(n,j,p,q)\rho_{pq} \quad , \tag{4.34}$$

where $w(n,j,p,q)$ represents the weighting factor for the pixel (p,q) combined with the ray $R_{n,j}$. In the simplest case $w(n,j,p,q)$ is equal to one if the center of the pixel lies within the ray $R_{n,j}$ and zero otherwise. Reconstruction by a simple summation can be described by

$$\tilde{\rho}_{pq} = \frac{1}{C} \sum_n \sum_j \frac{w(n,j,p,q)}{w_{nj}} P_{nj} \quad , \tag{4.35}$$

where $\tilde{\rho}_{pq}$ is an estimate of ρ_{pq}. We have used

$$w_{nj} = \sum_p \sum_q w^2(n,j,p,q) \quad . \tag{4.36}$$

With the simple assumption made above concerning the factors $w(n,j,p,q)$, w_{nj} is equal to the number of pixels whose centers are in the ray $R_{n,j}$. The constant C normalizes $\tilde{\rho}_{pq}$ appropriately.

Since ART is an iterative method, let $\rho_{pq}^{(\ell)}$ denote the estimated density after ℓ iterations. Using (4.34) we can calculate the ray sum

$$R_{nj}^{(\ell)} = \sum_p \sum_q w(n,j,p,q)\rho_{pq}^{(\ell)} \quad . \tag{4.37}$$

Now the density $\rho_{pq}^{(\ell)}$ will be related to the deviation between $R_{nj}^{(\ell)}$ and the measured value P_{nj}

$$\rho_{pq}^{(\ell+1)} = \rho_{pq}^{(\ell)} + \frac{w(n,j,p,q)}{w_{nj}}\left(P_{nj} - R_{nj}^{(\ell)}\right) \quad . \tag{4.38}$$

Starting from $\rho_{p,q}^{(0)} = 0$, all rays in a given projection are considered in turn, then the next projection is chosen and so on. The whole procedure (4.38) is repeated until a certain error criterion is satisfied.

One of the advantages of the algebraic methods is the possibility of incorporating constraints. In fact, if one knows that, for example, densities in an image
cannot be lower or higher than predetermined limits, one can replace "impossibly
corrected pixel values" (which are too low or too high) by the limiting values
(constrained ART methods).

The ART algorithm has been modified in many ways, by the use of statistically
motivated criteria for example. As one of the modifications, we discuss the SIRT
method [4.63] (Simultaneous Iterative Reconstruction Technique). The authors start
from the idea that not merely isolated single projections but the complete image
should be corrected in each cycle, using the error signals from all projections
simultaneously. Let us thus assume that these differences have been calculated
for all projections. From this difference projection body an "error reconstruction"
will be calculated by simple back-projection. This error image can be used for the
correction of the reconstructed image for the next cycle of the iterative procedure. Note that in this case corrections in the pixels along a ray $R_{n,j}$ are not
constant as in ART. It is obvious that this procedure again works only in approximation since back-projection is an imprecise method of reconstruction. However,
the errors in the "error back-projection" will become increasingly less important
if the procedure converges or if, in other words, the error image becomes small with
respect to the reconstructed image. It would obviously be possible to calculate
better approximations for the error image by using filtered projections. It is
tempting to contemplate an "iterative filtered back-projection" of the SIRT type
where the back-projections are replaced by their filtered versions. The convergence would be extremely fast—maybe so fast that an iterative scheme would become
senseless. Note in this connection that the use of unfiltered back-projection
in the algebraic direct space methods has the computational advantage that only
summations are necessary. From this point of view, a combination of Fourier
methods and algebraic iteration might be uninteresting. For the same reason the
multiplicative ART type reconstructions [4.62], which have also been proposed,
have hardly been used in practice. Recently it has been shown [4.64] that the influence of clutter and noise in ART can be considerably reduced simply by increasing the area of the reconstructed region (by a factor of about four) Extended
Field Iterative Reconstruction Technique (EFIRT)]. The corresponding projections
are also enlarged and filled with zeros in the new regions. The authors showed
that noise and clutter are spread over the extended field. They are, therefore,
"diluted" in the regions of interest. Note in this connection the discussion of
the use of extended fields in Fourier methods in this paper.

It is difficult to understand in detail the features of the algebraic iterative
techniques from general principles alone. Test calculations for different image
conditions are therefore of great interest. Such calculations have been performed
in the past in several papers. An interesting extended study [4.15] of a number of

algorithms (ART, SIRT, ILST[15]) has been made recently, comparing visually the various reconstructions of simple geometrical structures (points, arrangements of spheres, and hollow spheres of different size), using unconstrained and constrained versions of the algorithms. The results did not depend very much on the algorithm used except for the influence of noise. SIRT was found to be the least sensitive and ART the most sensitive. Of further interest are the results concerning the resolution achieved by the algebraic techniques. In order to check this question, the author has first reconstructed a single point at different positions and has calculated the power spectrum (which in the ideal case should be constant throughout the reciprocal space[16] using unconstrained versions of ART, SIRT, and ILST. Due to the linearity of these algorithms, a discussion in Fourier terms is obviously meaningful. The power spectrum was relatively constant with a sharp cutoff at a given resolution r^*_{max}. The size of this Fourier region did not depend on the position of the point, thus proving that the resolution is approximately uniform (as in the Fourier methods). In the next experiments the number of projection has been changed (from four to 24). r^*_{max} did increase roughly linearly with the number of projections. The resolution achieved with the ART and ILST algorithm is twice that given by (4.23). The SIRT resolution, however, increases less rapidly with the number of projections and with 24 projections it only reaches approximately the (4.23) resolution. The SIRT reconstruction thus tends to "smear" out the details somewhat — this is obviously also the reason why it suffers less from noise and clutter than the other two reconstruction algorithms.

This result is surprising at first sight. A connection between resolution and number of projections will be found if there are interpolation connections between the Ewald planes (or between the projections) of a certain kind (see the preceding sections). It is difficult to see how these connections arise in ART, for example, which treats all projections independently.

Let us discuss the case of back-projection as the simplest reconstruction method in direct space (see also Fig.4.16). It is obvious that, with proper matching of Δu and Δp with respect to the projection resolution, the smallest region to which all rays contribute has just the size of a projection resolution element irrespective of the number of projections. The Fourier transform of such a star is limited only by the reciprocal value of the projection resolution. However, as discussed in the section on Fourier methods, there is an approximately "smooth" region (over-

---

16 ILST (Iterative Least Square Technique) is an algorithm derived by GOITEIN [4.65] that minimizes the mean-square error between the calculated ray sums and measured projections.

15 These calculations have been made for single-axis tilting with $\alpha_{max} = 90°$. The reciprocal space is, therefore, two-dimensional.

lapping region) with an extent reaching at best (if partial overlap can be tolerated) twice the reciprocal resolution limit defined by (4.23). The size of the overlapping region will not be basically changed if Fourier filtering is applied. It can be assumed too that replacement of filtering by the ART algorithm will not influence the division of the Fourier space into an overlapping and a nonoverlapping region. In [4.15], the density distributions of the power spectra of reconstructed points have been reproduced as photographs taken from a visual display. One can recognize from these pictures that the division into the "smooth region" and the nonoverlapping regions does indeed take place. In the latter region, the power spectrum density will only be found along the reciprocal lines (especially pronounced in the reconstructions with only few (8, 12) projections). It is thus a matter of definition which limit is chosen as resolution limit; if we take the "smooth region limit", the connection with (4.23) is obvious.

It is easy to see from the power spectrum representations that the algorithms work rather like a Fourier filter function: the power spectrum of a point is approximately constant within the resolution range (overlapping range). However, closer inspection shows quite drastic changes within this region,[17] depending on the number of projections, the algorithm used, etc.; this demonstrates that the interaction between the crude representation of resolution elements by square pixels with sharp edges, the weighting function for the selection of pixels along a ray, and, last though not least, the algorithm have side effects which are difficult to analyze in detail in an iterative scheme.

There is no doubt that the algebraic reconstruction methods are powerful and basically correct. They are already in use with a good measure of success in medical tomography. We think, however, that for three-dimensional electron microscopy, methods that are based on Fourier transforms, on filtering, and perhaps on series expansions are more appropriate since the "optical" relations involving image point shapes, clutter, etc., are defined in a straightforward way. It is of importance in this connection that Fourier treatment of the projection must take place anyway since quantitative electron microscopy is only possible if two-dimensional image reconstruction is used for the removal of the "accidental" lens effects associated with parameters such as spherical aberration, axial astigmatism, and — at least to some extent — chromatic aberration. Furthermore, generalization to curved Ewald calottes — necessary for atomic resolution work — is easily possible, whereas the algebraic methods are restricted to real parallel projections. One

---

17 An interesting detail is worth mentioning: In ART and ILST reconstructions, the power spectrum varies only little up to the (4.23) limit. From (4.23) up to twice (4.23), the changes are much more pronounced. It is evidently the "partial" overlap in the latter regions which causes these irregularities and which is treated in a different way by the different algorithms.

should bear in mind that the optical measuring instrument "electron microscope" is in fact very different from the optical instrument "tomograph" and that the question of resolution is one of the fundamental questions in electron microscopy.

It is sometimes asserted that only the algebraic methods are suitable for the introduction of nonlinear constraints. This is, however, not true. On the contrary, phase determination in X-ray crystallography (which uses Fourier transform methods to a great extent) is in fact dependent on the utilization of constraints. It will later be shown that the "constraint" principle in the systematic form of crystallographic phase correction can easily be adapted to the problems of electron microscopy.

The algebraic methods are obviously not dependent in principle on the geometrical arrangement of the projections. It is therefore easy to generalize them, to conical tilting for example [4.15]. The main difference is that separation into a set of two-dimensional reconstruction discs with a plane sampling grid (see Fig.4.16) is no longer possible. The rays pass through the reconstruction space in oblique directions. One has, therefore, to replace the planar grid (with the area element "pixel") by a three-dimensional sampling lattice (the volume elements of which will be called "voxels"). The ray is a tube with an elliptic (or circular) cross-section. The selection rule for the voxels cut by the ray can be established as in the two-dimensional case. The results of the different test calculations described in [4.15] demonstrated the high quality of the reconstructions. As expected, anisotropic image points elongated along the optical axis are generated. For a conical angle of $\alpha = 45°$, the author found an elongation of 1.5:1 which is in good agreement with the theoretical value of 1.4:1 (for $\alpha = 50°$) in Fig.4.10. Conical tilting will certainly become — as already mentioned — the fundamental procedure in electron microscopy, since single-axis tilting with $\alpha_{max} = 90°$ is not possible for geometrical reasons. With respect to the choice between Fourier methods and algebraic methods the same arguments are valid as mentioned above for single-axis tilting. As already shown in a preceding section, adaptation of the Fourier methods to conical tilting is easily possible. The methods can evidently also be generalized to curved sampling surfaces (Ewald calottes).

4.3.7 Reconstruction of an "Infinite" Platelet with Restricted Tilting Angle

In the electron microscope, the object is usually deposited on a foil, which must be reconstructed three-dimensionally together with the object. Such a specimen cannot be tilted from $-90°$ to $+90°$ because the virtual thickness becomes extremely large near $\pm 90°$. The missing projections lead to "blind" regions in Fourier space. Three-dimensional reconstruction with the methods described hitherto is only possible if the unmeasured projections are replaced by zero. The consequences for the image point have already been discussed in Sect.3.3.4, using back-projection as an example.

The reconstruction body best adapted for the specimen should have the shape of a platelet with maximum thickness c and "infinite" extent in the other two directions. If, as is necessary for Cormack's method, a cylinder is used, another disadvantage appears. Let us assume that we want to extend the region for reconstruction by a factor of two, but only in one direction normal to the tilting axis. Thus the volume of the cylinder and also the number of image elements to be reconstructed must be increased by a factor of four. With the same number of projections, however, the information will only be doubled according to the suitably increased cutouts of the projections. The missing information leads to worse resolution and can only be matched by the measurement of further projections. A solution to this problem is given by the following reconstruction method which assumes restricted tilting angles and an "infinite" platelet as reconstruction body.

a) *One-Dimensional Whittaker-Shannon Treatment of Single-Axis Tilting*

Like all reconstruction methods for axial tilting, the description can be reduced to the x, z plane if the curvature of the Ewald calotte is negligible and if the tilting axis is parallel to the y axis. The cutout of the "infinite" platelet to be reconstructed has the dimensions a and c, respectively. The aim of the method is to evaluate all structure factors $F_{h_1 h_3}$ corresponding to the lattice points $(h_1/a, h_3/c)$ of the reciprocal space within a certain resolution limit. A Fourier transform then yields the reconstructed object. The calculation of the $F_{h_1 h_3}$ requires two steps which will be explained with the aid of Fig.4.17.

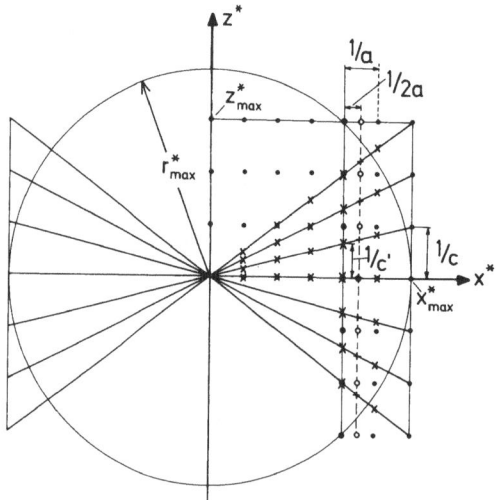

Fig.4.17. "Platelet" reconstruction represented in Fourier space: With appropriately chosen tilting angles equidistant lattices are generated by the intersections of the sampling lines with lines parallel to z*. The unknown structure factors (full circles) can easily be calculated from the measured values (crosses). If the reconstruction body is enlarged along x from a to 2a, only new lattice lines (open circles) have to be evaluated

We assume that all projections have been sampled with the sampling distance Δx. Considering the n-th projection with tilting angle α_n, a cutout with length $a \cdot \cos\alpha_n$ will be transformed by interpolation to a grid sample $\Delta x \cdot \cos\alpha_n$. As a consequence, the corresponding section in Fourier space will be stretched by the factor $1/\cos\alpha_n$. After transformation of all the projections, the sampling points in Fourier space are arranged in straight lines parallel to the z^* axis. We obtain exactly the structure factors $F_{h_1 n}$ corresponding to the points $(h_1/a, h_1 \tan\alpha_n/a)$, marked in Fig.4.17 by crosses. These points are defined as the intersection of the lattice lines $x^* = h_1/a$ with the section corresponding to the projections. Test calculations have shown that a simple cubic interpolation is sufficiently accurate for this step in direct space.

The second step must be performed in Fourier space: We have to calculate the desired $F_{h_1 h_3}$, marked in Fig.4.17 by circles, from the $F_{h_1 n}$. Since, for constant h_1, all the $F_{h_1 n}$ and $F_{h_1 h_3}$ are arranged on a straight line, the relation between these two sets of structure factors is given by a one-dimensional Whittaker-Shannon interpolation

$$F_{h_1 n} = \sum_{h_3} F_{h_1 h_3} \frac{\sin\pi\left(h_1 \dfrac{c \tan\alpha_n}{a} - h_3\right)}{\pi\left(h_1 \dfrac{c \tan\alpha_n}{a} - h_3\right)} . \qquad (4.39)$$

It is possible to avoid inversion of this linear system if the tilting angles are chosen according to the law

$$\alpha_n = \arctan\left(n \frac{2\Delta x}{c}\right) . \qquad (4.40)$$

The $F_{h_1 n}$ are then arranged equidistantly on each line $x^* = h_1/a$ with the sampling distance $2h_1 \Delta x/ac$. In particular, for $h_1 = a/2\Delta x$, the sampling distance is $1/c$. Thus we have already found $F_{h_1 h_3}$ on the lattice line $x^* = 1/2\Delta x$. For all the other lines $x^* = h_1/a$ we can apply the simple interpolation

$$F_{h_1 h_3} = \sum_n F_{h_1 n} \frac{\sin\pi\left(h_1 \dfrac{c'}{c} - n\right)}{\pi\left(h_1 \dfrac{c'}{c} - n\right)} . \qquad (4.41)$$

with the virtual thickness $c' = ac/2h_1 \Delta x$. As we have assumed the object to be zero outside the thickness c, this is also true for $c' > c$. Thus the use of (4.41) is always correct.

Obviously, there are no difficulties if the lateral extent of the object has to be enlarged. Let us assume that we increase a to 2a. The corresponding cutouts of the projections must be increased by the same factor. Since the sampling distance of each section in Fourier space is halved, the method described above leads to new lattice lines $x^* = h_1/2a$. Thus we obtain exactly the additional structure factors needed without any change in the number of projections or the limits of resolution.

As already mentioned in Sect.4.3.3d, it is impossible to define an isotropic resolution when the tilting range is restricted. One estimate of the resolution limits can be derived from the region in Fourier space which is covered with structure factors. We denote by x^*_{max} and z^*_{max} the largest dimensions of this region in the x^* and z^* directions, respectively. Assuming that N projections have been measured, we obtain from Fig.4.17

$$z^*_{max} = \frac{N}{2C} \quad , \tag{4.42a}$$

$$x^*_{max} = \frac{z^*_{max}}{\tan\alpha_{max}} = \frac{1}{2\Delta x} \quad . \tag{4.42b}$$

These values only describe a theoretical resolution, however. To determine the physical resolution of the reconstruction we have to consider the resolution limit r^*_{max} of the projections. If r^*_{max} is less than x^*_{max}, the reconstruction will be made with too high an accuracy, since the structure factors are practically zero outside the limit r^*_{max}. If r^*_{max} is larger than x^*_{max}, measured information will be thrown away. The most favorable relation is $x^*_{max} = r^*_{max}$, as shown in Fig.4.17. Assuming a given r^*_{max}, it is always possible to satisfy this condition by the use of appropriate tilting angles: We have to replace Δx by $1/2r^*_{max}$ in (4.40)

$$\alpha_n = \arctan\left(\frac{n}{cr^*_{max}}\right) \quad . \tag{4.43}$$

If the N projections can be measured with a larger resolution limit r^*_{max} the angles α_n and α_{max} will be decreased. From (4.42a), we see that the resolution along z remains unchanged because it is controlled only by the number N of projections and the thickness c of the specimen. Perpendicular to the z axis, however, the resolution increases proportionally to r^*_{max}.

b) Reconstruction by Interpolation in Projection Space

A new and interesting reconstruction method emerges if one rearranges the projections in a special manner. Like Cormack's scheme, the method works in the projection space using an interpolation involving the tilting angle α. So far, we have only used projection planes that are perpendicular to their projection directions and inclined to the reconstruction body at the tilting angle α. Now we transform all the projections in such a way that their planes are parallel to the x, y plane. As shown in Fig.4.18, this can be done by a stretching operation with the factor $1/\cos\alpha$. Since this transformation is only possible for $\alpha_{max} < 90°$, only single-axis tilting with restricted angle and conical tilting can be treated in this way. In the next step, the projections are arranged parallel to each other in a sequence corresponding to increasing tilting angle α. The distance between two neighboring projection planes is equal to the increment $\Delta\alpha$. In other words, the projections are sampling planes in a projection space using the coordinates of these planes and the tilting angle as cartesian coordinates, whereas a polar coordinate system is used in Cormack's scheme. As a consequence, the interpolation with respect to α cannot be performed on concentric circles about the origin but is now carried out on straight lines perpendicular to the projection planes. Obviously the same one-dimensional Fourier interpolation can be used.

As an example, the method has been applied to an object consisting of a single point. Figure 4.19a shows the plane D_0 containing all projections $P_{\bar{3}}$, $P_{\bar{2}}$, ..., P_3 of the point A. In Fig.4.19b these projections are rearranged in a sequence as described above. We can now interpolate along vertical lines such as Q_0 and Q_1. Since each of these lines contains only points with zero weight except for one with weight unity, the appropriate interpolation functions are those shown in Figs.4.19c

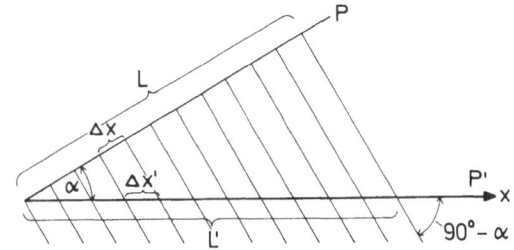

Fig.4.18. Projection with tilting angle α onto the x, y plane; all resolution elements are stretched by the factor $1/\cos\alpha$

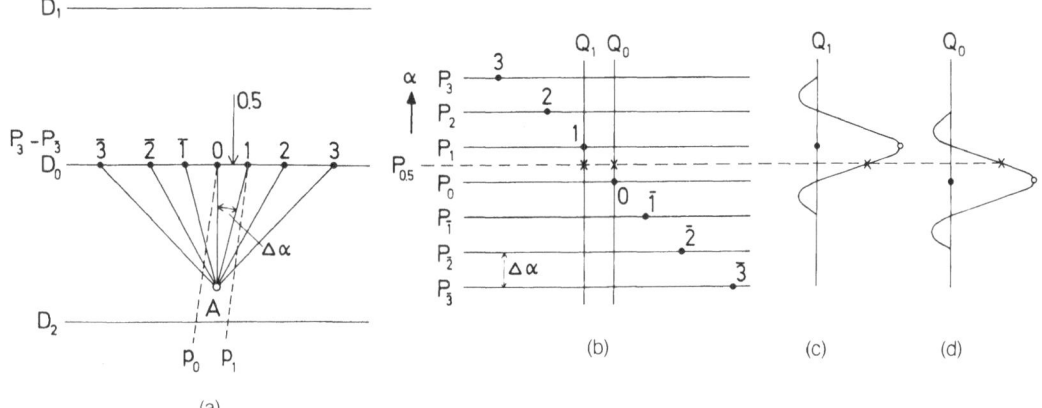

(a)

(b)

(c)

(d)

Fig.4.19a-d. "Platelet" reconstruction by interpolation in the projection space. (a) Projections of the point A onto the plane D_0. (b) Arrangement of the projections in the projection space with the cartesian coordinates x and α. (c) and (d) Interpolation along Q_0 and Q_1 leads to a double point on the projection $P_{0.5}$. According to (a) the projection body corresponding to $P_{0.5}$ consists of two parallel projection lines p_0 and p_1 which do not cross in A

and 4.19d. Let us now consider the projection plane $P_{0.5}$ in Fig.4.19b, interlacing
between P_0 and P_1 with $\alpha = 0.5\Delta\alpha$. The values on this plane as interpolated from Q_0
and Q_1 are marked by crosses. It can be seen that the point image has been split into
a double point by this interpolation. The consequences for the reconstruction can
be demonstrated by simple back-projection. If measured projections only are used,
we obtain a sharp point like A in Fig.4.19a. The addition of the interpolated projection $P_{0.5}$, however, blurs the reconstructed image point. The reason is that the
dashed lines p_0 and p_1 in Fig.4.19a, corresponding to the point back-projection
lines for $P_{0.5}$, do not intersect in A.

The resolution is thus not limited by the projection resolution, whatever method
is used for reconstruction starting from the interpolated projection space. In
Fig.4.19a, it has been demonstrated for the case of back-projection that the resolution limit is determined by the splitting of the correct back-projection line
into two parallel, somewhat incorrectly inclined lines with different weights.
Furthermore, the shape of the image point becomes dependent on the position in the
reconstruction body. The same result has been derived for the Cormack interpolation
in Sect.4.3.5a, where Fig.4.13 corresponds to Fig.4.19. The difference is that, in
Cormack's scheme, the image point shape is identical along circles around the
cylinder axis (for $\alpha_{max} = 90°$), whereas the point in the new scheme is identical
in planes parallel to D_0. In particular, in Cormack's scheme, only the points
along the cylinder axis are free of interpolation errors, unlike the interpolation
method of Fig.4.19b, where all points in the plane D_0 are error-free. It should be
mentioned that the stretching operation described above is responsible for this
favorable feature of the method. If unstretched projections are arranged and interpolated according to Fig.4.19, the lines or surfaces containing identical image
point shapes will have complicated curvatures.

There is no doubt that the new system — adapted on the one hand to restricted
α_{max} and on the other hand to the spcial shape of the object preparations — is in
principle better suited for electron microscopy of "infinite" platelets than a
system based on cylindrical interpolations. This is even more true if one works
with Cormack's system in the case $\alpha_{max} < 90°$, where the image point becomes ad-
ditionally dependent on the azimuth. Perhaps a system of orthogonal functions can
be found which will enable us to perform the reconstruction with the mathematical
elegance of the Cormack system.

c) *The Partially Defined "Unit Cell"*

We have neglected a problem arising from the definition of a unit cell in the
case of an "infinite" platelet. Let us suppose that all cutouts of the projections
used for reconstruction are correlated exactly. Nevertheless, for different tilting
angles different regions of the platelet have been imaged at the borders of the
cutouts. Figure 4.20a shows the case of a platelet reconstruction, where contributions

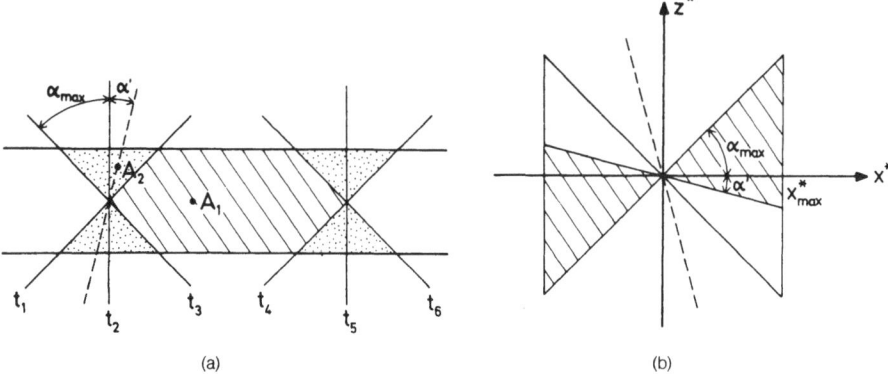

(a)　　　　　　　　　　　　　　　　　(b)

Fig.4.20a,b. The partially defined "unit cell". (a) Only the dashed region of the specimen can be imaged in the cutouts of all projections with tilting angles from $-\alpha_{max}$ to α_{max}. (b) Region in Fourier space accessible to the point A_2

from the dashed region of the specimen can be found in all projections within the tilting range from $-\alpha_{max}$ to α_{max}. The dotted region, however, has been imaged only in some of the projections. As a consequence, the unit cell defined by the lines t_2 and t_5 will be reconstructed incorrectly in some parts.

Nevertheless, it is possible to define formally a unit cell in full analogy to X-ray crystallography but with atoms somewhat modified: We replace the atomic sacttering factor f_j of each atom by

$$f'_j = \begin{cases} f_j & \text{for} \quad \underline{r}^* \text{ inside } Q \\ 0 & \text{for} \quad \underline{r}^* \text{ outside } Q \end{cases} \tag{4.44}$$

where Q is the reciprocal space region accessible to the experiment. Q depends on the position of the atom. For an atom at the point A_1 in Fig.4.20a, Q is defined in Fig.4.20b by the full angular range from $-\alpha_{max}$ to α_{max}, since A_1 will be imaged in all projections. An atom at A_2, however, can be seen only in projections with the tilting angle between $\alpha'(-\alpha_{max} < \alpha')$ and α_{max}. Thus Q is restricted to the dashed region in Fig.4.20b.

The advantage of (4.44) is that the distortions of the atoms can be understood immediately from general Fourier principles. The undistorted image of an atom has to be convolved with the Fourier transform of the function that is equal to one inside the region Q and zero outside. All atoms in the dashed region in Fig.4.20a will be reconstructed with the distortions defined by the full angular range from $-\alpha_{max}$ to α_{max}, whereas atoms in the dotted regions in Fig.4.20a will show stronger distortions. We can predict for the atom at A_2 that the resolution becomes worse in the direction of the dashed line in Fig.4.20b, which in Fig.4.20a corresponds to the line bisecting the angle between t_3 and the projection line through A_2 for the angle α'. Perpendicular to this direction, however, the resolution is equal to that of the point A_1. Thus we conclude that artefacts arising from the partially defined unit cell mainly occur in the triangles limited by the lines t_1, t_2, t_3 and t_4, t_5, t_6, respectively.

4.3.8 Determination of a Common Origin of the Projections

Since the radiation dose for the significant determination of an image detail, e.g., of a mark, is nearly the same as for a complete three-dimensional structure determination with the same significance level [4.3], it is obviously not possible to determine the positions of marks with sufficient significance if the dose necessary for the structure determination has to be kept as low as possible in order to minimize radiation damage. In this case, the reconstruction has to be made from noisy projections which are later combined to give the significant three-dimensional structure. Thus the unambiguous determination of a common origin of the projections by point marks is not possible. Note in this context that the errors of such a determination must be smaller than the resolution limit of the reconstruction. Some time ago, it was shown [4.30,43] that a new type of correlation functions —namely cross-correlation functions between slightly inclined projections —can be used for the definition of a common origin. The interesting point is that a correlation method is based on all features of a structure and not only on few selected points. It has been shown in another context that the significance of the cross-correlation maximum between two identical point structures increases with the number of points [4.37].

Let us consider an image divided into M image elements, where U_j is the signal of the j-th image element. The cross-correlation peak CCF_{max} of two images with the signals $U_j^{(1)}$ and $U_j^{(2)}$ is then given by

$$CCF_{max} = \sum_{j=1}^{M} U_j^{(1)} U_j^{(2)} \quad . \tag{4.45}$$

Now we consider $U_j^{(1)}$ and $U_j^{(2)}$ to be random variables with the same expectation values ρ_j and the same variances σ_j^2. In other words: The signals are two different and statistically independent realizations of an "ideal" image ρ_j. These assumptions require that the micrographs have been taken under identical conditions, without any modification of the object. The variance of the product $U_j^{(1)} U_j^{(2)}$ is then given by $2\rho_j^2\sigma_j^2 + \sigma_j^4$. Since the image elements are mutually independent, we can calculate the variance σ_c^2 of the cross-correlation peak

$$\sigma_c^2 = \sum_{j=1}^{M} \left(2\rho_j^2\sigma_j^2 + \sigma_j^4 \right) \quad . \tag{4.46}$$

If we replace the signal CCF_{max} by its expectation value $\sum_{j=1}^{M} \rho_j^2$, the signal-to-noise ratio P_c of CCF_{max} can be expressed by

$$P_c = \sum_{j=1}^{M} \rho_j^2 \Big/ \sqrt{\sum_{j=1}^{M} \left(2\rho_j^2\sigma_j^2 + \sigma_j^4\right)} \quad .$$ (4.47)

Assuming a bright-field image, the noise is nearly equal for all image elements, $\sigma_j = \sigma$, which leads to

$$P_c = \sqrt{M} \frac{\dfrac{1}{\sigma^2 M}\sum_{j=1}^{M}\rho_j^2}{\sqrt{\dfrac{2}{\sigma^2 M}\sum_{j=1}^{M}\rho_j^2 + 1}} \quad .$$ (4.48)

As already mentioned in Sect.4.2.3, it is convenient to transform bright-field images into dimensionless contrast distributions. In this case,

$$C = \sqrt{\sum_{j=1}^{M}\rho_j^2/M}$$ (4.49)

is equal to the root-mean-square contrast and can be considered as the average signal per image element. Writing $p = C/\sigma$ for the signal-to-noise ratio of one image element we obtain

$$P_c = \sqrt{M} \frac{p^2}{\sqrt{2p^2+1}} \quad .$$ (4.50)

Thus the signal-to-noise ratio of CCF_{max} is very much larger than that of one image element. Even with $p \ll 1$ a sufficiently large P_c is possible if the number M of image elements is not too small. In this case, (4.50) can be simplified to $P_c = \sqrt{M}\, p^2$.

In connection with trace structure analysis [4.37], the signal-to-noise ratio of the correlation peak of two three-dimensional point structures (three-dimensional dark-field images under ideal imaging conditions) has been derived. In this case the number M' of points corresponds to the number M of image elements and p is equal to the signal-to-noise ratio of the occupancy of one point by its dark-field electrons. For equal atoms, P_c can be estimated from (4.50) by replacing M by M'. If one calculates the variance σ_c^2 in the linear approximation using the Gaussian error propagation law, the term σ_j^4 in (4.46) and therefore the number 1 in the denominator of (4.50) will be negligible. Thus we obtain $P_c = p\sqrt{M'/2}$ which corresponds to the result in [4.37] (see also [4.38]).

The definition of a common origin can be explained using a two-dimensional point object as an example. We choose the y axis as tilting axis and assume N points with the weights U_j to be distributed in the x, z plane with the coordinates (x_j, z_j). The density $\rho(x,z)$ of the object is

$$\rho(x,z) = \sum_{j=1}^{N} U_j \delta(x - x_j)\delta(z - z_j)$$ (4.51)

where $\delta(x)$ denotes the Dirac delta function. From this we calculate the projections $P(x,\alpha)$ on the x axis inclined to the z axis at an angle α,

$$P(x,\alpha) = \sum_{j=1}^{N} U_j \delta(x - x_j + z_j \tan\alpha) \quad . \tag{4.52}$$

The correlation of two projections with the inclination angles α_1 and α_2 leads to

$$CCF(x) = \sum_{j=1}^{N} U_j^2 \delta[x - x_0 - z_j(\tan\alpha_2 - \tan\alpha_1)] \tag{4.53}$$

$$+ \sum_{j=1}^{N} \sum_{\substack{i=1 \\ i \neq j}}^{N} U_i U_j \delta(x - x_0 + x_i - x_j - z_i \tan\alpha_2 + z_j \tan\alpha_1) \quad ,$$

where x_0 is an unknown translation of the second projection relative to the first. For the case $\alpha_2 = \alpha_1$, (4.53) is equal to the autocorrelation function. This function shows one strong and sharp peak with the weight $\sum_{j=1}^{N} U_j^2$ at x_0 and a background distributed over the whole image. This background does not exceed the maximum of the products $U_i U_j$. If α_2 becomes different from α_1 the peak will be broadened more and more. For a large difference $\alpha_2 - \alpha_1$ the peak will disappear in the background. If $\alpha_2 - \alpha_1$ is not too large, however, the peak can be recognized very well.

Let us consider the center of gravity of the peak, defined by

$$x_s(\alpha_1,\alpha_2) = \frac{\int x\, CCF_{max}(x)dx}{\int CCF_{max}(x)dx} \quad , \tag{4.54}$$

where $CCF_{max}(x)$ contains only the quadratic terms of (4.53) contributing to the peak. We obtain

$$x_s(\alpha_1,\alpha_2) = \frac{1}{G} \sum_{j=1}^{N} U_j^2 [x_0 + z_j(\tan\alpha_2 - \tan\alpha_1)] \quad , \tag{4.55}$$

with $G = \sum_{j=1}^{N} U_j^2$. Now we calculate the projection $P_q(x,\alpha)$ of the object (4.51) but with squared weights U_j^2 instead of U_j.

$$P_q(x,\alpha) = \sum_{j=1}^{N} U_j^2 \delta(x - x_j + z_j \tan\alpha) \quad . \tag{4.56}$$

The center of gravity $x_{qs}(\alpha) = \int xP_q(x,\alpha)dx / \int P_q(x,\alpha)dx$ of this projection is

$$x_{qs}(\alpha) = \frac{1}{G} \sum_{j=1}^{N} U_j^2(x_j - z_j \tan\alpha) \quad . \tag{4.57}$$

Since the second projection again contains the unknown shift x_0, it is easy to see that the relation

$$x_s(\alpha_1,\alpha_2) = x_{qs}(\alpha_2) - x_{qs}(\alpha_1) \tag{4.58}$$

is valid. In other words, the center of gravity of the cross-correlation peak of two projections is equal to the difference between the centers of gravity of these projections taken from the same structure but with squared weights.

The application of relation (4.58) for the definition of a common origin is shown in Fig.4.21. S is the center of gravity of the structure with squared weights. For the projection $P(x,\alpha_1)$, the origin 0_1 can be chosen arbitrarily. The origin 0_2 of the projection $P(x,\alpha_2)$ will be shifted until the center of gravity (4.58) of the cross-correlation function is situated at the origin of the corresponding plane. According to (4.58), the distances 0_1S_1 and 0_2S_2 are then equal in both projections. We can now shift $P(x,\alpha_1)$ and $P(x,\alpha_2)$ along these directions α_1 and α_2 until 0_1 and 0_2 coincide with 0. Thus a common origin is defined, situated in the line parallel to the x axis through S. Obviously this construction applied to further projections leads to the same origin 0.

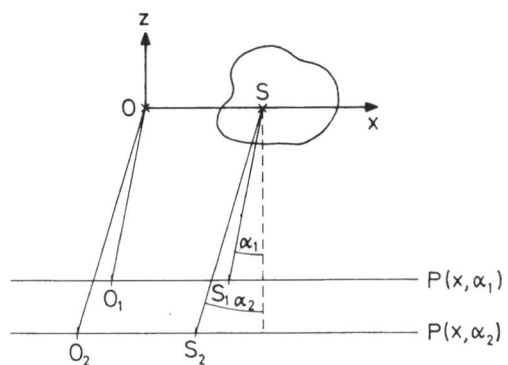

Fig.4.21. Construction of a common origin for two projections with different tilting angles α_1 and α_2. S is the center of gravity of the squared structure: its projection points are S_1 and S_2. By back-projection of 0_1 and 0_2, the common origin 0 can be found

In the proof of (4.58) as well as in Fig.4.21, inclined projections have been used. This is no restriction, however, because each orthogonal projection can be transformed to the corresponding inclined projection by multiplication with the factor $1/\cos\alpha$. The application of the method to real two-dimensional projections of a single-axis tilt series requires that the direction of the tilting axis be known. In the most cases, this direction can be determined from the experiment. For conical tilting a similar but more complicated procedure for the determination of a common origin is possible.

The usefulness of the method for real micrographs is limited by the accuracy with which the center of gravity $x_s(\alpha_1,\alpha_2)$ can be determined. As well as the radiation dose, the fluctuations of the background, neglected above, are of interest. Since the background also extends below the correlation peak, it may slightly displace its position. This position error does not occur if two identical ($\alpha_1 = \alpha_2$) noise-free projections are correlated. In this case the centrosymmetry of the correlation background can cause only small shape modifications of the peak. Since not only the correlation peak but also the side peaks undergo shape changes which increase with $\Delta\alpha = \alpha_2 - \alpha_1$, we might suppose that at small $\Delta\alpha$ the deviation from the centrosymmetry of the background and thus its influence on the peak position will be small. There is a second argument for a low background. Since the projections are recorded as contrast distributions with zero mean level they show positive and negative "peaks". Both sorts of peaks will be treated in the same manner by the squaring operations described above but the heavy atoms will become more pronounced whereas the averaged peak height of the light atoms will appear considerably reduced. Therefore, the background of the correlation function also remains low.

4.4 Aspects for the Future

4.4.1 The "Atom" Constraint

We have already used the finite body constraint which permits interpolation between successive projections. There are, however, further constraints well known in crystallography. It is useful to investigate whether these constraints can be exploited for three-dimensional electron microscopy.

The direct methods in crystallography can usually solve the phase problem for structures up to about 100 atoms (not counting hydrogen atoms), and are so successful that the structure determination of complicated organic molecules can be done nearly routinely today. These methods are based on the fact that the electron density is positive and peaked at atomic sites. If nothing is known of the structure (except the unit cell contents), probability relations between the phases of the structure factors can be utilized which are derived from the postulate that the integral of the third power of the electron density ρ should have the maximal possible value (ρ^3 criterion). It is easy to see that this postulate means that the correct structure shows an electron density condensed to singular peaks. One has to use the integral of ρ^3 and not of ρ^2 as the criterion since in the latter case negative areas of a wrong structure would increase and not decrease the integral. Exactly the same criterion can now be used for the refinement of the origin: by systematic variation of the relative parameters between the projections, the "best peaked" structure (with maximal ρ^3 integral) can be found. This procedure needs atomic resolution. It might, however, also be applicable at lower resolutions if for other reasons a peaked structure is expected (for example if the specimen is an amorphous arrangement of very small crystallites).

In another way, the "atom" constraint is used in the crystallographic method of "phase correction" [4.66-69]. In this case a first approximation to the structure — although with substantial phase errors — has already been determined. The method leads to a structure with minimal phase errors by an iterative scheme. Since errors in defining the origin are also reflected as phase errors of the structure factors, these methods might also be used for the correction of a structure with a somewhat incorrectly determined origin. One could then calculate the set of projections and check by correlations whether the phase correction has indeed produced a refinement of the origin. In the final step, however, one should use the experimentally determined phases (after refinement of the origin). It is one of the advantages of the electron microscopical method that amplitudes *and* phases can be measured.

There is a further possibility of using the atom constraint, which has already been tested some time ago in computer simulations [4.17]: if one keeps the measured structure factor amplitudes and phases fixed and uses a variant of phase correction, one can calculate the missing structure factors in the blind regions of reciprocal space. It is, however, not clear whether deviations from the ideal peak structure caused by averaging over a substantial part of unstable structures might lead to difficulties.

We see that there are many possible applications of the atom constraint in electron microscopy. However, the atom constraint is only one possible constraint; bond lengths, bond angles, and constraints in the chemical composition could also be used in sophisticated pattern recognition approaches. This is a great advantage and we note that these constraints can also be used as a check of the consistency of experimental results. The application of constraints can be carried even further. Their utilization in the determination of the dynamic structure of radiation-damaged specimens leads to the ideas of trace structure analysis [4.37,38].

4.4.2 The Use of a STEM as a Diffractometer

In the last few years, another type of electron microscope, the STEM (Scanning Transmission Electron Microscope) has been developed to a high technical standard and is now also commercially available. In contrast to the surface scanning microscope (resolution $d_{min} \sim 10\,nm$) it can also be used for medium ($d_{min} = 0.5 - 2\,nm$) and high ($d_{min} < 0.3\,nm$) resolution work. It is thus of interest to discuss the usefulness of these instruments for three-dimensional imaging. It has been shown that a STEM can be regarded to some extent as a device with "inverted" optical features [4.70,71]. It works in the usual mode as a dark-field microscope with incoherent illumination. Such an instrument cannot be regarded as a diffractometer in the usual sense, since there is not a unique primary beam direction S_0 as there is in Fig.4.1. However, as long as we can work with the plane approximation of the

Ewald sphere and as long as the lens is sufficiently ideal, the dark-field image resembles the square of the projection of the image amplitude ψ_s (anomalous scattering neglected). The illuminating aperture α_{ill} (Fig.4.22a) (and not the detection aperture α_d) determines the resolution of the image. In a CTEM (see Fig.4.22b[18]), the detection aperture α_d of the imaging lens L_{im} determines the resolution. If we "construct" a STEM, which has in addition the imaging lens system of a CTEM as shown in Fig.4.22b the image in the nonscanning mode is a spot with a diameter defined by α_{ill}; it will exhibit a dark-field fine structure with a resolution defined by α_d. The intensity distribution at the annular detector which replaces this lens in a normal STEM, corresponds to the Fourier transform square of the specimen inside the illuminated spot. For a three-dimensional body, the illuminated region is a cylinder along z, assuming that the plane approximation of the Ewald sphere is adequate, and the image density corresponds to the projection of the squared image amplitude ψ_s (although strictly speaking only if the plane approximation remains valid for the Ewald calotte up to α_d!). In a STEM, however, the scattered rays will not be utilized for the generation of a sharp dark-field image within the probe diameter but simply collected on the annular detector. It can be shown that this integral over the intensity distribution corresponds to the zero peak of the autocorrelation function of ψ_s or in other words to the integral (average) over the dark-field image within the probe (which is a reasonable result).

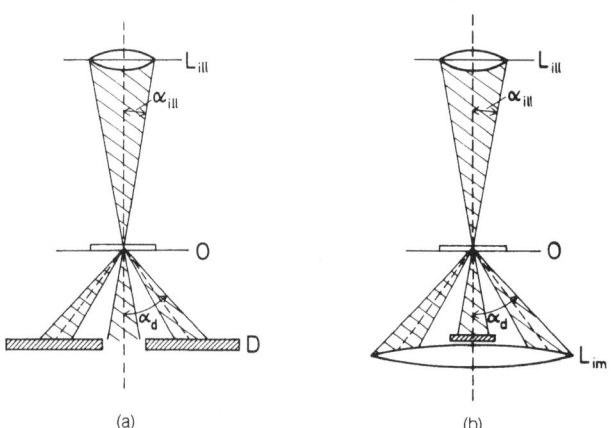

(a)   (b)

Fig.4.22a,b. Imaging with the STEM. (a) L_{ill} illuminating lens, α_{ill} illuminating aperture, O object, α_d detector aperture, D detector. (b) STEM with the additional facility for conventional imaging

---

18 The lower part of Fig.4.22b is similar to Fig.4.1a, where φ_{max} and L correspond to α_d and L_{im}, respectively.

As shown in an earlier paper [4.72], one could retrieve the information most fully by measuring the diffractogram in the detector plane for each image element instead of the integrated diffraction intensity. This can be done by replacing the annular integrating detector by a position-sensitive detector. A Fourier transformation of the corresponding amplitudes (not intensities!) then leads to the fine structure within the spot (replacement of the lens L_{im} by reconstruction from the diffracted rays). It has been shown in the same paper that the phases of the diffracted beams —which are lost since only the squared amplitude (intensities) can be measured—could then be determined by a method which we later called ptychography. Unfortunately, for amorphous specimens, the evaluation procedure leads to nonlinear equations, which are difficult to solve. However, for crystals this method is appropriate—especially if (as shown in [4.73]) the primary beam is split by a double-slit aperture or a biprism into two inclined beams (see also [4.74]).

Ptychography is a lensless method for the determination of amplitudes and phases in the diffraction plane rather like holography. Let us now arrange a position-sensitive detector inside the (sufficiently wide) illumination cone. In this case, a diffraction diagram will again be projected onto the detector. However, the primary beam waves interfere with the scattered waves, acting as reference waves in the sense of holography. Divided detectors in the primary beam cone were first proposed in [4.75,76].

In order to understand the "matrix detector holographic STEM", we first discuss the well-known setting of the STEM for phase contrast [4.70]. In this case, the annular dark-field detector in Fig.4.22a is replaced by a small detector D_1 in Fig.4.23a. Each "primary beam" inside the illumination cone of the probe lens L_{ill} will produce a "scattered beam" in D_1. All these scattered beams will interfere since the beams within the probe aperture are coherent in the STEM. This interference will take place in exactly the same way as in the corresponding image point of a CTEM. The important point is that Fig.4.23a again describes a diffractometer; now, however, we are concerned with "all primary waves producing a single scattered wave of defined orientation" instead of "all scattered waves produced by a single primary wave of defined orientation".

A STEM image taken in this way can be treated like a CTEM bright-field image, it can be separated by Fourier transformation into the corresponding scattered waves, and it can be corrected by two-dimensional image reconstruction. One can show that the scattering in such an "inverted" diffractometer can be described in a generally similar way to the scattering in a CTEM with coherent illumination. There is a surface (similar to the Ewald sphere surface) in reciprocal space which contains all Fourier coefficients contributing to a single scattered ray. Figure 4.23b demonstrates that the origins of all the Ewald spheres belonging to the cone of primary beam directions of the probe lens lie on the surface of a sphere (dashed in Fig.4.23b). Since only the scattered wave in direction $\underline{s}$ will contribute to the image, the $\underline{r}^*$

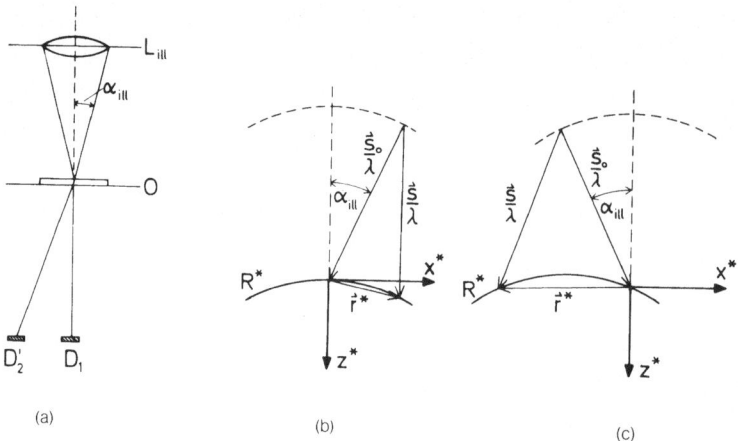

(a)

(b)

(c)

Fig.4.23a-c. The STEM as a diffractometer. (a) STEM with positive-sensitive detectors D_1 and D_2. (b) Representation in the reciprocal space: D_1 measures the intensity arising by superposition of all waves scattered into the same direction $\underline{s}$ but
with different primary directions $\underline{s}_0$. In other words, an integral over the structure factors on the calotte R^* is measured. (c) Like (b), but for the detector D_2

sphere R^* in Fig.4.23b will be recorded. We again have a pure case since no de-
trimental overlap of different scattering directions will occur.[19] Let us now shift
the detector off the optic axis into an arbitrary position inside the illumination
cone (D_2 in Fig.4.23a). Figure 4.23c shows that, in this case too, a spherical surface will be selected in reciprocal space. This calotte is shifted relative to the
calotte in Fig.4.23b. In this mode, the STEM works like a CTEM phase contrast microscope with inclined illumination. It is now easy to see that a STEM provided with
a matrix of N detectors corresponds to a CTEM which delivers N phase contrast images
taken with N different illumination directions. In contrast to the corresponding
CTEM, however, the STEM delivers these images simultaneously. The images are not
completely equivalent, since the corresponding diffraction surfaces in reciprocal
space (the R^* calottes in Fig.4.23b) vary their positions with the position of the
detector element. In the case of a nonnegligible curvature of the Ewald sphere
they occupy a certain volume of the reciprocal space (see Fig.4.23c). This has the
interesting consequence that the collected data can be used for three-dimensional
reconstruction although with a very anisotropic image point (see [4.71]).

It is of interest that there is an alternative way of interpretating the data
of the matrix detector STEM. Let us remember that the first proposal in holography
[4.77] was to provide a point source which illuminates the specimen and which delivers at the same time a coherent reference wave background for the scattered rays.

---

19 Note that the strict equivalence with the CTEM case is only valid if all primary
   beams in the probe cone have the same intensity. This point is, however, of little
   importance, since the influence of different intensities can easily be corrected.

The illumination conditions in a defocused STEM correspond to the Gabor conditions (negligible curvature of the Ewald sphere assumed). Therefore, the density distribution in the small area illuminated by the primary beam can be determined by Gabor holography and the complete scanned region can be regenerated from these small holographic reconstructions. As usual in holography with an on-axis reference beam, the twin image will overlap the correct image. However, in the STEM case the image field can be very small (small defocus assumed). The defocused twin image is larger than the correct image. It will, therefore, only overlap the correct image. (We assume that only the image field of the correct image will be reconstructed.) It is interesting that holography in the STEM discriminates at least partially against the twin image in spite of the fact that the primary beam is used as reference beam. Unlike the first reconstruction method, this second method could also be used on-line.

There is no experimental experience with matrix STEMs. We have discussed these possibilities in some detail in order to show that a STEM can also be used as a diffractometer in the crystallographic sense, although in a somewhat complicated form. It is obvious that the replacement of one STEM image by a great number of STEM "images" leads to substantial problems in data collection and data handling. It remains to be seen whether the experimental advantages — notably the straightforward electronic recording of the data and the strict restriction of the illumination to the image field — will outweigh these complications.

References

4.1   W. Hoppe, D. Typke: "Three-Dimensional Reconstruction of Aperiodic Objects in Electron Microscopy", in *Advances in Structure Research by Diffraction Method*, Vol.7, ed. by W. Hoppe, R. Mason (Pergamon Press, Oxford, and Vieweg, Braunschweig 1978) pp.137-190
4.2   W. Hoppe, P. Bussler, A. Feltynowski, N. Hunsmann, A. Hirt: "Some Experience with Computerized Image Reconstruction Methods", in *Image Processing and Computer-Aided Design in Electron Optics*, ed. by P.W. Hawkes (Academic Press, London 1973) pp.91-126
4.3   R. Hegerl, W. Hoppe: Z. Naturforsch. *31a*, 1717-1721 (1976)
4.4   A. Jones: J. Appl. Cryst. *11*, 268 (1978)
4.5   W. Hoppe: Acta Cryst. A*26*, 414-426 (1970)
4.6   W. Hoppe: Naturwiss. *61*, 239-249 (1974)
4.7   J. Kieffer: Radiology *33*, 560-585 (1939)
4.8   P. Edholm: The tomogram. Its formation and content. Acta Radiol., Suppl. *193* (1960)
4.9   R.N. Bracewell, A.C. Riddle: Astrophys. J. *150*, 427-434 (1967)
4.10 D.W. Sweeney, C.M. Vest: Appl. Opt. *11*, 205-207 (1972)
4.11 R.G. Hart: Science *159*, 1464-1467 (1968)
4.12 D.J. DeRosier, A. Klug: Nature (London) *217*, 130-134 (1968)
4.13 W. Hoppe, R. Langer, G. Knesch, Ch. Poppe: Naturwiss. *55*, 333-336 (1968)
4.14 J.G. Colsher: Comput. Graphics Image Process. *6*, 513-537 (1977)
4.15 J.G. Colsher: Thesis, University of California (1976)
4.16 R.A. Crowther, A. Klug: Ann. Rev. Biochem. *44*, 161-182 (1975)

4.17 W. Hoppe: Ber. Bunsenges. Phys. Chem. *74*, 1090-1100 (1970);
      Phil. Trans. Roy. Soc. London *261*, 71-94 (1971)
4.18 J. Frank, P. Bussler, R. Langer, W. Hoppe: Ber. Bunsenges. Phys. Chem. *74*,
      1105-1115 (1970)
4.19 R. Langer, J. Frank, A. Feltynowski, W. Hoppe: Ber. Bunsenges. Phys. Chem.
      74, 1120-1126 (1970)
4.20 W. Hoppe: Manchester (1972) pp.612-617
4.21 W. Hoppe: Naturwiss. *48*, 736-737 (1961)
4.22 W. Hoppe: Optik *20*, 599-606 (1963)
4.23 G. Möllenstedt, R. Speidel, W. Hoppe, R. Langer, K.-H. Katerbau, F. Thon:
      Rome (1968) Vol.1, pp.125-126
4.24 W. Hoppe, H. Wenzl, H.J. Schramm: Jerusalem (1976) Vol.2, pp.58-60
4.25 W. Hoppe, H. Wenzl, H.J. Schramm: Hoppe-Seyler's Z. Physiol. Chem. *358*,
      1069-1076 (1977)
4.26 D.F. Parsons: Science *186*, 407-414 (1974)
4.27 R.M. Glaeser: J. Ultrastruct. Res. *36*, 466-482 (1971)
4.28 R. Henderson, P.N.T. Unwin: Nature (London) *257*, 28-32 (1975)
4.29 W. Hoppe, J. Gassmann, N. Hunsmann, H.J. Schramm, M. Sturm: Hoppe-Seyler's Z.
      Physiol. Chem. *355*, 1483-1487 (1974)
4.30 W. Hoppe, H.J. Schramm, M. Sturm, N. Hunsmann, J. Gassmann: Z. Naturforsch.
      31a, 645-655 (1976)
4.31 W. Hoppe, H.J. Schramm, M. Sturm, N. Hunsmann, J. Gassmann: Z. Naturforsch.
      31a, 1370-1379 (1976)
4.32 W. Hoppe, H.J. Schramm, M. Sturm, N. Hunsmann, J. Gassmann: Z. Naturforsch.
      31a, 1380-1390 (1976)
4.33 W. Hoppe, B. Grill: Ultramicroscopy *2*, 153-168 (1977)
4.34 J. Frank: Jerusalem (1976) Vol.1, pp.273-274
4.35 J. Frank: Ultramicroscopy *1*, 159-162 (1975)
4.36 J. Frank: "Optical Use of Image Information Using Signal Detection and Aver-
      aging Techniques", in *Short Wavelength Microscopy*, Ann. N.Y. Acad. Sci. *306*,
      112-120 (1978)
4.37 W. Hoppe: Z. Naturforsch. *30a*, 1188-1199 (1975)
4.38 W. Hoppe: "Trace Structure Analysis", in *Short Wavelength Microscopy*, Ann.
      N.Y. Acad. Sci. *306*, 121-144 (1978)
4.39 O. Scherzer: Optik *33*, 501-516 (1971)
4.40 H. Rose: Optik *33*, 1-24 (1971); Optik *34*, 285-311 (1971)
4.41 E. Plies, W. Hoppe: Optik *46*, 75-92 (1976)
4.42 W. Hoppe, R. Guckenberger: Z. Naturforsch. *29a*, 1931-1932 (1974)
4.43 W. Hoppe: Naturwiss. *61*, 534-536 (1974)
4.44 P. Schiske: Rome (1968) Vol.1, pp.145-146
4.45 W. Hoppe, R. Langer, F. Thon: Optik *30*, 538-545 (1970)
4.46 W. Hoppe: Z. Naturforsch. *26a*, 1155-1168 (1971)
4.47 W. Hoppe: Optik *29*, 617-621 (1969)
4.48 C.E. Shannon, W. Weaver: *The Mathematical Theory of Communication* (University
      of Illinois Press 1975)
4.49 J.L. Harris: J. Opt. Soc. Am. *54*, 931-936 (1964)
4.50 R.A. Crowther, D.J. DeRosier, A. Klug: Proc. Roy. Soc. Lond. A*317*, 319-340
      (1970)
4.51 V. Knauer: Diplomarbeit, TU München (1976)
4.52 W. Hoppe: Z. Naturforsch. *27a*, 919-929 (1972)
4.53 D. Typke, W. Hoppe, W. Sessler, M. Burger: Jerusalem (1976) Vol.1, pp.334-335
4.54 W. Hoppe, N. Hunsmann, H.J. Schramm, M. Sturm, B. Grill, J. Gassmann:
      Jerusalem (1976) Vol.1, pp.8-13
4.55 M. Radermacher, W. Hoppe: Toronto (1978) Vol.1, pp.218-219
4.56 R.A. Crowther, L.A. Amos, A. Klug: Manchester (1972) pp.593-597
4.57 A.M. Cormack: J. Appl. Phys. *35*, 2908-2913 (1964)
4.58 P.R. Smith, T.M. Peters, R.H.T. Bates: J. Phys. A*6*, 361-382 (1973)
4.59 E. Zeitler: Optik *39*, 396-415 (1974)
4.60 A. Klug, R.A. Crowther: Nature *238*, 435-440 (1972)
4.61 N. Hunsmann: Thesis, TU München (1975)
4.62 R. Gordon, R. Bender, G.T. Herman: J. Theor. Biol. *29*, 471-481 (1970)

4.63 P.F.C. Gilbert: J. Theor. Biol. *36*, 105-117 (1972)
4.64 R.A. Crowther, A. Klug: Nature *251*, 490-492 (1974)
4.65 M. Goitein: Nucl. Instrum. Methods *101*, 509-518 (1971)
4.66 W. Hoppe, R. Huber, J. Gassmann: Acta Cryst. *16*, A4 (1963)
4.67 W. Hoppe, J. Gassmann: Acta Cryst. B*24*, 97-107 (1968)
4.68 W. Hoppe, J. Gassmann, K. Zechmeister: "Some Automatic Procedures for the Solution of Crystal Structures with Direct Methods and Phase Correction", in *Crystallographic Computing II*, ed. by F.R. Ahmed (Munksgaard, Copenhagen 1970) pp.26-36
4.69 J. Gassmann: "Improvement and Extension of Approximate Phase Sets in Structure Determination", in *Crystallographic Computing II*, ed. by F.R. Ahmed (Munksgaard, Copenhagen 1976) pp.144-154
4.70 E. Zeitler, M.G.R. Thomson: Optik *31*, 258-280 (1970)
4.71 O. Scherzer: Ber. Bunsenges. Phys. Chem. *74*, 1154-1167 (1970)
4.72 W. Hoppe: Acta Cryst. A*25*, 508-514 (1969)
4.73 W. Hoppe: Acta Cryst. A*25*, 495-501 (1969)
4.74 R. Hegerl, W. Hoppe: Ber. Bunsenges. Phys. Chem. *74*, 1148-1154 (1970)
4.75 H. Rose: Optik *39*, 416-436 (1974)
4.76 L.H. Veneklasen: Optik *44*, 447-468 (1975)
4.77 D. Gabor: Nature *161*, 777-778 (1948)

5. The Role of Correlation Techniques in Computer Image Processing

J. Frank

With 23 Figures

Many electron microscopists are familiar with the analysis of electron micrographs by using optical diffraction [5.1]. The optical diffractogram, which is essentially the absolute square of the Fourier transform of the optical density on the plate, reveals information on periodicities [5.2], astigmatism, defocus [5.3], drift [5.4] illumination [5.5-7] and defocus spread [5.8].

We shall here be concerned with another method of analysis which is less known but has a number of useful applications: the statistical analysis of images in real space by correlation functions.

While the *autocorrelation function* (ACF) is simply the inverse Fourier transform of the optical diffractogram and contains the same information, the *cross-correlation function* (CCF) introduces a new feature, the statistical comparison of two images.

The ACF has been known in X-ray crystallography for a long time under the name of Patterson function, and is here an important tool for structure determination [5.9]. While this function is routinely interpreted in terms of structural features in X-ray crystallography, the statistical interpretation is important in the analysis of images of nonperiodic objects. The main concepts and theorems relating to correlation functions have been developed in the field of electrical engineering [5.10] and have found their way into statistical optics [5.11,12].

The correlation functions were introduced into electron microscopy [5.13,14] when it became clear from optical diffraction experiments [5.15] that different electron micrographs of the same specimen can be aligned with high accuracy despite their apparent differences due to contamination, radiation damage, electron and photographic noise, and changes in defocus.

The present survey of the computer applications of correlation functions in electron microscopy goes back to a symposium lecture for the 1974 meeting of the Electron Microscopy Society of America [5.16]. A brief introduction is found in [5.17], and in a writeup of the Kontron seminar on Biomolecular Electron Microscopy [5.18].

5.1 Correlation Functions

5.1.1 The Cross-Correlation Function

For definition of the cross-correlation function (Fig.5.1), we consider two images that are scanned on a square grid and represented by a set of optical density measurements $i_1(r_{jk})$ and $i_2(r_{jk})$, respectively. The images are shifted against each other by a vector r_{pq} contained in the sampling grid, and the product of the data in the overlapping area is averaged over the whole area of overlap. The resulting number is the value of the CCF at the argument position r_{pq}

$$\Phi(r_{pq}) = \frac{1}{N'} \sum_j \sum_k i_1(r_{jk} - r_{pq}) i_2(r_{jk}) \quad . \tag{5.1}$$

Here N' denotes the number of points in the overlap region. Variable transformation $r'_{jk} = r_{jk} - r_{pq}$ leads to a common alternative formulation

$$\Phi(r_{pq}) = \frac{1}{N'} \sum_j \sum_k i_1(r'_{jk}) i_2(r'_{jk} + r_{pq}) \quad . \tag{5.2}$$

The whole procedure when repeated for all displacement vectors contained in the sampling grid results in a two-dimensional function that can be displayed like a picture, using gray tones (Fig.5.2). If the two images contain similar features with the same orientation, then there exists a particular shift vector r_{pq} for which complete overlap of these features occurs, producing a high value of the cross-correlation function at the corresponding position, visible as a bright peak in a "picture" display of the function. The location of the peak relative to the origin thus marks the translation vector by which one image has to be shifted against the other for exact alignment. The cross-correlation function therefore plays an important role in image processing as a tool for alignment [5.17].

Implied in the definition of the correlation functions is the stationarity of the images. An image is said to be *stationary* if the statistics (characterized by mean, variance, and higher moments) of a selected image patch do not change with its position. The patch used for a test of stationarity has to be large enough for the statistical description to be meaningful. Electron micrographs of amorphous films frequently used for supporting biological specimens have this property but are no longer stationary in a strict sense if they contain extended objects.

5.1.2 The Autocorrelation Function

One obtains the autocorrelation function (ACF) if one computes the CCF of an image with itself, i.e., $i_1(r) \equiv i_2(r)$ in (5.1). Obviously, exact alignment always occurs at zero displacement so the ACF has a peak at the origin. In this position, not only object-related features of the image (the "signal") but also noise features

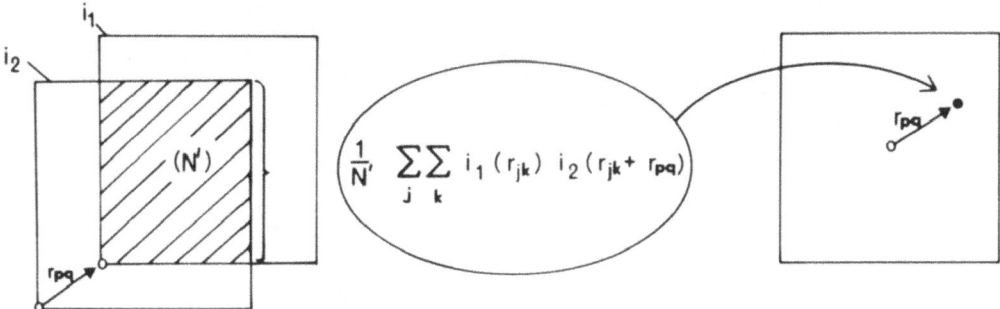

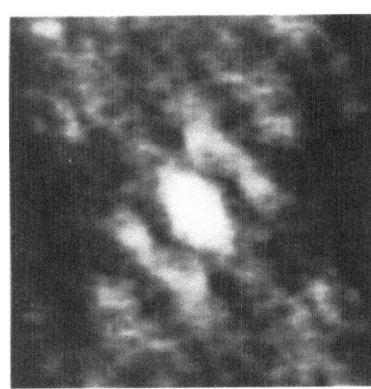

Fig.5.1. Definition of the cross-correlation function. For each shift vector r_{pq} sum up over products of overlapping points N' and place result into the cross-correlation function at the position r_{pq}

Fig.5.2. Cross-correlation function of two electron micrographs of carbon film

come to an exact overlap with themselves and thus contribute to the peak (Fig.5.3a-c). Since noise patches from *different* areas of the image are not correlated, the contribution from noise to the image ACF is confined to the vicinity of the origin. The radius of this peak is often referred to as the "noise autocorrelation radius". If one wishes to study the ACF of the signal part alone, one has to cross correlate two electron micrographs of the same object taken under the same conditions, as these contain different realizations of the noise superposed on the same signal component. Since in such an experiment the different noise realizations are not statistically related to each other, they are cancelled out in *any* relative position by the averaging implicit in the defining formula (5.1).

A comparison between the CCF of two successive electron micrographs of the same object (carbon film) and the ACF of one of the micrographs indeed shows the occurrence of the central noise peak in the ACF as the major difference (Fig.5.3c,d). In the case of a periodic image, the ACF shows a peak repeated on the periodic lattice because a periodic image is invariant under translation by any vector contained in the lattice (Fig.5.4). Only the peak at the origin shows the additional noise contribution for zero shift. In contrast, the CCF of two different patches of the same periodic image shows no such contribution. (For an ideal periodic object, two nonoverlapping patches of the same image are just as "similar" to each other as are corresponding patches of two different images of the same object area.)

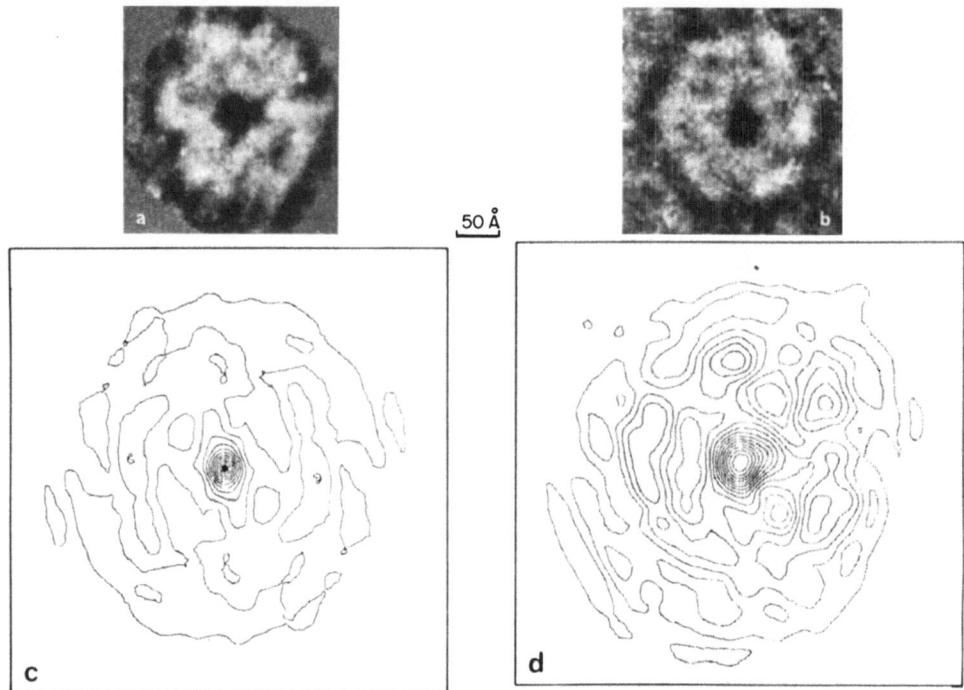

Fig.5.3. (a,b) Two negatively stained glutamine synthetase particles [5.19], (c) cross-correlation function of the two particles, (d) autocorrelation function of the particle in (a)

A conceptual understanding of the ACF can be obtained as follows. Consider an electron micrograph showing lattice fringes. We make two copies on film, and superpose these on a white background or on a viewing screen. As we slide one film with respect to the other in the direction normal to the fringes, the resulting transparency becomes alternately light and dark, showing enhancement or cancellation of the periodicities. Insofar as we perceive a visual impression of the *overall* changes of the resulting transparency we are in fact observing the ACF of the film as a function of the displacement in a particular direction.

In general, a maximum of the ACF in a position $\pm r_m$ indicates that features separated by $\pm r_m$ in the image are highly correlated. Let us take the example of a motif that appears in three positions in the image. Exact overlap of the motif with itself occurs for three pairs of difference vectors where the two vectors of each pair are only distinguished by sign. The ACF of this image therefore consists of six repeats of the motif ACF arranged on the end points of the difference vectors (cf Fig.5.5).

Besides its role in other applications to be mentioned later, the ACF is a very sensitive tool for detecting the presence of weak periodic features in the image [5.23-25]. AEBI et al. [5.26,27] have used it to verify the noiselike character of the background structure in computer filtrations. As a true noise function, it

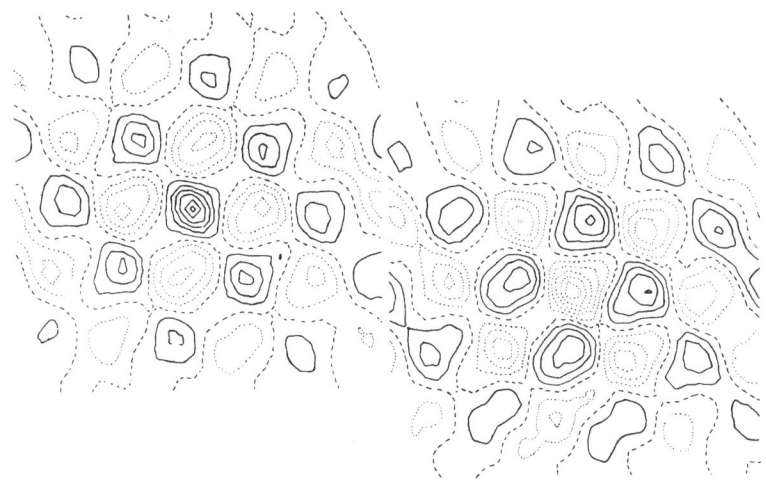

Fig.5.4. ACF (left) and CCF (right) of a published [5.21] electron micrograph of sodium faujasite showing lattice fringes. The two contour diagrams are mounted in such a way that their similarity is emphasized. Note the pronounced noise contribution on the central ACF peak which is missing in the CCF

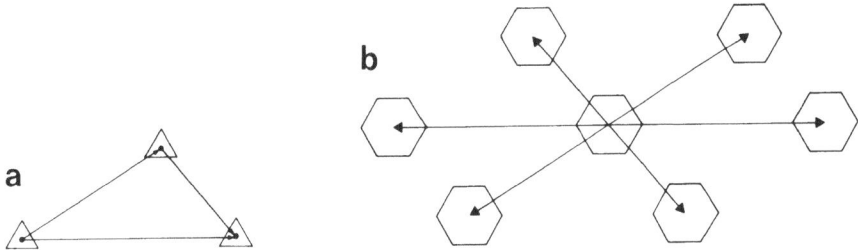

Fig.5.5, (a) A structure consisting of three repeats of a simple motif, (b) its autocorrelation function. [5.22] (By courtesy of North-Holland Publishing Company)

would only produce a central peak in the ACF. Any additional structure visible in the ACF indicates the presence of a (in this case periodic) signal component.

5.1.3 Correlation and Similarity

The CCF of two pictures can be interpreted in terms of "similarity" if we understand this term in a very narrow sense, permitting "rigid body" movements but excluding any geometric distortions. To see this, one may write down the normalized Euclidean distance

$$E(\underline{r}_{pq}) = \left\{ \frac{\sum_j \sum_k [i_1(\underline{r}_{jk}) - i_2(\underline{r}_{jk} + \underline{r}_{pq})]^2}{[\sum_j \sum_k i_1^2(\underline{r}_{jk}) \sum_j \sum_k i_2^2(\underline{r}_{jk})]^{\frac{1}{2}}} \right\}^{\frac{1}{2}} \tag{5.3}$$

(normalized averaged squared difference), which becomes zero for a certain shift vector r_{mn} if the images are identical and assumes a minimum for that vector if they are statistically related. Since stationarity implies that the statistics are invariant under translation, the "image energy" $\sum_j \sum_k [i_2(r_{jk} + r_{pq})]^2$ is not dependent on r_{pq} for stationary or quasi-stationary images, and the behavior of $E(r_{pq})$ is entirely determined by the cross-correlation term

$$\Phi(r_{pq}) = \sum_j \sum_k i_1(r_{jk}) i_2(r_{jk} + r_{pq}) \quad . \tag{5.4}$$

A large cross-correlation value at a certain shift vector r_{mn} indicates that the Euclidean distance between the images is minimized in this position, i.e., that the images are similar.

Apart from the normalization, the Euclidean distance (5.3) corresponds to the measure of dissimilarity [5.28] which can be used to study structural changes quantitatively [5.29]. AL-ALI [5.20] used the dissimilarity measure to study the overall similarity between micrographs of a defocus series.

The eye perceives as similar images that are generated from some original image by convolution with a large class of functions.[1] These functions have a small lateral range (of the order of the smallest resolved distance in the original image) and, generally, centrosymmetry in common. Experience has shown that a significant correlation peak at the correct position can be found in the CCF of two such related images although it appears with reduced height, increased width, and sometimes with reversed sign (cf Sect.5.3.2).

5.2 Computation

The correlation functions are conveniently computed from the digital representation of images by using the well-known convolution theorem. The discrete Fourier transforms of the images are

$$F_1(k_{lm}) = \sum_{j=0}^{M-1} \sum_{k=0}^{N-1} i_1(r_{jk}) \exp[2\pi i (x_j u_l + y_k v_m)] \quad , \tag{5.5}$$

where $r_{jk} = (x_j, y_k) = (j\Delta x, k\Delta y)$ and $k_{lm} = (u_l, v_m)$ with $u_l = 1/M\Delta x$ and $v_m = m/N\Delta y$ are vectors on cartesian grids in the image and the Fourier plane, respectively.

---

1 The class of all operations that preserve visual similarity is of course much wider, including scale transformations and geometric distortions which are not considered here

The discrete Fourier transform of the cross-correlation function is obtained as the conjugate product of the image transforms

$$F_c(\underline{k}_{lm}) = F_1^*(\underline{k}_{lm})F_2(\underline{k}_{lm}) \quad .$$

(5.6)

From this the CCF is obtained by inverse Fourier transformation

$$\Phi_{12}(\underline{r}_{pq}) = \sum_{l=0}^{M-1} \sum_{m=0}^{N-1} F_c(\underline{k}_{lm})\exp[-2\pi i(x_p u_l + y_q v_m)] \quad .$$

(5.7)

Computation of the CCF matrix therefore involves three Fourier transformations and one scalar matrix multiplication. Fast Fourier transform algorithms make this method of computation much faster than the direct evaluation of (5.1).

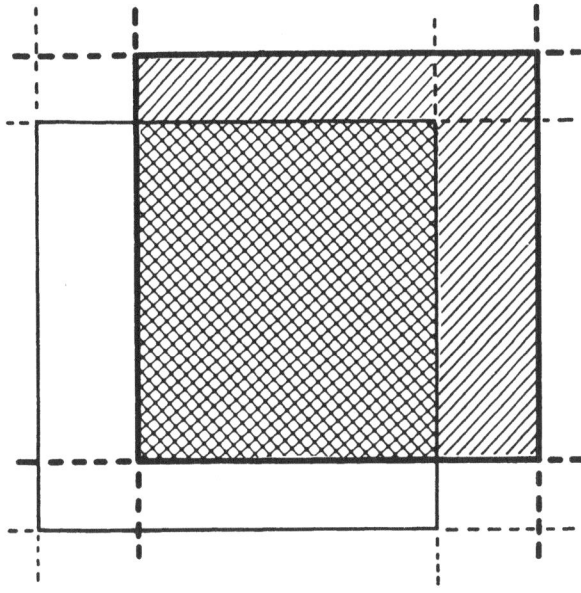

Fig.5.6. Explanation of the wraparound artefact due to discrete Fourier sum representation of the images. Area of desired overlap is crosshatched, area of undesired overlap shaded

However, there is an important difference between the CCF so obtained and the CCF previously defined. The discrete finite Fourier series does not represent the original image but, rather an infinite two-dimensional lattice on which this image is repeated. Therefore, the Fourier method of computation effectively cross correlates two infinite lattices. It is seen from Fig.5.6 that the resulting CCF contains contributions from "false" overlaps between left and right as well as upper and lower parts of the original images. Since the infinite two-dimensional periodic lattice is equivalent to a torus, this phenomenon is sometimes referred to as "wraparound artefact." In order to avoid artefacts from these false overlaps, one has to augment the image with a frame containing values equal to its average. The width of this frame has to be as large as the largest correlation shift one is interested in. In order to obtain the full artefact-free correlation matrix, the frame widths have to be equal to half the image dimensions in both directions, respectively. In this case, the linear dimensions of the augmented image are twice the linear dimensions of the original image. For instance, a 64 × 128 image matrix has to be embedded into a 128 × 256 matrix. Figure 5.7 shows an example of original and augmented images as well as their correlation functions. The artefact from the wraparound overlap of left and right and top and bottom parts of the particle is noticeable in the ACF of the nonaugmented image, but is missing in the ACF of the augmented one.

Because of the invariance of the ACF under translation of the image, the position of the frame with respect to the original image is irrelevant. The same is true for the CCF as long as the relative position between frame and original image is the same for both images.

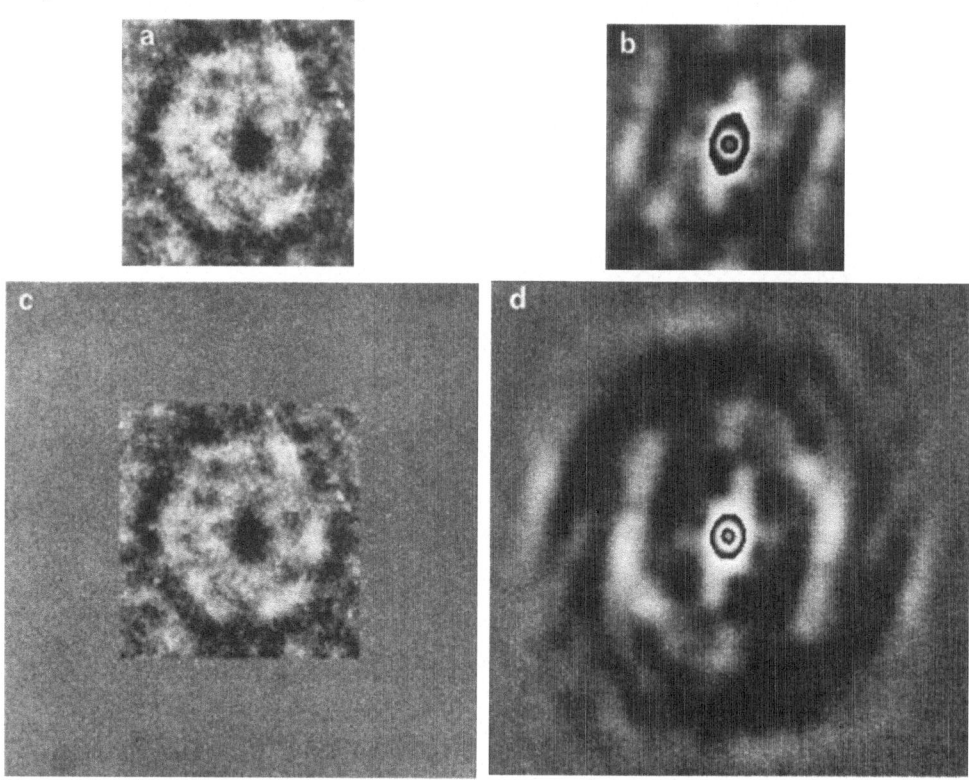

Fig.5.7a-d. Demonstration of the wraparound artefact. (a) Micrograph of a stained glutamine synthetase particle represented by a 64 × 64 image array [5.19]; (b) ACF of (a) obtained by using the Fourier theorem, displayed with bit clipping; (c) the image in (a) floated into a 128 × 128 array; (d) ACF of (c) obtained by using the Fourier theorem, displayed with bit clipping. Only (d) is the valid autocorrelation function of the image (a). Differences between (d) and (b) are due to the overlapping of adjacent images in the discrete Fourier sum representation

5.3 Some Important Theorems

5.3.1 CCFs of Images Containing Signal and Noise

Let us consider the CCF of two images, each consisting of two additive parts, i.e., a signal part $s(\underline{r})$ and a noise part $n(\underline{r})$

$$i_1(\underline{r}) = s_1(\underline{r}) + n_1(\underline{r})$$

$$i_2(\underline{r}) = s_2(\underline{r} + n_2)(\underline{r}) \quad . \tag{5.8}$$

Denoting correlation by the symbol $\otimes$ one obtains by application of (5.1)

$$i_1(\underline{r}) \otimes i_2(\underline{r}) = s_1(\underline{r}) \otimes s_2(\underline{r}) + s_1(\underline{r}) \otimes n_2(\underline{r})$$

$$+ n_1(\underline{r}) \otimes s_2(\underline{r}) + n_1(\underline{r}) \otimes n_2(\underline{r}) \quad , \tag{5.9}$$

where the construction of the four terms apparently follows the same rules as multiplication. However, it is easy to verify from (5.1) that correlation, unlike multiplication and convolution, is not commutative,

$$i_1(\underline{r}) \otimes i_2(\underline{r}) \neq i_2(\underline{r}) \otimes i_1(\underline{r}) \quad . \tag{5.10}$$

Frequently the signal and noise parts are not correlated, and the same holds for the two noise functions. In these cases, (5.9) reduces to

$$i_1(\underline{r}) \otimes i_2(\underline{r}) = s_1(\underline{r}) \otimes s_2(\underline{r}) \quad . \tag{5.11}$$

In the special case where $s_1(r) = s_2(r) = s(r)$ (i.e., the images are only different in the realizations n_1 and n_2 of the noise process), the cross-correlation function of the two images is equal to the auto-correlation function of the signal *common to both images*. However, note that the auto-correlation function of each image contains an additional noise autocorrelation term

$$i_1(\underline{r}) \otimes i_1(\underline{r}) = i_2(\underline{r}) \otimes i_2(\underline{r}) = s(\underline{r}) \otimes s(\underline{r}) + n_1(\underline{r}) \otimes n_1(\underline{r}) \tag{5.12}$$

since any noise realization is correlated with itself. For a noise function, the correlation extends only over very short distances so the additional function consists only of a sharp peak at the origin. (Since n_1 and n_2 are realizations of the same noise process, they have of course the same ACF.)

5.3.2 The CCF of Blurred Signals

The lack of definition of fine detail in blurred images is reflected in an increased width of their correlation function peaks. Defocused images in electron
microscopy are blurred representations of the object's projection. To obtain a
quantitative description, we have to use the Fourier transform relationship

$$F_c(\underline{k}) = F_1^*(\underline{k})F_2(\underline{k}) \quad , \tag{5.13}$$

where $F_c(\underline{k})$ is the Fourier transform of the CCF and $F_1(\underline{k}),F_2(\underline{k})$ are the Fourier
transforms of the blurred images. Let us assume that the images come from the
same signal $s(\underline{r})$ with transform $S(\underline{k})$ and that the blurring is due to point spread
functions $h_1(\underline{r}),h_2(\underline{r})$ with transforms (transfer functions) $H_1(\underline{k})$ and $H_2(\underline{k})$.
According to the convolution theorem

$$F_1(\underline{k}) = H_1(\underline{k})S(\underline{k})$$

$$\tag{5.14}$$

$$F_2(\underline{k}) = H_2(\underline{k})S(\underline{k})$$

$$F_c(\underline{k}) = F_1^*(\underline{k})F_2(\underline{k}) = S^*(\underline{k})S(\underline{k})H_1^*(\underline{k})H_2(\underline{k}) \quad . \tag{5.15}$$

The corresponding real space relationship is

$$C(\underline{r}) = [s(\underline{r}) \otimes s(\underline{r})] * [h_1(\underline{r}) \otimes h_2(\underline{r})] \quad . \tag{5.16}$$

We obtain the result that the ACF of the unblurred signal appears convolved with
the CCF of the point spread functions.

Because of the wide range of shapes assumed by the point spread functions for
different defocus values [5.30], the CCF of two point spread functions also shows
a large variety of forms [5.20,31,32]. The central peak may be positive or negative,
or may sometimes be entirely absent (cf Fig.5.8). According to (5.7) and (5.16),
the central value of the CCF peak may be obtained by integrating the transfer func-
tion product $H_1^*(\underline{k})H_2(\underline{k})$ over the resolution domain. Height and polarity of the
peak are therefore determined by size and polarity of the balance between regions
("lobes") where the transfer functions have the same sign and regions where they
have different sign. The peak vanishes only if the positive contributions happen
to match the negative contributions exactly.

In this case, a change of the integration domain will destroy the balance and
make the peak appear. Figure 5.9 shows the integral $\int_B H_1^*(\underline{k})H_2(\underline{k})d\underline{k}$ calculated
for two phase contrast transfer functions [5.13]. As the circular integration
domain B is increased, the integral reverses sign.

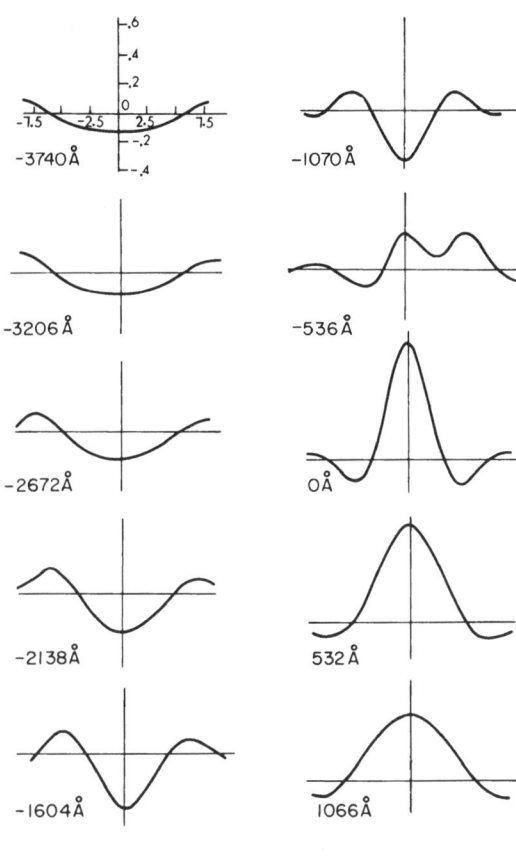

Fig.5.8. Profiles of CCF peaks between a Scherzer focus electron micrograph and 10 micrographs of a defocus series. The figures at the bottoms of the profiles give the defocus difference $\delta\Delta z$ in Ångströms. The peak height becomes maximum and the peak width becomes minimum for zero defocus difference ($\delta\Delta z = 0$). [5.32] (By courtesy of the author)

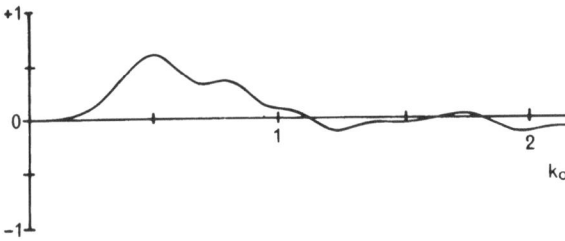

Fig.5.9. The heights of the CCF peak of two bright-field electron micrographs as a function of the radius of the integration domain in reciprocal space. [5.13]

A negative correlation peak (cf [5.32,33]) indicates contrast reversal between the two electron micrographs investigated, a phenomenon which is predicted by phase contrast transfer theory from the relative signs and position of the lobes of the transfer functions, and which is also frequently observed in experiments (e.g., [5.34]).

The CCF halfwidth depends on the positions of "destructive" and "constructive" overlaps of the lobes of the different transfer functions within the spatial frequency domain. For phase contrast electron microscopy where different defocus values introduce the largest changes in the high-resolution range of the transfer function we have the general rule that the peak halfwidth increases with increasing defocus difference [5.20,31,32].

In the special case where one tries to find repeats of a particle in the same micrograph (cf Sect.5.6) one has $h_1(\underline{r}) \equiv h_2(\underline{r})$ which means according to (5.15) that the cross-correlation peak in self-detection and matched filtering applications is always positive.[2]

To appreciate the relationship (5.16) one has to realize that the "signal" in electron microscopic applications is the projection of the atomic potential. For a random structure, the ACF consists of a single peak whose width is of the order of the width of the atomic potential plus small off-origin peaks with small weight due to overlaps between different atoms of the structure. This extremely sharp peak would be visible in the CCF for an ideal electron microscope. According to (5.16) the blurring due to wave aberrations in the electron microscope introduces statistical dependence between image points over much longer distances than originally present in the object: each object point gives rise to a finite disk in the image with a size ranging from 2 to 5 Å, depending on the resolution of the instrument.

The short-range image correlation introduced by the electron microscope and the interpretation of the peak shape in terms of electron optical aberrations are the subjects of a later section (Sect.5.6) of this chapter.

5.3.3 Some Thoughts on Signal, Noise, and Correlation

Frequently we have a situation where the image contains superposed signal and noise portions. An example is an electron micrograph of stained virus particles lying on a carbon film. Both particles and film form the object which is imaged by the electron microscope. In addition to this, the image contains shot noise from fluctuations of the grain distribution in the photographic emulsion. What is signal and what noise in this situation? It is quite clear that the virus particle is signal, as the subject of study, and equally clear that shot noise and photographic noise carry no useful information. However, the carbon film seems to have signal or noise character, depending on the observer's point of view. To render an old German saying[3] in free translation "someone's signal is someone else's noise".

The noise character is obvious to anyone who is trying to interpret the image of the virus particle. The term "structural noise" is in fact frequently used for the part of the image that is due to the supporting film structure. The signal character, however, is a consequence of the fact that the film is a unique struc-

---

2 Unless, of course, the micrograph is of a tilted specimen where a varying defocus produces a varying transfer function. This complication will occur in the three-dimensional reconstruction of individual particles from a tilted low-dose micrograph, cf concluding remarks in [5.35]

3 In the Low German version: *Dem enen sin Uhl is dem andern sin Nachtigall.*

ture that yields a *reproducible* image on the photoplate. It is the reproducible part, or the part common to both images, that produces a significant value of the correlation function and hence appears as a signal, as opposed to the irreproducible noise part.

We can go one step further and do an experiment where even the photographic noise appears as part of the signal. We simply have to make two copies of an electron micrograph. Besides specimen and supporting film structure, the two copies have the specific photographic grain distribution of the original in common! (This is easily demonstrated by an optical correlation experiment: sandwich the two copies in the image plane of the optical diffractometer. When the images are nearly in register, a pattern of Young's fringes appears [5.36] whose extension goes far beyond the electron optical resolution limit.) By the copying experiment, we have made the noise portion of the original image "reproducible" since a replica of it now appears in each copy.

5.4 Determination of Relative Positions

5.4.1 Translation

It was already pointed out that the CCF is used for alignment of images containing a common signal part $s(\underline{r})$ which appears in different positions separated by an unknown vector $\underline{r}_s$. The CCF of two such images shows a sharp peak at $-\underline{r}_s$ and thus indicates both direction and amount of the displacement. The accuracy of position determination depends on the width of the peak, but is better than this by a factor of at least five, because the centrosymmetry of the peak which is theoretically expected and experimentally observed allows the accurate determination of its center. Since the width of the peak is of the order of the instrument resolution, an accuracy of 1 Å or better in the determination of position can be achieved. Formally this can be seen if one represents the shifted signal by convolution of the unshifted signal with a δ-function positioned at $\underline{r}_s$

$$i_1(\underline{r}) = s(\underline{r}) + n_1(\underline{r})$$

$$i_2(\underline{r}) = s(\underline{r} - \underline{r}_s) + n_2(\underline{r}) = s(\underline{r}) * \delta(\underline{r} - \underline{r}_s) + n_2(\underline{r}) \quad . \tag{5.17}$$

The cross-correlation function of these images is

$$i_1(\underline{r}) \otimes i_2(\underline{r}) = s(\underline{r}) \otimes [s(\underline{r}) * \delta(\underline{r} - \underline{r}_s)] \tag{5.18}$$

$$+ \text{ uncorrelated terms,}$$

which can be rewritten as

$$i_1(\underline{r}) \otimes i_2(\underline{r}) = [s(\underline{r}) \otimes s(\underline{r})] * \delta(\underline{r} - \underline{r}_s) \qquad (5.19)$$

and shows the autocorrelation function positioned at $\underline{r}_s$.

It was mentioned in an earlier section (Sect.5.3.2) that a defocus change does not in general destroy the correlation between two electron micrographs of the same specimen [5.20,31]. This fact is of great practical importance for the alignment of electron micrographs with different defocus. Examples are restoration [5.37,38] and phase determination [5.39] from a bright-field defocus series.

Figure 5.10 shows central portions of cross-correlation functions of a carbon film defocus series. An image at $\Delta z = 3600$ Å was used as reference. The peak appears in various off-center positions indicating direction and size of the displacements. As expected from the discussion of the CCF of blurred signals (Sect. 5.3.2), the peak width increases and its height decreases with increasing defocus difference.

$\Delta z_1 \approx \quad 3600$ Å

$\Delta z_2 \approx \quad 3000$ Å $\qquad\qquad$ 2400 Å $\qquad\qquad$ 1800 Å $\qquad\qquad$ 1200 Å

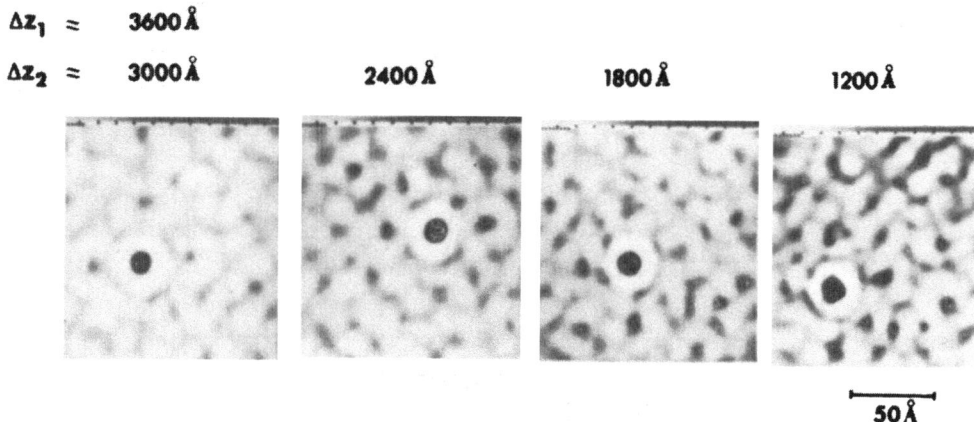

50 Å

Fig.5.10. Central area of the CCFs of a carbon film defocus series showing increasing peak widths with increasing defocus difference

The Fourier relationship $F_c(\underline{k}) = F_1^*(\underline{k})F_2(\underline{k})$ mentioned earlier shows that structural changes of an object between two exposures do not destroy the correlation between the micrographs as long as these changes affect only part (in all practical cases the high-resolution part) of the Fourier transform. Structural changes caused by radiation damage can therefore be investigated by digital comparison of images in an exposure series [5.29].

The decrease in the size of the Fourier domain contributing to the CCF signal causes a decrease in peak height and simultaneously an increase of the peak width. As one would expect, the resulting peak width is of the same order as the smallest structural details preserved during the structural change. An instructive example for the changes to the CCF peak due to radiation damage is found in [5.40]; cf Fig.5.11.

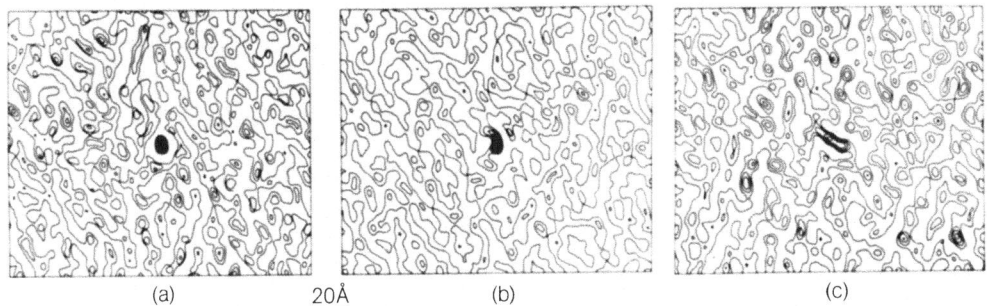

Fig.5.11a-c. Structural changes of a carbon film due to radiation damage as reflected in the cross-correlation function. The contour plots show cross-correlation
functions between a reference image and a series of successive exposures taken with
increasing preilluminations. (a) 2.5 C/cm^2, (b) 13 C/cm^2, (c) 63 C/cm^2. Equal contour line distance in (a) and (b) halved in (c). Relative heights of the correlation
peaks (a) 7810, (b) 2910, (c) 910. [5.40] (By courtesy of North-Holland Publishing
Company)

The difference image method [5.13-15] is based on the fact that the supporting
film structure remains essentially unchanged when a specimen is applied to it. The
cross-correlation function of the micrographs before and after preparation is
dominated by the ACF of the supporting film structure, with the additional specimen
structure visible in the second image appearing as a "random" feature, which does
not change the position of the CCF peak (in this case, signal and noise have completely exchanged their roles, cf Sect.5.3.3).

In a first extensive application of this idea, HOPPE et al. [5.41-43] have
studied the changes in the high-resolution structure of carbon, graphite, and Al$_s$O$_3$
films that occur over an extended period of time as radiation damage and contamination proceed (Fig.5.12).

These experimental studies are currently being complemented by theoretical work
based on a statistical model (e.g., random walk) of the structural changes [5.44-46].

A further development of the difference image method is the so-called trace
structure analysis [5.47-49], the analysis of changes of a structure over some
period of time based on a series of electron micrographs.

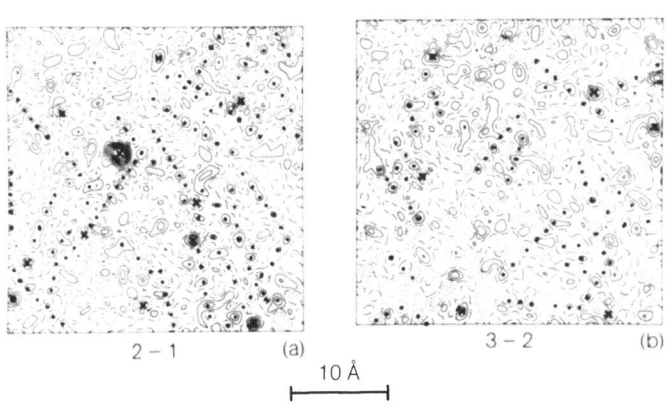

Fig.5.12a,b. Difference
images between three suc-
cessive exposures of a
thin (~50 Å) Al$_2$O$_3$ foil.
Resolution limit 0.33 Å^{-1},
defocus 800 Å. Illumi-
nation dose increasing
in equal steps of
1.2 C/cm^2. (a) 2-1, (b)
3-2. The dots indicate
negative and positive
peaks that seem to be
preferentially arranged
in linear periodic ar-
rays. [5.40] (By courtesy
of North-Holland Pub-
lishing Company)

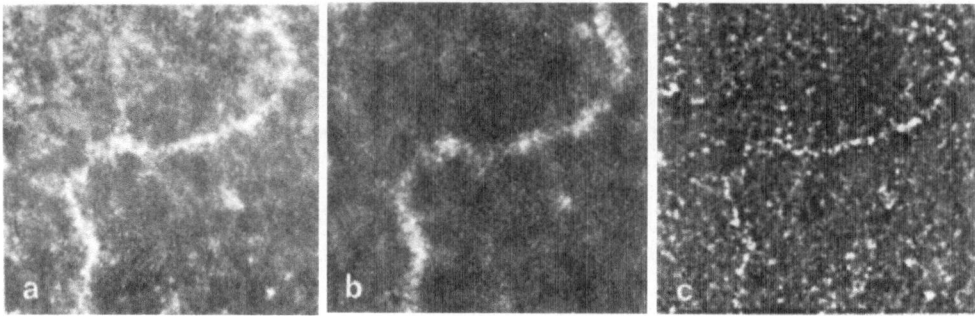

<u>Fig.5.13a-c.</u> Digitally processed micrographs of uranyl-stained SV40-DNA on a carbon film. (a) Best focus, (b) inelastic image, (c) difference image (a-b), showing elastic contribution. All images are composed of 256 × 256 pixels and the side dimension of each micrograph is 1280 Å (5 Å/pixel). [5.50] (By courtesy of the Institute of Physics, Great Britain)

A recent application of high-resolution translational alignment is the elimination of the "inelastic image" from bright- or dark-field electron micrographs by subtraction of two images that are taken with different defocus values [5.50-52]; cf Fig.5.13.

5.4.2 Alignment of Projections

It is of great significance for three-dimensional reconstruction of amorphous objects that different views of the object are correlated; the correlation signal can be used to determine the common origin of the views.

HUNSMANN et al. [5.53] reported that projections with up to 20° difference could be related to each other by cross-correlation.

This technique was first applied by HOPPE et al. in the three-dimensional reconstruction of fatty acid synthetase [5.54]. A detailed discussion of the common origin determination using cross-correlation is found in [5.40,43,55].

The statistical dependence between adjacent projections can be seen as a consequence of the boundedness of the object [5.56]. FRANK [5.57] showed that the behavior of the CCF peak height and width (in the direction normal to the tilt axis) as a function of angle between the views can be explained in terms of the "area of influence" around each projection Fourier coefficient. The size of this area corresponds to the extension of the shape factor associated with the object's boundary. The results of this analysis are in agreement with earlier computations by MATTHEWS [5.58].

While the peak width in the direction normal to the tilt axis quickly increases with increasing tilt angle, the width in the direction *parallel* to the tilt axis is expected to remain the same: the different sections of the three-dimensional Fourier transform which correspond to different projections intersect each other in the so-called common lines. Irrespective of the angle between the projections,

these lines carry the same Fourier coefficients, and should therefore give rise to
a (one-dimensional) correlation signal that can be used to align projections in the
direction of the tilt axis. In practice, of course, the peak decreases with increasing angle [5.43,57] and becomes obscured by noise for large angles. The concept of common lines is known from three-dimensional reconstruction of objects
with symmetry [5.59] where they are used to combine Fourier information from different views.

5.4.3 Centering of a Centrosymmetric Particle

An elegant application of the correlation technique is the centering of a motif
(in most applications a molecule or virus particle) that exhibits centrosymmetry
[5.35]. Such a centering is necessary for real space superposition [5.35] or for
rotational filtering [5.60]. The method described here can be used in place of
the least-squares residual method of CROWTHER and AMOS [5.60].

Let us assume that an image $i(\underline{r})$ contains a centrosymmetric motif $m(\underline{r}) = m(-\underline{r})$
in an arbitrary position

$$i(\underline{r}) = b(\underline{r}) + m(\underline{r} + \underline{r}_0) \quad , \tag{5.19a}$$

where $b(\underline{r})$ stands for background image which may be random and does not exhibit
any symmetries. The unknown offset vector is denoted by $\underline{r}_0$.

A rotation of the image by 180° has the effect that all argument vectors change
their sign,

$$i(-\underline{r}) = b(-\underline{r}) + m(-\underline{r} - \underline{r}_0) \quad . \tag{5.19b}$$

The cross correlation between original and 180° rotated image gives the following result

$$i(\underline{r}) \otimes i(-\underline{r}) = b(\underline{r}) \otimes b(-\underline{r}) + m(\underline{r} + \underline{r}_0) \otimes m(-\underline{r} - \underline{r}_0) + \text{mixed terms} \quad . \tag{5.19c}$$

Since the background image has no centrosymmetry, we can ignore the first term on
the right-hand side of (5.19c). The second term may be written in the form [cf (5.17)]

$$[m(\underline{r}) * \delta(\underline{r} + \underline{r}_0)] \otimes [m(-\underline{r}) * \delta(\underline{r} - \underline{r}_0)] = [m(\underline{r}) \otimes m(-\underline{r})] * [\delta(\underline{r} + \underline{r}_0) \otimes \delta(\underline{r} - \underline{r}_0)]$$

$$= [m(\underline{r}) \otimes m(\underline{r})] * \delta(\underline{r} + 2\underline{r}_0) \quad . \tag{5.19d}$$

Due to the centrosymmetry of the motif, the ACF of the motif appears displaced
from the origin of the CCF by twice the unknown vector.

In the computing procedure based on this algorithm, the displaced ACF peak is located, and the resulting vector divided in half, sign reversed, and applied to the original image

$$i_{centered}(\underline{r}) = i(\underline{r} - \underline{r}_0) = b(\underline{r} - \underline{r}_0) + m(\underline{r}) \quad . \tag{5.19e}$$

5.4.4 Determination of Relative Orientation

If the images are not only shifted but also rotated with respect to each other, the cross-correlation technique cannot be used without modification. Since both angle and center of rotation are unknown, it is not in general possible to align the images by simply rotating them against each other to maximize the averaged cross product. Three techniques have been developed to deal with this problem:

1) Iterative search technique [5.61]. The rotational misalignment is frequently so small that the cross-correlation peak can still be found although it appears reduced in height and increased in width. Computation of the CCF therefore yields a first estimate of the shift vector. After shifting one image with respect to the other, one proceeds by searching for a rotation angle that maximizes the averaged cross product. The angle so found is taken as the first estimate of the rotation angle, and is used to align the images. This procedure is repeated until angle and shift vector converge to stable values. SAXTON [5.61] reports that for small angles ($< 5^\circ$) a few cycles lead to satisfactory results.

This procedure becomes difficult to handle for large images because of storage and computing time problems. On the other hand, the use of a small portion of such images does not give sufficiently accurate rotation angles.

2) Triangulation techniques [5.62]. Small portions near the edge of the large images are used to determine a set of offset vectors (Fig.5.14). While two offset vectors are in principle sufficient for determination of rotation and translation of the large images, the accuracy of these parameters can be enhanced by using more than two vectors and a least-squares evaluation. As in the first technique, the rotation angle has to be restricted to a few degrees.

These first two approaches are appropriate for determining the orientation between images that are prealigned by eye on the microdensitometer or, in direct image readout applications such as STEM, come prealigned from the instrument. They are, however, not suited for motifs that have completely random positions and orientations such as a set of virus particles on a carbon film. The third technique solves this problem.

3) Use of translation-invariant functions [5.14]. A technique related to the Patterson search methods in X-ray crystallography [5.63] has been generalized by LANGER et al. [5.14] for application in electron microscopy.

Both autocorrelation function and power spectrum, which are translation-invariant for stationary images, contain characteristic oriented features related to aniso-

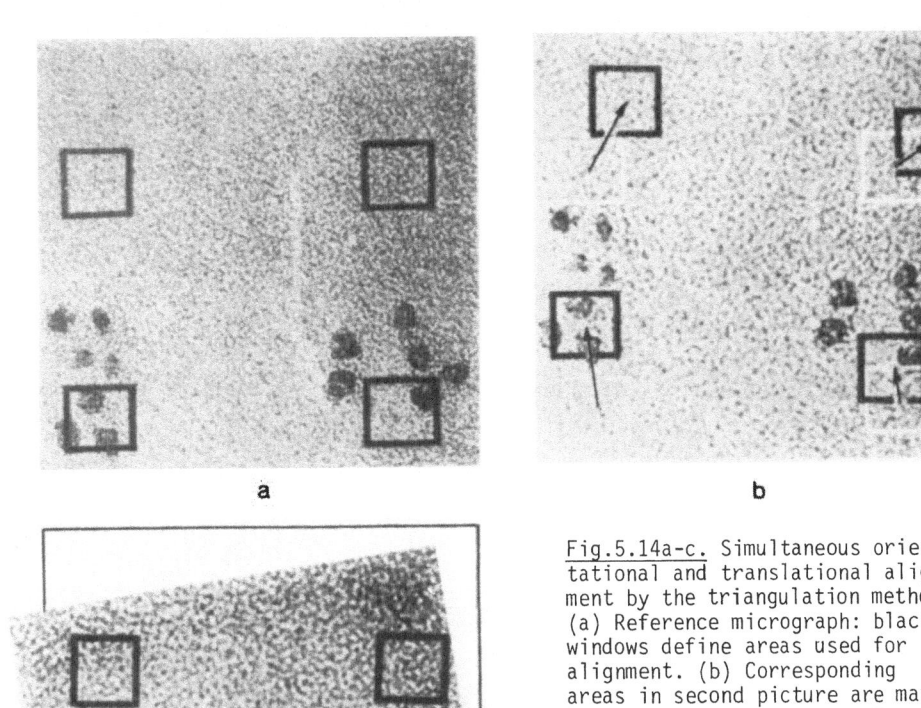

a

b

Fig.5.14a-c. Simultaneous orien-
tational and translational alignment by the triangulation method.
(a) Reference micrograph: black
windows define areas used for
alignment. (b) Corresponding
areas in second picture are marked
with white windows. The CCFs
between corresponding areas show
that a perfect match occurs for
the offset vectors indicated (the
areas on the left-hand side have
obviously to be larger than marked
by the author because the CCF
method of alignment requires at
least partial overlap). (c) The
offset vectors are used to determine the effective shift and
rotation to be applied to the second picture for exact alignment.
[5.62] (By courtesy of Claitor Publishing Division)

c

tropic features of the image. To obtain the relative orientation of two patterns
lying in arbitrary positions in a larger image, one has first to cut out the image
patches fully containing these patterns, and compute their Fourier transform. In
one variant of the method, the absolute square of the transform (i.e., the estimate
of the power spectrum) is used as test function; in the other one the inverse
transform of this (i.e., the estimate of the ACF) is used instead. The test functions are rotated with respect to each other by small angle increments between
$+ \pi/2$ and $- \pi/2$, and in each position their product is summed up

$$P_{12}(\varphi_k) = \sum_i T_1(R_i,\varphi_i)T_2(R_i,\varphi_i + \varphi_k)w(R_i) \quad . \tag{5.20}$$

The use of weights $w(R_i)$ makes it possible to emphasize the contribution from particular correlation distances or spatial frequency zones R_i, respectively.

The plot of the resulting values of P_{12} against the test rotation angle shows a high peak at the matching orientation angle of the images (Fig.5.15).

A recent implementation of this technique by SAXTON [5.64] makes use of one-dimensional correlation (using the fast Fourier transform routine) along zones of the test functions. A display of the zonal correlation functions vs the zonal position R_i for the case of the power spectrum test function [5.61] shows the contribution of different spatial frequency zones to the anisotropy of the pattern (Fig.5.16).

A detailed exploration of the orientation search algorithm is found in SAXTON's thesis [5.61]. Unfortunately, most of his studies are done with micrographs of magnetic stripe domains which are not typical of electron microscopic objects as they exhibit an unusual degree of anisotropy. A systematic study of the performance of the translation-invariant search technique for various biological objects is so far missing.

Figure 5.17 is a block diagram of the combined rotational and translational alignment procedure that may be used in the computation of a difference image (cf Sect.5.4.1) or in the computation of the average over a large number of randomly oriented particles (cf Sect.5.5). First, the ACFs of the two pictures are computed and the matching angle determined by using angular cross correlation. One of the pictures is then rotated by the angle found. Since the pictures now have the same orientation, cross-correlation can be employed to find the relative position vector. This vector is finally used to shift the second picture into exact registration with the first one.

So far not mentioned in this list are systematic searching methods. Here one image is systematically rotated and shifted with respect to the other one until a best match occurs in the sense of the minimum Euclidean distance criterion (5.3). In Fourier space, the corresponding phase shift operations are performed on the Fourier coefficients to be compared. The Euclidean distance criterion is easily recognized in the "minimum power loss criterion" of SMITH et al. [5.65], where the so-called $D(Z,k)$ transform coefficients of two helical particles are compared with each other for various phase offsets, and in AMOS's "least square residual" criterion [5.66], where such a comparison is made between the helical transforms of two particles.

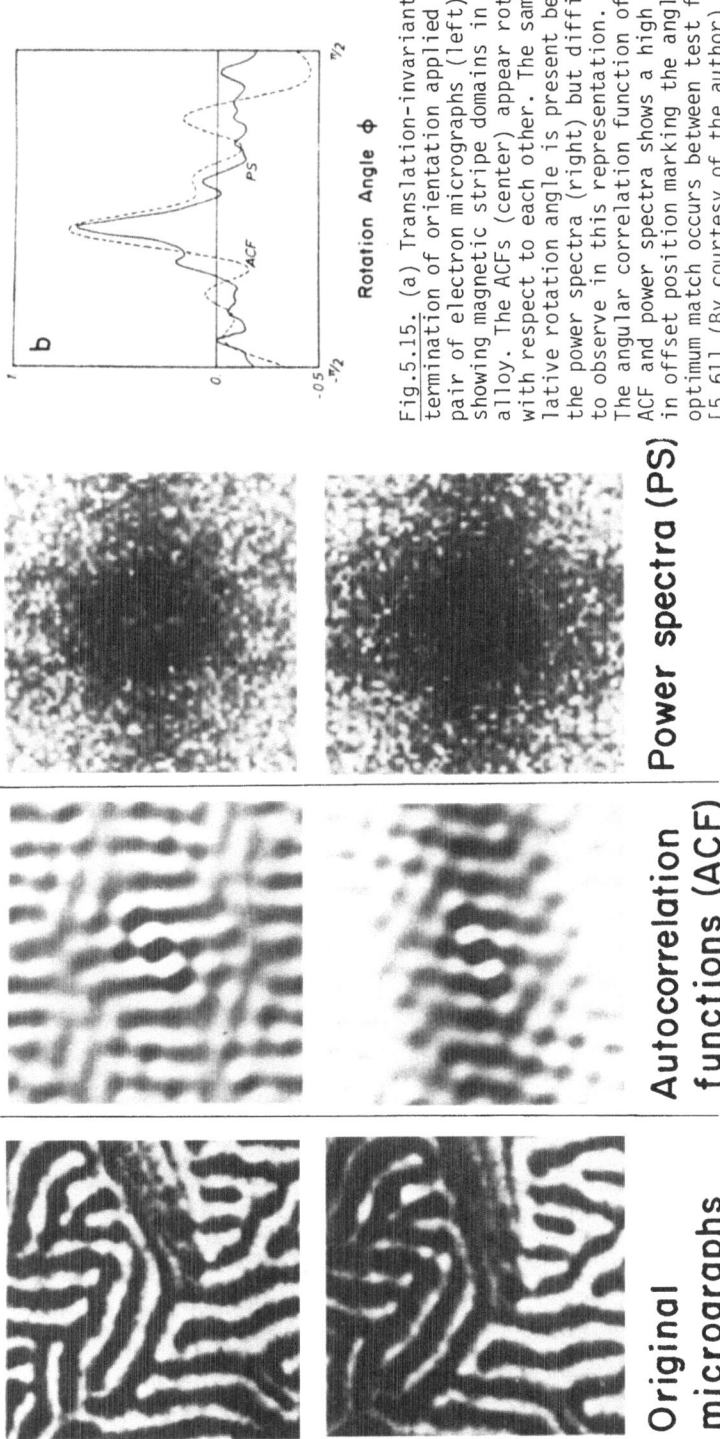

Power spectra (PS)

Autocorrelation functions (ACF)

Original micrographs

Rotation Angle φ

Fig.5.15. (a) Translation-invariant determination of orientation applied to a pair of electron micrographs (left) showing magnetic stripe domains in permalloy. The ACFs (center) appear rotated with respect to each other. The same relative rotation angle is present between the power spectra (right) but difficult to observe in this representation. (b) The angular correlation function of both ACF and power spectra shows a high peak in offset position marking the angle where optimum match occurs between test functions. [5.61] (By courtesy of the author)

208

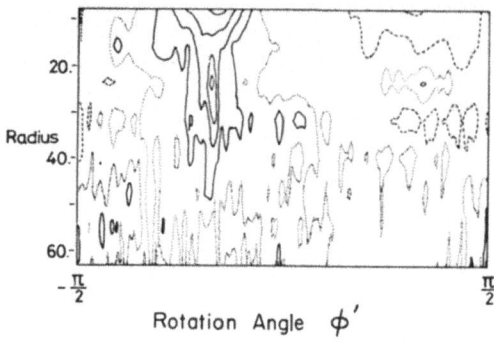

Fig.5.16. The radial dependence of angular correlation in the case of the power spectrum as test function. The radius is proportional to the spatial frequency; low-resolution side is at the top, high-resolution side at the bottom of the diagram. Positive contours are solid, negative ones are broken, and the zero contour is dotted. The elongated peak on the left-hand side marking the matching angle shows that the orientation match of low-resolution features is consistent with that of high-resolution features of the images. [5.61] (By courtesy of the author)

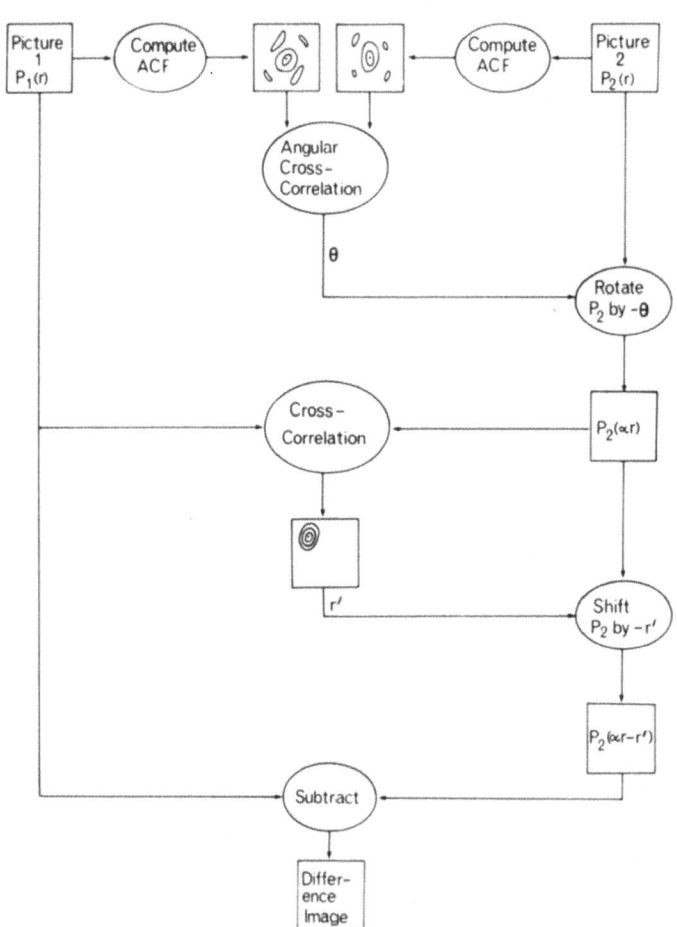

Fig.5.17. Block diagram showing sequence of alignment procedures for difference image method. [For expediency, the rotated picture is denoted as $P_2(\alpha r)$ where α is the two-dimensional rotation matrix associated with a rotation by the angle θ]

5.5 Matched Filtering

The translation search technique described in the last section is the basis for a
pattern recognition method due to VANDER LUGT, which is known under the name
"matched filtering" [5.67]. If combined with the translation invariant orientation
search technique, matched filtering allows one to find a known pattern that is
buried in noise with arbitrary orientation and position. In contrast to general
pattern recognition methods (e.g., for recognition of handwritten characters
[5.68]), this method is restricted to translations and rotations of the "rigid"
structure of the pattern, excluding any geometric distortions. Matched filtering
is thus appropriate for detection of biological objects with defined morphology or
macromolecules with defined structure, and with preferential attachment to the supporting grid.

A look at the Euclidean distance (5.3) introduced earlier makes it easy to under-
stand this method. Let $s(\underline{r})$ be the pattern to be detected and $p(\underline{r})$ the pattern used
to detect this. Among all possible functions $p(\underline{r})$, the function $s(\underline{r})$ itself has the
smallest Euclidean distance and the largest cross-correlation value with respect
to $s(\underline{r})$. This remains true if we introduce additive uncorrelated noise $s'(\underline{r})$
$= s(\underline{r}) + n(\underline{r})$; the smallest Euclidean distance is still assumed for $p(\underline{r}) = s(\underline{r})$.

The name "matched filtering" can be understood from the optical implementation
by VANDER LUGT [5.67]: The light-optically generated Fourier transform of the transparency containing the pattern to be sought is passed through (i.e., "multiplied
with") a complex filter and is then back-transformed. The resulting intensity distribution is according to (5.13) equal to the CCF between the pattern and the inverse transform of the filter used. VANDER LUGT showed that among all possible
complex filters, the complex conjugate of the pattern transform maximizes the detection peak in the CCF.

The orientation search described in the previous section is essentially a one-
dimensional matched filtering procedure: The test functions (ACF or power spectrum
estimates) of two identical patterns show the highest possible angular correlation
peak at the matching angle.

Matched filtering has found only a few applications in biological electron microscopy. These applications have mostly to do with automation of mass screening for
clinical purposes (e.g., search for virus particles) rather than the detection of
low-contrast objects. Heavy metal staining usually provides sufficient contrast
and makes elaborate detection methods unnecessary.

The recent development of specimen-preserving preparation and low dose electron
microscopy [5.69,70], however, has created new interest in methods for the detection of low-contrast signals embedded in noise [5.22,71-74].

The noise-free signal can be reconstructed by averaging over a large number of
noisy repeats of the signal. UNWIN and HENDERSON [5.69] used this principle to

reduce the dose and to eliminate the need for staining. For a periodic signal, the shift vectors between the identical, equally orientated repeats are simply found by inspection of the Fourier transform. In fact, in the normal approach where averaging is done by filtering of the Fourier transform, the shift vectors do not explicitly enter into the computation.

In contrast to this, the averaging of randomly orientated and positioned objects requires translation and orientation parameters (i.e., three independent parameters if rotation is restricted to the plane of the supporting film) to be accurately known for each repeat. Since an a priori model of the object structure with the required resolution is not available in most cases, matched filtering cannot be applied in its original form. However, as the noisy repeats all contain the common signal, any one of them can be used as prototype function for the others ("self-detection principle", cf [5.74]). The same applies to the orientation search: the test functions for two repeats contain a common function due to the signal, which produces a maximum response of the angular correlation function at the matching angle.

Recently FRANK [5.72], and SAXTON and FRANK [5.74] studied the dose-dependent performance of self-detection (i.e., between two noisy repeats) in quantum noise limited situations. In the case of self-detection, the height of the correlation peak was found to depend on dose n, resolution distance d, contrast C, and diameter of the particle D. Requiring that the peak be three times as large as the standard deviation of the noise background in the correlation function, the authors found the criterion

$$n \geq \frac{3}{C^2 dD} \quad , \tag{5.21}$$

which calls for an electron dose that is smaller by a factor of 8D/d than the dose required for direct visual interpretation [5.75]

$$n \geq \frac{25}{C^2 d^2} \quad . \tag{5.22}$$

In the case of the spherical virus particle used to test the method, a dose reduction down to a value of 1 el/Å proved to be feasible in the model computation (cf Fig.5.18). Although a number of simplifying assumptions were made, these calculations suggested that, contrary to popular belief, low-exposure averaging techniques are not restricted to periodic specimens.

Fig.5.18. Detection of four repeats of a spherical virus particle (top) in a simulated low exposure electron micrograph. Column (a) low-dose versions of the model image; (b) low-dose versions of a single particle; (c) cross-correlation functions of these two arrays; and (d) perspective displays of the cross-correlation functions. The numbers on the left indicate the exposures used (electrons per picture element). The exposures correspond to object doses of 2.6, 2.1, 1.6, 1.15, and 0.6 electrons/Å^2. [5.74] (By courtesy of North-Holland Publishing Company)

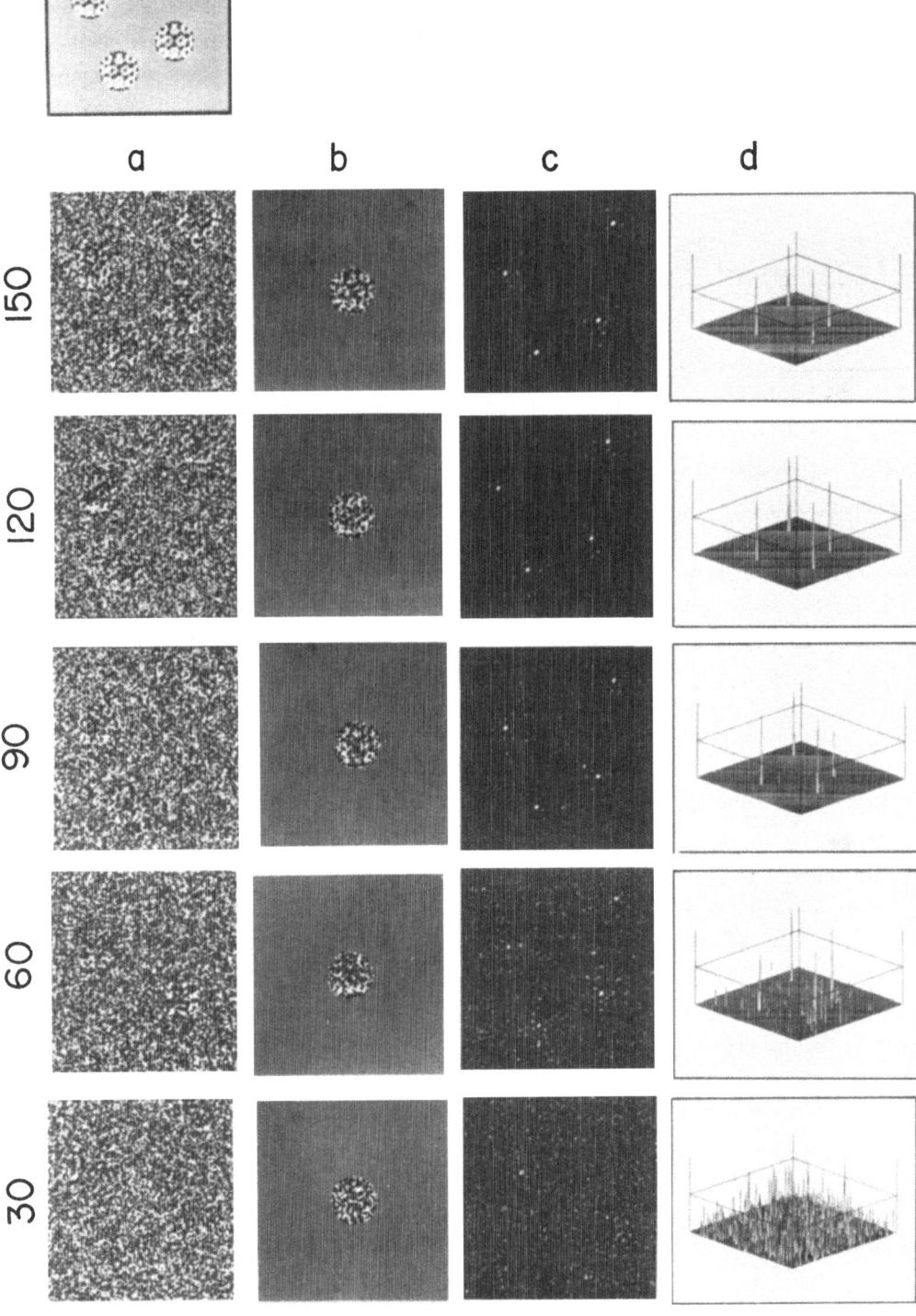

We should mention that similar results were obtained by HOPPE et al. [5.49,76] who considered the correlation contribution of each of the atoms that make up the structure.

Meanwhile low-dose averaging of individual particles based on the alignment algorithms described above has been developed into a practical procedure [5.77,78]. First averages of glutamine synthetase molecules [5.19,35,79] show an amount of detail previously not achieved with other methods of investigation.

5.6 Characterization of Instrument Conditions

The peak shape of the image ACF shows a pronounced dependence on the instrumental factors. The finite width of the peak reflects small range correlations in the object as well as correlations introduced by the image forming system [5.80]. Since the object's ACF is usually much narrower than the range of the correlation due to the instrument, the ACF peak of the image can be used to characterize the instrument conditions.

When linear transfer theory is applicable to the description of the imaging process, and a suitable object with random structure is chosen, the ACF peak of the image closely approximates the ACF of the point spread function: According to the theorem in Sect.5.3.2, the ACF of the image Φ_{ii} is the convolution product of the ACF of the object with that of the point spread function,

$$\Phi_{ii}(\underline{r}) = \Phi_{oo}(\underline{r}) * \Phi_{hh}(\underline{r}) \quad . \tag{5.23}$$

With an object ACF resembling a delta function, this reduces to $\Phi_{ii}(r) \approx \Phi_{hh}(r)$.

Classical resolution criteria based on the instrument's response to two object points at variable distance become meaningless in the domain of atomic dimensions: resolution so defined cannot be tested because there is no device that would allow us to keep two single atoms at a defined distance suspended in vacuum.

Equation (5.23) shows an alternative way of obtaining information on the imaging system. The width of the ACF of the point spread function may be related to the resolution distance of the classical two-point resolution criterion in the following way:

Consider a hypothetical experiment where two points are imaged by the electron microscope. The resulting image shows two partly overlapped point spread functions. The ACF of this image will indicate whether or not the information on the separation of the two points has been passed on to the image, by the presence or absence of distinct side maxima on either side of the central maximum. It was pointed out before that the ACF of a single point spread function can be obtained by measurement. Since the ACF of two points separated by Δr is a sequence of three δ functions at $-\Delta r$, 0, Δr with weights 1/2, 1, 1/2, the ACF of the two-point image can be constructed by placing the ACF of the point spread function on top of itself at $-\Delta r$, 0, Δr with weights 1/2, 1, 1/2. The two points may then be considered resolved in the image if the two side maxima become distinctly visible on either side of the main maximum of the resulting function. We may therefore say that by this method the resolution test is taken outside the instrument. This criterion can be expected to give a good estimate of the resolution if the point spread function is mainly concentrated in a single maximum. Application to the case of incoherent diffraction-limited imaging indeed shows good agreement between the resolution distances obtained by the Rayleigh criterion with that obtained by the correlation criterion [5.80].

The method of resolution estimation proposed in [5.80] has been applied by AL-ALI [5.32] to a defocus series of carbon film micrographs. For each electron micrograph, the author computed the ACF and superposed the peak on top of itself at various postions $-\Delta r$, 0, $+\Delta r$ with weights 1/2, 1, 1/2. The result is shown for four particular displacements (see Fig.5.19). For $\Delta r = 0$ displacement, the original peak shape is seen. For 5 Å displacement, bumps become visible on both sides of the main maximum, for 6.25 Å these bumps turn into horizontal shoulders, and for 8 Å we have clearly separated side maxima. Since the appearance of a horizontal slope turns out to be very sensitive to slight variations of the distance, it may be used as a criterion to define a heuristic "resolution distance".

The distance found in this way is somewhat higher than the halfwidth of the peak (cf Fig.5.20). Both quantities have a pronounced defocus dependence, with

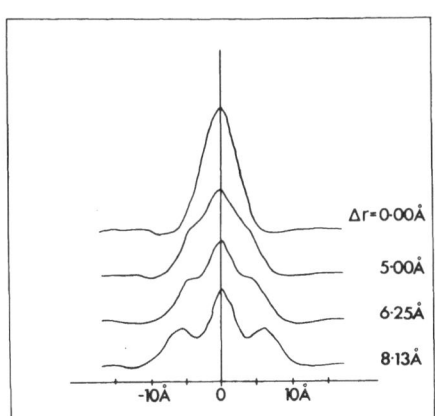

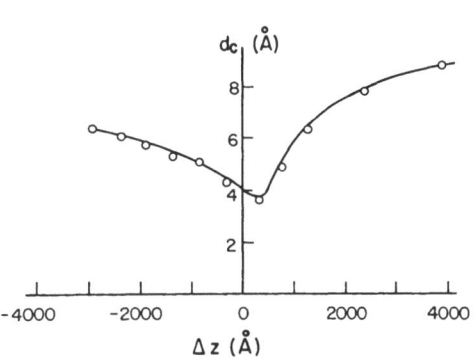

Fig.5.19. ACF of two object points at variable distance constructed by super-position of experimentally obtained ACFs. For a distance of $\Delta r = 5$ Å, horizontal slopes begin to appear on either side of the maximum

Fig.5.20. Resolution distances d_c as a function of defocus Δz obtained by using the ACF superposition and horizontal slope criterion (cf Fig.5.19). [5.32] (By courtesy of the author)

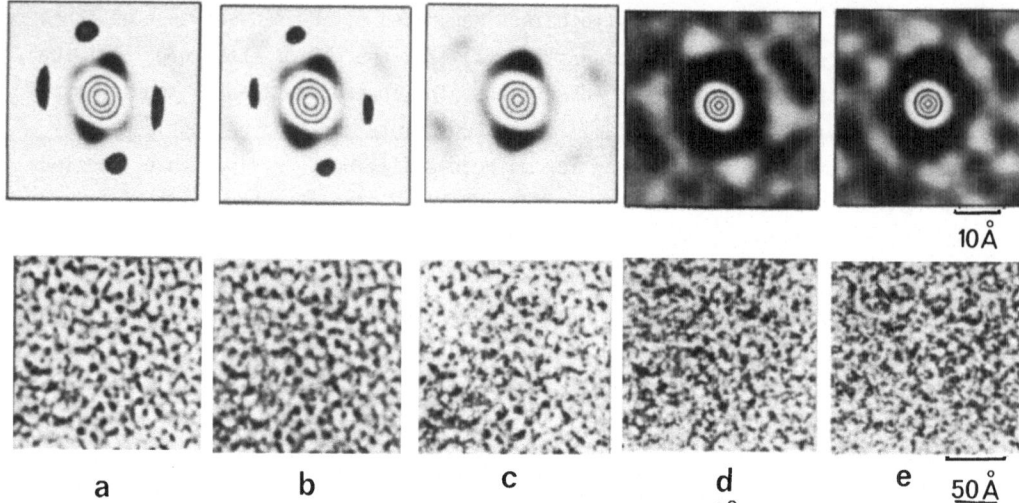

a   b   c   d   e   50 Å

<u>Fig.5.21a-e.</u> Electron micrograph of a carbon film at 270 el/Å^2 dose with ACF (a) simulated low-exposure electron micrographs with reduction factors (b) 1, (c) 5, (d) 10, and (e) 20, respectively, and their ACFs. [5.81] (By courtesy of the Institute of Physics, Great Britain)

the minimum occurring somewhere between the Gaussian ($\Delta z = 0$) and the Scherzer ($\Delta z = 800$ Å) focus.

Since both focus and astigmatism show up in the shape of the ACF peak, this peak could be used for controlled focusing and stigmating of instruments that are connected on-line to the computer [5.81]. The advantages of this method over the use of the Fourier transform [5.82,83] are that only a small number of coefficients in the peak region need to be computed and that the analysis of the important parameters of the peak shape, width, and deviation from circular symmetry is readily incorporated into a computer program.

FRANK [5.81] used a computer simulation to study the ACF peak of a 128 × 128 array of a carbon film micrograph as a function of electron dose (cf Fig.5.21). It is seen that the exposure can be reduced by a factor of 10 against normal exposure (270 el/Å^2) before the peak loses its elliptic shape and shows noise dominance. (Here the display of the ACF was done by using bit clipping to produce a contour effect.)

5.7 Signal-to-Noise Ratio Measurement

5.7.1 Theory

The signal-to-noise ratio is a figure that gives the relative amount of significant information compared with noise in the image, and is thus a measure of information content. Unfortunately, the definition of the S/N ratio differs in different fields. In electrical engineering and statistical optics it is defined as the energy (variance) ratio between signal and noise, while in the electron microscopical literature the square root of this is mostly considered. We shall here make use of the first definition because the method of measurement used is closely related to methods developed in electrical engineering.

In electron microscopy, the S/N ratio is a very convenient parameter for describing the useful information contained in an electron micrograph or alternatively, for a given object and dose, the performance of the instrument by which it is obtained.

We now investigate the connection between the signal-to-noise ratio and the size of the cross-correlation peak. Consider the following experiment: a stable object is twice imaged with the same electron optical conditions. (The same experiment was previously described as a way of observing the ACF of a signal embedded in noise, cf Sect.5.1.2.) The two images resulting from this experiment contain the same signal. The noise functions, which are assumed to be additive to the signal and uncorrelated with the signal and with each other belong, to the same statistical ensemble characterized by a noise power spectrum.

Conceptually it is clear that the CCF peak increases with increasing signal-to-noise ratio since only the signal contributes to the sum in (5.1). Examination of the defining formulae for both quantities reveals a very simple relationship which can be used for measuring the S/N ratio of electron micrographs in situations where estimates of signal and noise variances cannot be separately obtained.

The cross-correlation coefficient of the two images is defined as

$$\rho = \frac{\Phi_{12}(0) - \mu_1\mu_2}{\sigma_1\sigma_2} \quad , \tag{5.24}$$

where $\Phi_{12}(0)$ is the peak height of the CCF of the two images in aligned position[4], and $\mu_{1,2}$ and $\sigma_{1,2}$ their respective means and rms values. From this, the S/N ratio can be obtained as

$$\alpha = \frac{\rho}{1-\rho} \quad . \tag{5.25}$$

This equation immediately follows from (5.24) if one assumes that the noise is additive and uncorrelated to the signal. A formula for estimating α derived by BERSHAD and ROCKMORE [5.84] for discretely sampled images approaches (5.25) rapidly as the number of samples becomes large.

5.7.2 Measurement

An interesting application of this measurement technique is the study of the defocus dependence of the S/N ratio since the defocus is a free parameter in the

---

4 In nonaligned position, the peak height would have to be corrected for the overlap of uncorrelated portions due to the infinite image repeat inherent to the digital Fourier representation.

electron microscopical experiment. Although the existence of a focus position with maximum contrast and information transfer had been known since Scherzer's famous paper on the theoretical resolution limit of the electron microscope [5.85], the information content of a phase contrast micrograph and its defocus dependence had never been quantitatively studied. FRANK and AL-ALI [5.86,87] investigated a carbon film series and found an agreement between experimental S/N ratio and predictions from phase contrast transfer theory.

For a stable object and constant exposure and recording conditions throughout the whole defocus pair series, one should expect the S/N ratio measured from (5.25) to be proportional to the signal variance (apart from some signal-dependence of the shot noise which may become noticeable for high contrast). This quantity can be calculated by using Parseval's theorem which states that the signal variance is equal to the integrated power spectrum of the signal,

$$\text{var}(s) = \int P(\underline{k})d\underline{k} \quad . \tag{5.26}$$

The signal power spectrum is the scattering intensity $S(\underline{k})$ multiplied with the absolute square of the contrast transfer function $H(\underline{k})$

$$P(\underline{k}) = S(\underline{k})|H(\underline{k})|^2 \quad . \tag{5.27}$$

These functions are known, or can easily be determined from experiments: the scattering intensity from electron diffraction data and the transfer function from optical diffraction data — so that the value of the variance integral can be com-

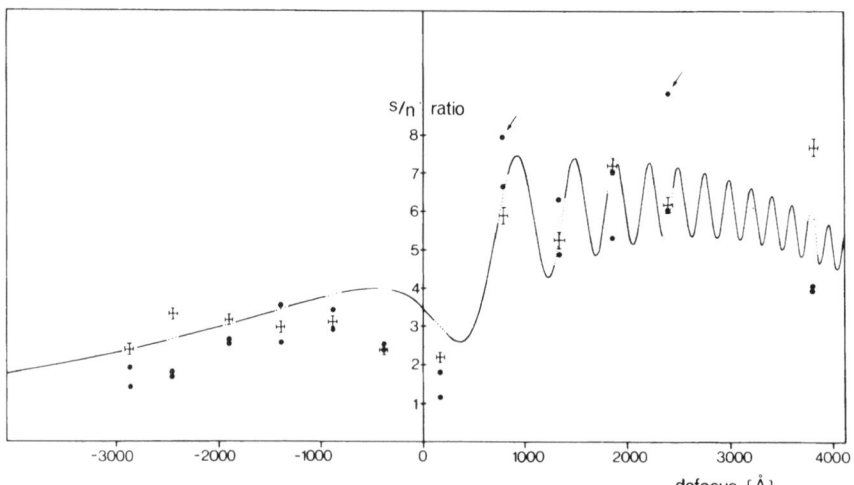

Fig.5.22. Theoretical S/N ratio of electron micrographs of a thin carbon film as a function of defocus (solid line) and experimental S/N ratios obtained by cross-correlation measurements (dots). [5.87] (By courtesy of Macmillan Journals Ltd., London)

puted. The transfer function consists of two factors: the transfer function for coherent illumination and an envelope factor describing the influence of finite source size [5.5] and energy spread [5.8,88]. The coherent transfer function can be determined from the positions of the contrast bands in the optical diffractogram by using THON's method [5.3]. The parameters of source and energy spread can be measured by optical diffraction [5.8]. With these parameters, the effective transfer function can be computed in the defocus range of interest, and from this the resulting [cf (5.26,27)] signal variance. Figure 5.22 shows the experimental S/N ratio values obtained from correlation measurements, together with the theoretical variance curve obtained with electron diffraction data for carbon. The main features of the theoretical curve are a jump from negative to positive defocus, and large rapid oscillations on the positive defocus side.

The match between theoretical curve and experimental data, obtained with a single scaling factor, is apparent from Fig.5.22. Even though the error range in the experimental S/N ratio values makes it impossible to corroborate individual points of the oscillating section of the curve, the important features of the theoretical curve are seen to be supported: constant S/N ratio in the overfocus region and high fluctuating S/N ratio values in the underfocus region.

A more detailed investigation of the defocus dependence of the S/N ratio with the method applied in [5.87] would be very tedious, involving the evaluation of dozens of electron micrographs. Using a more elegant approach, KÜBLER and DOWNING [5.89] measured the entire curve from a single pair of electron micrographs. They examined the one-dimensional CCF of the two micrographs obtained with a tilted specimen where the correlation function is calculated row by row in the direction parallel to the tilting axis. The resulting S/N ratio curve is in agreement with the results of [5.87]. In particular, a sequence of maxima can be distinguished at the normalized defocus values $\Delta \tilde{z} = 1$, $\sqrt{3}$, $\sqrt{5}$, etc., as expected from theory.

In contrast to FRANK and AL-ALI's [5.87] and KÜBLER and DOWNING's [5.89] results, HOPPE and GUCKENBERGER's experiments [5.33] with optically obtained correlation functions did not indicate an appreciable defocus dependence of the CCF peak. This discrepancy must be attributed to the poor quantitative performance of optical methods. Another possible explanation is to be found in the behavior of the theoretical S/N curve for predominantly defocus-spread limited transfer functions, which do not show a large defocus dependence.

Why are the experimental S/N ratio values so small? To understand this, one has to see that the signal and noise variances measured by the cross-correlation method are relative to the resolution of the digital representation. For a "white" noise spectrum, and a monotonically decreasing signal spectrum, the ratio between signal and noise variance decreases with increasing resolution. In the experiment described in [5.87], the resolution was set at the highest resolution observed in the series, at a point where the Fourier contribution of the signal becomes barely perceptible [5.8]. The S/N figures in Fig.5.22 relate to a resolution of 2.5 Å.

Another reason, so far not investigated in the context of S/N ratio measurements, lies in the fact that the object undergoes some changes between the two exposures (cf [5.42]). Since a stable object structure is assumed in this method of measurement, any changing part of the object will be interpreted as part of the noise, resulting in an underestimation of the S/N ratio.

It should be mentioned that the theoretical defocus behavior of the signal rms (i.e., the square root of the variance considered so far) has also been studied by AEBI [5.26], who used a computer-generated carbon film. While the dip of the curve near the Gaussian focus is very pronounced, the image element size chosen (5 Å) and the coarse sampling of the defocus range make it impossible to discern the other features of the curve found in [5.87].

5.7.3 Consequences for Phase Contrast Microscopy

Some important conclusions can be drawn from the result of the S/N ratio measurement in Fig.5.22:

a) In high-resolution operation, with typical conditions of illumination (source halfwidth 5×10^{-4} rad) and energy spread (ΔE between 1 and 2 eV), the gain in S/N ratio is between 2 and 3 if one goes from overfocus to underfocus.

b) It is known that for direct interpretation of a picture, Scherzer's focus (the position of the first maximum) is most preferable because of lack of distortions. However, from an information theoretical point of view, the next three maxima in Fig.5.22 are equally acceptable. If we consider subsequent interpretation by image processing, these defocus positions may in fact be preferable, because the information is here more concentrated at high spatial frequencies.

c) If we had a device that would compute the variance on-line from the image on the screen, the distinct positions of the maxima on the positive defocus side could be made visible. Such a device could be used for absolute focus selection in the underfocus range most important for practical work.

5.7.4 Generalized Signal-to-Noise Ratio Measurement

The method of S/N ratio measurement described in the previous sections can be generalized in the following way: In Fig.5.23 the concept of information channels in the combined electron microscope/data collection system is outlined: In general, one makes a repeated experiment in order to obtain information on a certain signal embedded in noise; the experiments can be represented by different *information channels* with identical signal but different uncorrelated noise function, the latter characterized by the same power spectrum.

These channels are connected at the branching point with the single channel that carries the information common to all repeated experiments. By placing the branching point at different positions in the electron microscope, one can collect statistical information on the effect of each step of image formation and collec-

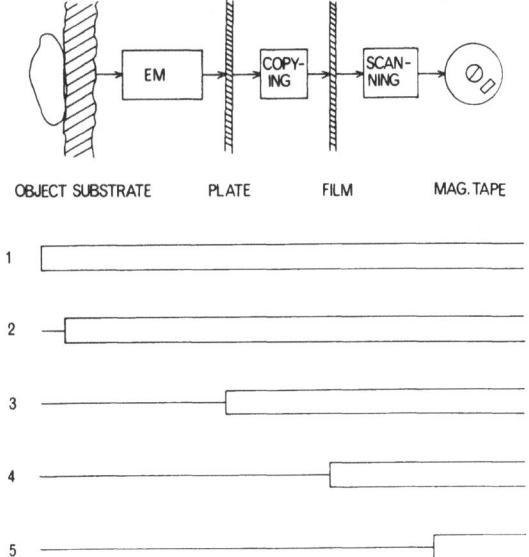

Fig.5.23. The concept of information channels associated with reproducible experiments. For explanation, see text

tion. Each fork in Fig.5.23 represents such an experiment: In (1), the two images of different object particles are compared with each other, to check the reproducibility of specimen preparation; in experiment (2), which is closely related to the image difference method [5.14], a specimen is placed on different supporting films; the fork (3) corresponds to the experiments described in [5.87] where the ability of the electron microscope and its recording devices to reproduce object details is tested; the next two experiments have to do with the subsequent processing of the information: experiment (4) is a test for faithful representation of the electron micrograph on film if the copying step should be necessary; and in (5) the reproducibility of digital densitometry can be checked. By correlation analysis, the signal-to-noise ratio can be measured for each of these experiments, and the assessment of different methods of preparation, imaging, and recording may be based on this criterion.

5.8 Conclusions

The present review attempted to show correlation techniques as important tools in the field of structural electron microscopy.

Both the extension of low-dose imaging techniques to nonperiodic biological specimens and the extension of three-dimensional reconstruction methods to objects without symmetry make signal detection methods indispensable. Moreover, apart from direct applications of correlation techniques for structure determination, the assessment of the instrument performance in terms of resolution and signal-to-noise ratio can be accomplished by using correlation functions.

Even though the computer cannot compete with optical correlations for speed and simplicity of operation, its accuracy and flexibility make it superior to use.

Acknowledgement. I would like to thank Dr. Harald Rose for a critical reading of the manuscript. I am grateful to the authors of various publications cited here, for kindly giving me permission to reproduce their illustrations.

References

5.1   B.V. Johansen: In *Principles and Techniques of Electron Microscopy*, Vol. 5, ed. by M.A. Hayat (Van Nostrand Reinhold Company, New York 1975) pp. 114-173
5.2   A. Klug, J.F. Berger: J. Mol. Biol. *10*, 565-569 (1964)
5.3   F. Thon: Z. Naturforsch. *21a*, 476-478 (1966)
5.4   J. Frank: Optik *30*, 171-180 (1969)
5.5   J. Frank: Optik *38*, 519-536 (1973)
5.6   A. Beorchia, P. Bonhomme: Optik *39*, 437-442 (1974)
5.7   W. Hoppe, D. Köstler, P. Sieber: Z. Naturforsch. *29a*, 1933-1934 (1974)
5.8   J. Frank: Optik *44*, 379-391 (1976)
5.9   R. Hosemann, S.N. Bagchi: *Direct Analysis of Diffraction by Matter* (North-Holland, Amsterdam 1962)
5.10 W.B. Davenport, W. Root: *Random Signals and Noise* (McGraw-Hill, New York 1958)
5.11 J.W. Goodman: *Introduction to Fourier Optics* (McGraw-Hill, New York 1968)
5.12 E.L. O'Neill: *Introduction to Statistical Optics* (Addison-Wesley, Reading, Mass. 1963)
5.13 J. Frank: Ph.D. Thesis, Technische Hochschule München (1970)
5.14 R. Langer, J. Frank, A. Feltynowski, W. Hoppe: Ber. Bunsenges. Phys. Chem. *74*, 1120-1126 (1970)
5.15 W. Hoppe, R. Langer, J. Frank, A. Feltynowski: Naturwiss. *56*, 267-272 (1969)
5.16 J. Frank: In *Proc. 32nd Annual Meeting EMSA, St. Louis*, ed. by J. Arceneaux (Claitor, Baton Rouge 1974) pp. 336-337
5.17 J. Frank: In *Advanced Techniques in Biological Electron Microscopy*, ed. by J. Koehler (Springer, Heidelberg 1973) pp. 215-274
5.18 J. Frank: In *Biomolecular Electron Microscopy*, ed. by A.K. Kleinschmidt (Kontron GmbH, Eching 1975)
5.19 J. Frank, W. Goldfarb, M. Kessel, D. Eisenberg, T.S. Baker: Biophys. J. *21*, 89a (1978)
5.20 L. Al-Ali: In *Developments in Electron Microscopy and Analysis*, ed. by J.A. Venables (Academic Press, London 1975) pp. 225-228
5.21 J.W. Menter: Adv. Phys. *7*, 299-348 (1958)
5.22 J. Frank: Ultramicroscopy *1*, 159-162 (1975)
5.23 D.R. Ensor, C.G. Jensen, J.A. Fillery, R.J.K. Baker: Toronto (1978), Vol. II, pp.32-33
5.24 A.M. Fiskin, M. Beer: Science *159*, 1111 (1968)
5.25 S.D. Golladay: In *Proc. 35th Annual Meeting EMSA, Boston*, ed. by G.W. Bailey (Claitor, Baton Rouge 1974) pp. 82-83
5.26 U. Aebi: Ph.D. Thesis, University of Basel (1976)
5.27 U. Aebi, P.R. Smith, J. Dubochet, C. Henry, E. Kellenberger: J. Supramol. Struct. *1*, 498-522 (1973)
5.28 E. Linfoot: J. Opt. Soc. Am. *46*, 740-752 (1956)
5.29 J. Frank: J. Phys. D: Appl. Phys. *7*, L75-78 (1974)

5.30 A. Engel, J.W. Wiggins, D.C. Woodruff: J. Appl. Phys. *45*, 2739-2747 (1974)
5.31 J. Frank: Manchester (1972) pp. 622-623
5.32 L. Al-Ali: Ph.D. Thesis, University of Cambridge (1976)
5.33 W. Hoppe, R. Guckenberger: Z. Naturforsch. *29a*, 1931-1932 (1974)
5.34 J.R. Parsons, H.M. Johnson, C.W. Hoelke, R.R. Hosbons: Phil. Mag. *27*, 1359-1368 (1973)
5.35 J. Frank, W. Goldfarb, M. Kessel, D. Eisenberg, T.S. Baker: Ultramicroscopy *3*, 283-290 (1978)
5.36 J. Frank, P.H. Bussler, R. Langer, W. Hoppe: Ber. Bunsenges. Phys. Chem. *74*, 1105-1115 (1970)
5.37 J. Frank: Biophys. J. *12*, 484-511 (1972)
5.38 P. Bussler, A. Feltynowski, W. Hoppe: Manchester (1972) pp. 626-627
5.39 D.L. Misell: J. Phys. D: Appl. Phys. *6*, L6-9 (1973)
5.40 W. Hoppe, B. Grill: Ultramicroscopy *2*, 153-168 (1977)
5.41 A. Feltinowski, W. Hoppe: Canberra (1974), Vol. I, pp. 206-207
5.42 W. Hoppe, P. Bussler, A. Feltynowski, N. Hunsmann, A. Hirt: In *Image Processing and Computer-Aided Design in Electron Optics*, ed. by P.W. Hawkes (Academic Press, London 1973) pp. 91-126
5.43 W. Hoppe, H.J. Schramm, M. Sturm, N. Hunsmann, J. Gassmann: Z. Naturforsch. *31a*, 645-655 (1976)
5.44 M. Eckert: Thesis, Technische Hochschule München (1976)
5.45 R. Hegerl, A. Feltynowski, B. Grill: Toronto (1978), Vol. I, pp. 214-215
5.46 R. Hegerl, W. Hoppe: Z. Naturforsch. *31a*, 1717-1721 (1976)
5.47 W. Hoppe: Naturwiss. *61*, 239-249 (1974)
5.48 W. Hoppe: Z. Naturforsch. *30a*, 1188-1199 (1975)
5.49 W. Hoppe: Ann. N.Y. Acad. Sci. *306*, 121-144 (1978)
5.50 W. Krakow, K.B. Welles, B.M. Siegel: J. Phys. D: Appl. Phys. *9*, 175-181 (1976)
5.51 K. Welles, W. Krakow, B.M. Siegel: Canberra (1974), Vol. I, pp. 320-321
5.52 D.L. Misell, R.E. Burge: J. Microsc. *103*, 195-202 (1975)
5.53 N. Hunsmann, P. Bussler, W. Hoppe: Manchester (1972) pp. 654-655
5.54 W. Hoppe, J. Gassmann, N. Hunsmann, H.J. Schramm, M. Sturm: Hoppe-Seyler's Z. Physiol. Chem. *355*, 1483-1487 (1974)
5.55 W. Hoppe: Naturwiss. *61*, 534-536 (1974)
5.56 W. Hoppe: Optik *29*, 617-621 (1969)
5.57 J. Frank: In Topical Meeting on Image Processing for 2-D and 3-D Reconstruction from Projections: "Theory and Practice in Medicine and the Physical Sciences", Stanford, California (1975)
5.58 R.M.C. Matthews: Master's Thesis, The University of Texas in Austin (1972)
5.59 R.A. Crowther, D.J. DeRosier, A. Klug: Proc. R. Soc. London *A317*, 319-340, (1970)
5.60 R.A. Crowther, L. Amos: J. Mol. Biol. *60*, 123-130 (1971)
5.61 W.O. Saxton: Ph.D. Thesis, University of Cambridge, England (1974)
5.62 K. Welles: In *Proc. 33rd Annual Meeting EMSA, Las Vegas, 1975*, ed. by G.W. Bailey (Claitor, Baton Rouge 1975) pp. 198-199
5.63 W. Hoppe: Z. Elektrochemie *61*, 1076-1083 (1957)
5.64 W.O. Saxton: Canberra (1974), Vol. I, pp. 314-315
5.65 P.R. Smith, U. Aebi, R. Josephs, M. Kessel: J. Mol. Biol. *106*, 243-275 (1976)
5.66 L.A. Amos: J. Mol. Biol. *99*, 65-73 (1975)
5.67 A. Vander Lugt: IEEE Trans. IT-*10*, 139 (1964)
5.68 J. Duvernoy: J. Opt. Soc. Am. *65*, 1331-1335 (1975)
5.69 P.N.T. Unwin, R. Henderson: J. Mol. Biol. *94*, 425-440 (1975)
5.70 R. Henderson, P.N.T. Unwin: Nature *257*, 28-32 (1975)
5.71 I.A.M. Kuo, R.M. Glaeser: Ultramicroscopy *1*, 53-66 (1975)
5.72 J. Frank: Jerusalem (1976), Vol. I, pp. 273-274
5.73 J. Frank: Ann. N.Y. Acad. Sci. *306*, 112-120 (1978)
5.74 W.O. Saxton, J. Frank: Ultramicroscopy *2*, 219-227 (1977)
5.75 A. Rose: Adv. Electron *1*, 131-166 (1948)
5.76 W. Hoppe, R. Hegerl, R. Guckenberger: Z. Naturforsch. *33a*, 857 (1978)
5.77 J. Frank: Toronto (1978), Vol. III, pp. 87-93
5.78 J. Frank, B. Shimkin: Toronto (1978), Vol. I, pp. 210-211
5.79 J. Frank, W. Goldfarb, M. Kessel: Toronto (1978), Vol. II, pp. 8-9

5.80   J. Frank: Optik *43*, 25-34 (1975)
5.81   J. Frank: J. Phys. E: Sci. Instr. *8*, 582-587 (1975)
5.82   V. Witt, H.P. Englmeier, W. Hoppe: Manchester (1972) pp. 632-633
5.83   O. Kübler, R. Waser: Optik *37*, 425-438 (1973)
5.84   N.J. Bershad, A.J. Rockmore: IEEE Trans. IT-*20*, 112-113 (1974)
5.85   O. Scherzer: J. Appl. Phys. *20*, 20-29 (1949)
5.86   J. Frank, L. Al-Ali: In *Developments in Electron Microscopy and Analysis*, ed.
       by J.A. Venables (Academic Press, London 1975) pp. 229-232
5.87   J. Frank, L. Al-Ali: Nature *256*, 376-379 (1975)
5.88   K.J. Hanszen, L. Trepte: Optik *32*, 519-538 (1971)
5.89   O. Kübler, K.H. Downing: Private communication (1977)

6. Holographic Methods in Electron Microscopy

R. H. Wade

With 12 Figures

The aim of this chapter is to describe the two-stage imaging process common to all holographic procedures. The general aspects of hologram formation are first described. The different types of holography are then treated and the zone plate representation is used to show the essential unity of these apparently disparate techniques. Experimental work on electron holography from the early 1950s onwards is then reviewed in some detail. Finally contrast transfer theory and holography are shown to be complementary descriptions of the same imaging process.

6.1 Historical Background

The diffraction limited resolution δ_R of an image forming-system is given to within a multiplicative constant by an expression of the form $\delta_R = \lambda/\alpha$ where λ is the wavelength of the irradiation and α is the limiting angular aperture. This leads to a best resolution of around 5×10^{-5} cm for a microscope operating in the visible region of the electromagnetic spectrum.

In 1926 BUSCH [6.1] showed that any axially symmetric electromagnetic field acts as a lens for electrons. De BROGLIE [6.2] had already shown that a wavelength $\lambda = h/p$ could be associated with the motion of electrons of momentum p; h is Planck's constant. Since an electron accelerated through a potential of 50 keV has an associated wavelength of roughly 5×10^{-10} cm, a wavelength limited resolution below atomic dimensions should be attainable in a microscope using electron irradiation. Encouraged by this prospect Knoll and Ruska began to construct the first electron microscope in 1928, and others soon followed; a brief account of this period is given by GABOR [6.3].

As early as 1936, SCHERZER [6.4] proved theoretically that spherical aberration cannot be eliminated from electron lenses since axially symmetric electron optical components always have a converging lens action. The spherical aberration limits the useful aperture of a lens so that the resolution depends essentially on the term $(C_s\lambda^3)^{\frac{1}{4}}$, where the spherical aberration constant C_s has the dimension of length.

However, the first period of steady progress in the experimentally attained resolution came virtually to an end only in 1947 when HILLIER and RAMBERG [6.5] showed that the residual astigmatism of the magnetic objective lens could be corrected using an electromagnetic stigmator. They obtained image resolutions essentially equal to the theoretical limit of around 5-10 Å, and since the resolution is proportional to the fourth root of the spherical aberration there then seemed little hope of obtaining further significant improvements in the resolution capabilities of electron microscopes.

In 1948, GABOR [6.6] published a paper which suggested a way around this difficulty. The proposition concerned a two-stage imaging procedure. It was first necessary to obtain an image which, although it might be a bad visual representation of the object, would contain all the information on the complex scattering amplitude of the object. This information was to be retained in the image by the device of adding a coherent background wave to act as a phase reference. GABOR showed in later articles [6.7,8] that the resulting interferogram contains all the necessary information required to reconstruct the original wave scattered by the object. For this reason he called this type of interference image a hologram from the Greek *holos* meaning complete. Since the reconstructed wave could be shown to be deformed by the phase distortions arising from the electron lens aberrations, GABOR proposed as the second stage of the new imaging procedure the use of the developed hologram as a transparency in a light-optical arrangement designed to correct the wave aberrations. The precision with which glass lenses could be manufactured would readily permit the light-optical aberrations to be correctly matched to those of the electron microscope. The light-optical experiments described in his articles showed clearly that these ideas were basically correct.

Experimental work on electron holography was carried out by HAINE et al. [6.9,10] over a period of three years or so beginning in 1950. A modified AEI EM2 was used. This work was not successful in obtaining improved or even equivalent resolution to that available in an ordinary micrograph. The primary limitation arose from the need to use a small illumination source in order to provide a background wave coherent with the wave scattered by the object. The long exposure times necessary to record the resulting very low electron intensity distributions meant that the practical resolution of the hologram was limited by mechanical vibrations, stray magnetic fields, specimen stage creep, and object contamination. In present-day instruments these problems have been essentially overcome. The advent of more efficient electron guns and in particular the field emission gun means that the spatial coherence conditions can be satisfied at much higher useful beam currents so that the exposure times necessary to record a high-resolution hologram can be drastically reduced. In addition the present generation of transmission electron microscopes allows image resolutions of around 3 Å to be readily obtained. Despite this, and as we shall see later, more recent attempts at electron holography have not been essentially more successful than the pioneering work of the early 1950s.

The aim of this chapter is to present in a reasonably digestible form the essential features of holography to a public of electron microscopists interested and active in image treatment. I have tried to present this material in a physically descriptive manner and for the sake of clarity I have not attempted to describe in detail all the many theoretico-speculative articles concerning electron holography. However, I have tried to make the list of references as complete as possible. The experimental work is described with, hopefully, no serious omissions. For background reading the books by DEVELIS and REYNOLDS [6.11], GOODMAN [6.12], and FRANÇON [6.13] can be warmly recommended. The key articles on the theory of electron holography are those by GABOR [6.6-8] and by HANSZEN [6.14-17], who with various collaborators develops the subject in terms of contrast transfer theory. This aspect of holography is also developed at length by WEINGÄRTNER et al. [6.18-21] with particular reference to the effect of partial spatial coherence. The series of papers by LEITH and UPATNIEKS [6.22-24] also make instructive reading. A recent review of electron holography with a very complete set of references appears as part of an article on partial coherence by HAWKES [6.25].

6.2 Holographic Schemes

6.2.1 The Generalized Hologram

The detection of any object depends quite naturally on the degree to which it scatters the illuminating irradiation. If the complex wave amplitude due to this scattering is U at an observation plane beyond the object, the visually observed or photographically recorded quantity in that plane is the intensity $I = |U|^2$. This square law detection has eliminated the phase information which was present in the complex wave amplitude. The addition of a reference wave U_r onto the object-dependent wave yields an intensity at the observation plane given by

$$I_i = |U_r + U|^2$$

$$= |U_r|^2 + |U|^2 + U_r U^* + U_r^* U \quad . \tag{6.1}$$

The first two terms on the RHS of (6.1) depend only on the intensities associated with the object and reference waves while the last two terms depend on both their amplitudes and relative phases.

In order to decide whether this amplitude and phase information is readily accessible we must first examine some of the characteristics of the photographic recording. The total light energy recorded by the photographic plate is $W = I\tau$ for an exposure time τ. After development the amplitude transmission t_a of the photographic negative depends on W in the manner shown in Fig.6.1. If the total exposure includes only small variations around a mean value W_0 the amplitude transmission of the plate will be linearly related to the exposure W according to the relation

$$t_a = t_b - \beta(W - W_0) \quad , \tag{6.2}$$

where β is the slope of the t_a/W curve at the point W_0. If we suppose that the reference wave intensity $|U_R|^2$ is uniform across the photographic recording plane then to within a constant exposure-time factor the amplitude transmittance of the photographic negative, which represents the holographic record of the object wave, is obtained by combining (6.1) and (6.2) to give

$$t_a = t_b - \beta(|U|^2 + U_r U^* + U_r^* U) \quad . \tag{6.3}$$

Equation (6.3) holds provided the fluctuations in the amplitude transmission remain in the linear region of the curve of Fig.6.1. The satisfaction of this condition will depend on the contrast of the original image intensity distribution, (6.1), defined as

contrast = (total intensity - background)/background

$$= \left| \frac{U}{U_r} \right|^2 + \left(\frac{U}{U_r} \right)^* + \frac{U}{U_r} \quad .$$

The contrast will be low if the condition

$$|U| \ll |U_r| \tag{6.4}$$

is satisfied. If the overall contrast is low the amplitude transmission of the plate will remain in the linear region of the t_a/W curve. When the hologram, whose amplitude transmittance is described by (6.3), is illuminated by the reference wave alone the amplitude of the transmitted light is given by

$$t_a \cdot U_r = t_b U_r - \beta |U_r|^2 \left(\frac{|U|^2}{U_r^*} + U + \frac{U_r}{U_r^*} U^* \right) \quad . \tag{6.5}$$

Since we have supposed that the intensity of the reference wave is uniform the third term on the RHS of (6.5) is seen to be equal to the original complex wave amplitude due to the object. The fourth term is proportional to the complex conjugate of the original object wave. The first term represents a uniform background wave and the contrast condition (6.4) ensures that the second term can be neglected.

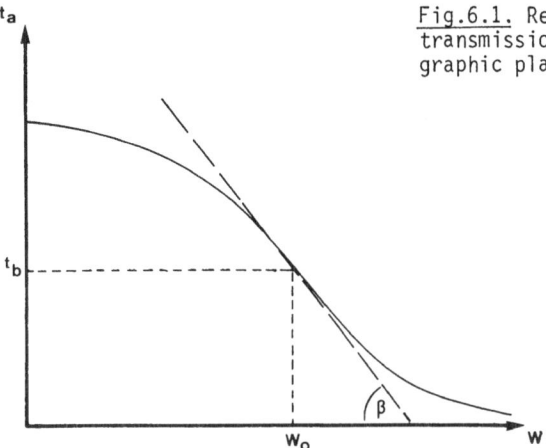

Fig.6.1. Relationship between the amplitude
transmission t_a and exposure W of a photographic plate

The third and fourth terms, then, represent the twin waves which occur in the reconstruction of all types of holograms and they both carry the same energy.

It can be concluded that an object-dependent wave can indeed be reconstructed
from the hologram. The practical viability of the holographic method will mainly
depend on whether the imaging wave can be efficiently separated from the background
and the conjugate waves.

We shall now go on to discuss specific holographic methods.

6.2.2 In-Line Fresnel and Fraunhofer Holograms

An incident parallel wave illuminates the object O in Fig.6.2a. The object must
have a high average transparency so that the transmitted wave $U_0(\underline{r})$ can be considered to consist of a strong plane wave unscattered component plus a weak
scattered component, i.e., $U_0(\underline{r}) = 1 + a_0(\underline{r})$ where $a_0(\underline{r}) \ll 1$. This will ensure that
the weak contrast condition (6.4) is satisfied. The photographic plate Pl records
the interference pattern generated by the directly transmitted and the scattered
wave components a distance z downbeam from the object. The wave amplitude $U_i(\underline{r}_i)$
at the plate is nothing other than the Fresnel diffraction amplitude associated
with the object U_0 and is given by the Fresnel-Kirchhoff diffraction integral which
can be expressed to within a multiplicative constant as

$$U_i(\underline{r}_i) = \int d\underline{r} \cdot U_0(\underline{r}) \cdot \exp\left[i \frac{\pi}{\lambda z} (\underline{r} - \underline{r}_i)^2 \right]$$

$$= 1 + a_i(\underline{r}_i) \quad , \tag{6.6}$$

where

$$a_i(\underline{r}_i) = \int d\underline{r} \, a_0(\underline{r}) \exp\left[i \frac{\pi}{\lambda z} (\underline{r} - \underline{r}_i)^2 \right] \tag{6.7}$$

and $d\underline{r} = dxdy$.

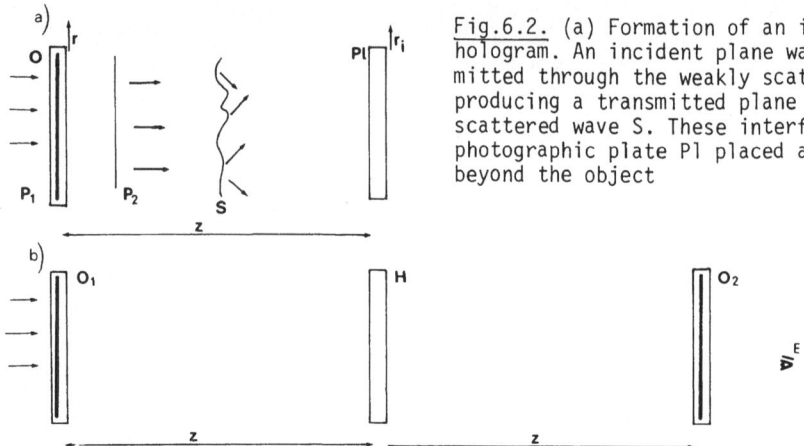

Fig.6.2. (a) Formation of an in-line Fresnel hologram. An incident plane wave P_1 is transmitted through the weakly scattering object O producing a transmitted plane wave P_2 and a scattered wave S. These interfere at the photographic plate Pl placed a distance z beyond the object

Fig.6.2. (b) Reconstruction of the scattered wave front occurs by illuminating the developed plate H by a plane wave. An observer at E sees the original object O_1 and its conjugate O_2 situated distances z before and beyond H, respectively

The photographic recording of the intensity $|U_i|^2$ constitutes the hologram formation step. The reconstruction stage is carried out by removing the object and replacing the developed photographic plate, the hologram, in its original position. If the illuminating plane wave is taken as having unit amplitude the amplitude transmission of the hologram is obtained from (6.5) by setting

reference wave: $U_r = 1$

object wave   : $U = a_i$

in accord with (6.6). This gives

$$t_a = t_b - \beta(|a_i|^2 + a_i + a_i^*) \quad . \tag{6.8}$$

An observer at E, Fig.6.2b, looking through the hologram sees two superposed twin images O_1 and O_2 situated at distances z on either side of the hologram. These images result, respectively, from the terms a_i and a_i^* of (6.8) as can be seen by changing the sign of z in (6.7). It is directly apparent from Fig.6.2b and can be shown quite simply from the equations that if the observer or an auxiliary optical system focuses on one or the other of the twin images the other image appears also with twice the primitive defocus. If a more general wave aberration perturbs the initial hologram a corrected image can always be obtained by using a reconstruction system with an adapted conjugate aberration; the superposed twin will then have twice the initial aberration.

This technique, similar to that originally proposed by GABOR, has become known as in-line Fresnel holography because the reference and scattered waves propagate

along the same axis to produce an out-of-focus or Fresnel image. The main disadvantages of the method are that it is restricted to weakly scattering objects
and that the reconstructed image is always degraded by the presence of the aberrated twin image.

The same arrangement as in Fig.6.2a is used to record in-line Fraunhofer holograms. In this case the individual scattering particles in the object must be
small enough to satisfy the Fraunhofer or far-field diffraction condition. If
the maximum particle size is such that

$$r_{max}^2 << \lambda z \tag{6.9}$$

the integral of (6.7) reduces to the form

$$a_i(r_i) \simeq \exp\left(\frac{i\pi r_i^2}{\lambda z}\right) \int dr \; a_0(t) \cdot \exp\left(\frac{-i2\pi rr_i}{\lambda z}\right) .$$

$$= \tilde{a}_0 \left(\frac{r_i}{\lambda z}\right) \cdot \exp\left(\frac{i\pi r_i^2}{\lambda z}\right) \tag{6.10}$$

The scattered wave amplitude at the recording plane is shown by (6.10) to consist of the Fourier transform $\tilde{a}_0$ of the object function multiplied by a quadratic
phase factor.

The inequality (6.9) can be satisfied for the individual particles while the
overall object surface is still in the Fresnel or near-field diffraction region.
This ensures that the hologram is produced by the interference of the background
plane wave with the scattered wave a_i of (6.10). The reconstructed focused image of
an individual particle is less perturbed by the superposed twin than in the Fresnel
case since (6.8) and (6.10) show that the twin must appear in the form of its
Fraunhofer diffraction image. That this is a broad function, and implicitly of low
contrast since both twins carry the same energy can be shown by considering the
simple example of a disc object of diameter d. Figure 6.3 shows that in the reconstruction stage an in-focus image appears at the plane 0_2 riding on the background
Fraunhofer image of diameter D produced by the twin situated in the plane 0_1. Since
an object of diameter d scatters essentially within an angle $\theta = \lambda/d$ the Fraunhofer
image of the twin will have a diameter in the plane 0_1 given by $D = 2z\theta = 2z\lambda/d$. If
$D >> d$ we can consider the in-focus image to be only slightly perturbed by the slowly
varying intensity distribution due to the twin. Substituting for D in this inequality
gives us the condition (6.9) to within a constant factor. The inequality (6.9)
therefore ensures that the initial hologram is formed in the Fraunhofer condition
and simultaneously that the reconstructed in-focus image will only be slightly perturbed by the twin image. If the particle density is high the overlapping twin contributions may well perturb the reconstructed image. It appears from this discussion

230

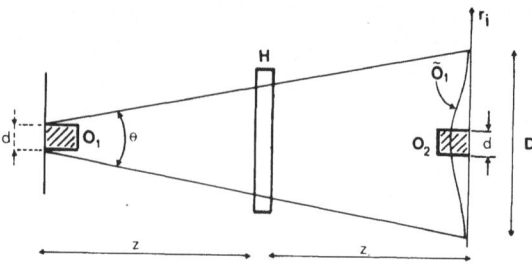

that the Fraunhofer hologram itself may lack contrast when the particle size is
very small.

6.2.3 Sideband Fresnel Holograms

A parallel incident wave illuminates the object, O, of Fig.6.4a. A prism deflects a part of the beam down through an angle α to overlap the wave scattered
by the object in the plane of the photographic plate P. In the absence of the object the coherent addition of the two inclined wave fronts produces a set of parallel two-beam interference fringes of separation λ/α in the recording plane. The
amplitude and phase modulations of the light transmitted by the object produce local
modulations of the intensity and position of the fringes. This can be seen by considering the intensity distribution produced by the interference of the object de-
pendent wave $U_i = a_0 \exp(i\varphi_0)$ and the tilted reference plane wave $U_r = a_r \exp(i\underline{k}_\alpha \cdot \underline{r}_i)$;
in this case (6.1) takes the form

$$I = a_r^2 + a_0^2 + 2a_0 a_r \cos(\underline{k}_\alpha \cdot \underline{r}_i - \varphi_0) \quad , \tag{6.11}$$

where $|\underline{k}_\alpha| = 2\pi\alpha/\lambda$.

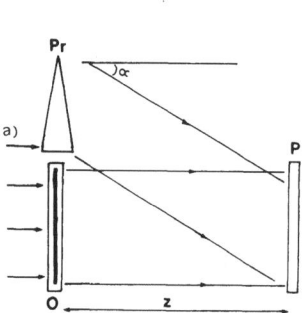
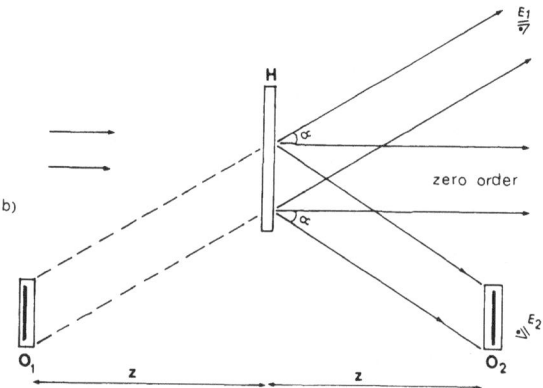

Fig.6.4. (a) Sideband Fresnel hologram. A prism Pr deflects part of the incident
plane wave through an angle α to interfere at the photographic plate P with the wave
scattered by the object O. (b) The reconstructed waves are deflected off at angles
$\pm\alpha$ to the zero-order beam directly transmitted by the hologram : observation at E_1
or E_2 permits the twin images to be viewed separately

The last term of (6.11) shows how the spatial carrier wave of frequency k_α has been amplitude and phase modulated by the object-dependent wave.

The reconstruction stage is carried out using a parallel illumination incident perpendicular to the hologram as schematized in Fig.6.4b. The amplitude transmission of the hologram is given by combining (6.2) and (6.11)

$$t_a = t_b a_r - a_r^2 \left\{ \frac{a_o^2}{a_r} + a_o \exp[i(\underline{k}_\alpha \cdot \underline{r}_i - \varphi_o)] + a_o \exp[-i(\underline{k}_\alpha \cdot \underline{r}_i - \varphi_o)] \right\} \quad .(6.12)$$

Equation (6.12) is correct provided $a_r \gg a_o$; the reference wave is now produced independently of the object wave so it may be simpler to adjust the amplitude a_r to ensure that the inequality is satisfied. The cosine term in (6.11) produces the two Fourier components which propagate at angles $\pm\alpha$ with respect to the directly transmitted beam at the reconstruction stage. These components are the last two terms on the RHS of (6.12) and it can be seen that each carries one of the twin images. When the reference beam angle is correctly chosen, the angular separation produced by the cosine grating ensures that the twin images are completely separated from each other and from the on-axis zero order beam (Fig.6.4b).

6.2.4 Fourier Transform Holograms

A parallel light beam illuminates the on-axis object O and an offset point reference scattering object R situated in the same plane, Fig.6.5a. The point reference can also be introduced by using an auxiliary lens. The size of the object and the reference scatterer and the separation z between the object and recording planes are such that the Fraunhofer condition (6.9) is satisfied. If the complex amplitude transmittance of the object and reference are U_o and U_r, respectively, (6.10) shows that sum of the wave amplitudes at the recording plane P is given by

$$A(r_i) = \exp\left(\frac{i2\pi r_i^2}{\lambda z}\right)\left[\tilde{U}_o\left(\frac{r_i}{\lambda z}\right) + \tilde{U}_r\left(\frac{r_i}{\lambda z}\right) \cdot \exp(i\underline{k}_\alpha \cdot \underline{r}_i)\right] \quad ,$$

where the tildes indicate the Fourier transforms as before and the phase factor multiplying the term U_r arises because the reference point is offset by the distance $\varepsilon = \alpha z$ with respect to the axis. The resulting intensity distribution at the photographic plate is

$$I(r_i) = |\tilde{U}_r|^2 + |\tilde{U}_o|^2 + \tilde{U}_r\tilde{U}_o^* \cdot \exp(i\underline{k}_\alpha \cdot \underline{r}_i) + \tilde{U}_r^*\tilde{U}_o \exp(-i\underline{k}_\alpha \cdot \underline{r}_i) \quad .$$

It is important to notice that the quadratic phase factor which occurs in the Fraunhofer holograms has been eliminated in the Fourier transform hologram and that this occurs because the object and the reference point are situated in the same plane.

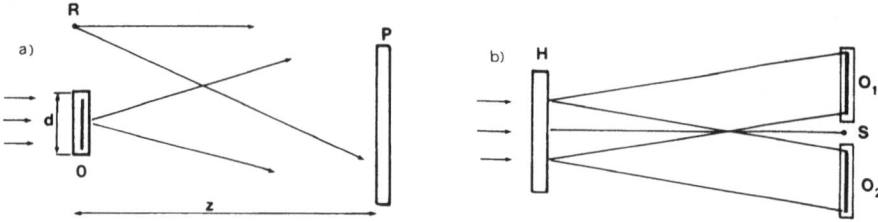

Fig.6.5. (a) Fourier transform hologram formed by interference between a point reference R situated in the same plane as the object 0. The defocus z satisfies the
Fraunhofer condition $z > d^2/\lambda$. (b) In the reconstruction the twin images are found
symmetrically on either side of the central spot S due to the undeflected beam

If the amplitude transmittance of the hologram satisfies the linearity condition
$|U_0| \ll |U_r|$ the reconstruction using a parallel illumination yields, in the plane
0 of Fig.6.5b, an amplitude distribution proportional to

$$A_R = U_r^* \otimes U_r^* + U_0^* \otimes U_0^* + U_r / U_0^*(\varepsilon + x) + U_r^* U_0(\varepsilon - x) \quad . \qquad (6.13)$$

The last two terms on the RHS of (6.13) represent the twin image terms which are
symmetrically placed on either side of the centered autocorrelation terms $U_r^* \otimes U_r^*$
and $U_0^* \otimes U_0^*$. Note that the defocus-dependent term does not appear in (6.13) and that
in general the aberrations introduced by the milieu between the object and the
recording plane are of no consequence since the spherical reference wave is sub-
mitted to the same aberrations as the object wave.

6.2.5 Single-Sideband Holograms

A weakly scattering object $U_0 = 1 + a_0$ is illuminated by a parallel beam of light.
A lens, L in Fig.6.6, is set to image the object plane near the photographic plate
P. The spatial frequency spectrum of the object $\tilde{a}_0(\underline{p})$ and the unscattered beam component $\delta(\underline{p})$, focused into the central spot, appears in the back focal plane of the
lens. A half plane aperture stops off one half of the object spectrum so that the
hologram is formed on the photographic plate by the interference between the
directly transmitted light and half of the object spectrum which is phase distorted
by the lens aberrations. The intensity distribution of the hologram is

$$I(\underline{r}_i) = 1 + \int_0^\infty \tilde{a}_0(\underline{p})T^*(\underline{p}) \cdot \exp-(i2\pi\underline{p} \cdot \underline{r}_i)d\underline{p} + \int_{-\infty}^0 \tilde{a}_0(\underline{p})T(\underline{p}) \cdot \exp(-i2\pi\underline{p} \cdot \underline{r}_i)d\underline{p} \quad (6.14)$$

provided a_0 is real and the lens aberrations are axially symmetric. The second
and third terms of (6.14) represent the twin image terms whose spectra are now
separated into the positive and negative spatial frequency regions. The function
$T(\underline{p}) = \exp[-i\gamma(\underline{p})]$ where $\gamma(\underline{p})$ represents the phase distortion of the imaging wave
due to the lens aberrations and includes the effects of defocus.

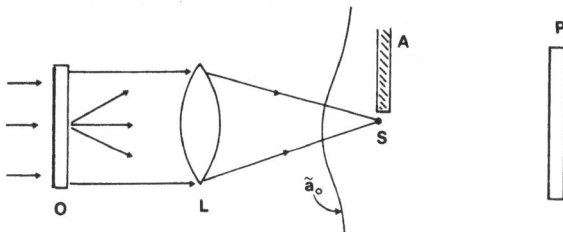

Fig.6.6. Single-sideband hologram. The directly transmitted wave from the object O is focused by the lens L into the central spot S. The scattered wave forms the distribution $\tilde{a}_0$. The aperture A cuts off half the scattered wave. The image forms on the plate P

The reconstruction stage uses the same setup as for the hologram formation; the object is now replaced by the hologram. The wave amplitude transmitted by the hologram is proportional to the intensity expressed by (6.14). If the half plane aperture position is inversed one of the twin images is eliminated and the other is phase distorted by the aberration term $T(\underline{p})$. The diffraction amplitudes contributing to the final image are proportional to $\tilde{I}_1(\underline{p})$ where

$$\tilde{I}_1(\underline{p}) = \delta(\underline{p}) + \tilde{a}_0(\underline{p})T(\underline{p})T^*(\underline{p}) \quad \text{for} \quad \underline{p} > 0$$

$$= \delta(\underline{p}) + \tilde{a}_0(\underline{p}) \quad .$$

The reconstructed image intensity is therefore to first order

$$I(\underline{r}_i) = 1 + \int_{-\infty}^{+\infty} \tilde{a}_0(\underline{p}) \exp(-i2\pi\underline{p} \cdot \underline{r}_i)d\underline{p}$$

$$= 1 + a_0(\underline{r}_i) \quad .$$

If the object function a_0 is complex it is necessary to take two holograms with inverted half plane aperture positions and by taking the sum and the difference of the reconstructed images the amplitude and phase components can be separated.

6.2.6 Zone Plate Interpretation

In the second article concerning his proposal for the holographic two-stage imaging process using "reconstructed wave fronts" GABOR observed that a hologram can be considered as a type of Fresnel zone plate. Shortly afterwards ROGERS showed that it is instructive to discuss the holographic imaging process in terms of these zone plates whose geometrical imaging properties are well known [6.26,27].

A circular zone plate which has the intensity transmission profile shown in Fig.6.7a, where the ring radii increase as the square roots of successive integers, produces a series of real and virtual point images along its axis when it is illuminated with a parallel light beam. It acts like a multiple focus lens. If the intensity transmission profile is modified to take the sinusoidal form of Fig.6.7b only two images, one real the other virtual, are observed along the axis at the distance $\pm d^2/\lambda$ from the plate illuminated with a parallel light beam; d is the radius of the first zone of the plate.

According to the transmission properties of photographic plates discussed in Sect.6.2.1 it will not be possible to produce photographically the zone plate profile of Fig.6.7b. However, the addition of a strong background intensity will reduce the contrast of the profile, a photograph of which will then have linear transmission/exposure characteristics. The transmission of light through this low-contrast sinusoidal zone plate will produce a directly transmitted beam in addition to the real and the virtual images.

Now consider the experiment schematized in Fig.6.8a in which a plane wave front P interferes with the diverging spherical wave front S emitted by a point scatterer O. Provided that the plane wave amplitude is greater than that of the spherical wave the developed photographic recording of the interferogram will have the amplitude transmission shown in Fig.6.8b. We remark that the same profile would have resulted from the interference of the plane wave with a spherical wave converging to the point O' in Fig.6.8a. When the photographic plate is illuminated with a parallel incident wave it will act as a sinusoidal zone plate and produce a real and a virtual image of the point scatterer O. These images are the foci of the two oppositely curved spherical waves which could have produced the zone plate. The experiment described represents the elementary holographic recording and wave front reconstruction process since any scattering object can be considered to consist of a sum of point scatterers.

If we schematize the formation of the elementary hologram, Fig.6.9a, as the interference between a spherical wave S and a plane wave P we see that in the reconstitution the two spherical waves S and S^* are generated. Since these are centered on and propagate along the zone plate axis they always appear superposed. This corresponds to the case of in-line Fresnel and Fraunhofer holograms.

The Fresnel (Fraunhofer) sideband method uses a tilted plane wave reference beam P, Fig.6.9b, and this produces an offset incomplete zone plate by interference with the spherical wave S. In the reconstruction the regenerated partial spherical waves are centered on the axis of the asymmetric zone plate but as shown in the figure propagate in different directions and are thereby separated so the images O and O^* can be viewed separately from the positions E_1 and E_2.

In the single-sideband method, Fig.6.9c, only half the spherical wave S generated by the point scatterer is allowed to contribute to the hologram formation by interference with the axial plane wave P. In the reconstruction stage the two half spherical waves are generated with foci on the zone plate axis but radiating along the directions indicated by the arrows. The images O and O^* can be viewed separately from E_1 and E_2 as in the Fresnel sideband method.

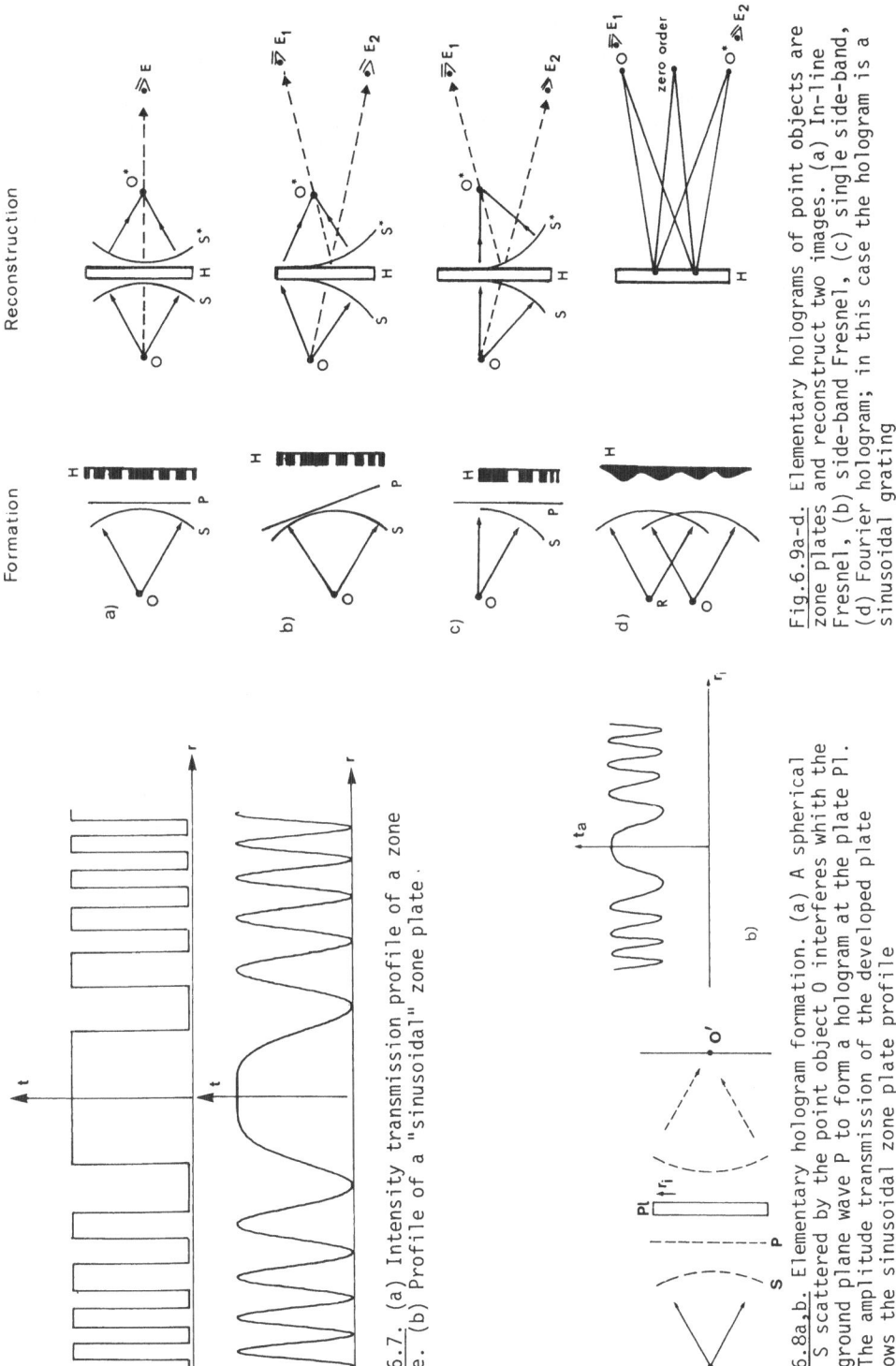

Fig.6.7: (a) Intensity transmission profile of a zone plate. (b) Profile of a "sinusoidal" zone plate.

Fig.6.8a,b. Elementary hologram formation. (a) A spherical wave S scattered by the point object O interferes whith the background plane wave P to form a hologram at the plate Pl. (b) The amplitude transmission of the developed plate follows the sinusoidal zone plate profile

Fig.6.9a-d. Elementary holograms of point objects are zone plates and reconstruct two images. (a) In-line Fresnel, (b) side-band Fresnel, (c) single side-band, (d) Fourier hologram; in this case the hologram is a sinusoidal grating

In the case of the Fourier transform hologram the reference point and the point scatterer lie in the same plane. They produce two spherical waves, S_R and S respectively, which generate sinusoidal fringes by interference at the plate. In the reconstruction the grating-like hologram produces a directly transmitted beam and two laterally separated spectral components each carrying one of the twin images. The images are focused at infinity, or in the focal plane of a lens and are symmetrically disposed on either side of the axial directly transmitted wave.

The Fourier transform holograms do not contain the zone plate term which is common to the other holograms and in consequence the resolution of the photographic plate used to record the hologram does not play the same role as for the other holographic procedures. When the sinusoidal fringes are too fine to be recorded by the plate only the field of view is limited. In the case of the zone plate type holograms the fineness of the circular zone pattern which can be recorded limits the resolution of the reconstructed image.

6.3 Experimental Electron Holography

The earliest experiments on electron holography arose directly from Gabor's proposed two-stage imaging process using reconstructed wave fronts. His colleagues DYSON, HAINE and MULVEY in the Associated Electrical Industries Research Laboratory used a modified EM3 microscope to investigate the "projection" and the "transmission" holographic methods. These correspond, respectively, to the in-line Fraunhofer and Fresnel methods which have been described in Sect.6.2. The first method was that originally proposed by GABOR [6.6,7] and the second, a modification of the first, was proposed by HAINE and DYSON [6.9] (see also [6.8]).

We should remark that the best resolution obtainable in a standard electron micrograph was of the order 5-10 Å at the period when this work was carried out.

In the experiments on the projection method all the four lenses of the electron microscope were used to form a highly demagnified image of the 60 keV electron source quite close to the object. The holographic image was formed directly onto a photographic plate some 30 cm below the object. The image resolution in this method depends both on the plate resolution and, because of a strong chromatic effect, on the radius of the field of view. The electrical stability of the microscope and the estimated source size were such as to allow a predicted resolution of 5 Å for a 200 s exposure time, but this implied using a mere 0.3 cm field of view on the plate. The best images showed a resolution worse than 50 Å.

The resolution in the transmission method was expected to be constant over the field of view since there is no field limitation due to chromatic instabilities. The instrument was operated as a classical electron microscope apart from improved

electrical stability and a small aperture, adjustable down to 2 μm square, in front
of the gun. Initially a diffraction resolution of the order of 20-30 Å was obtained
in the holograms themselves. The principal limiting factors were the high voltage
stability, stray magnetic fields, mechanical vibration, specimen contamination
and drift. It was found possible to reduce a number of these perturbing factors to
allow a diffraction resolution of around 6 Å as measured from the resolvable Fresnel
fringes around test objects consisting of zinc oxide crystals or carbon black. The
light-optical reconstruction apparatus used an incoherent source (the laser had yet
to be invented) and incorporated means of introducing variable spherical aberration
by using two identical biconvex lenses separated by twice their focal length. The
spherical aberration of the illuminating wave varied with the distance between the
source and the lens system [6.28]. The full diffraction resolution was never realized in the reconstructed images. A rather full geometrical optics treatment was
made of the hologram formation particularly as concerns the source coherence,
image intensity, exposure time, and instrumental stabilities.

In 1956, HIBI attempted to repeat these experiments using a pointed filament as
an electron source [6.29].

TONOMURA et al. [6.30,31] carried out some experiments on Fraunhofer electron
holograms in 1968. Their test object was small gold particles of diameter $d \simeq 100$ Å
supported on a carbon film. The images were recorded at a sufficiently large defocusing distance z to satisfy the Fraunhofer condition $z \gg d^2/\lambda_e$, where λ_e is the
wavelength of the electron beam. In their experiment $z = 0.19$ cm, $\lambda_e = 3.7 \times 10^{-10}$ cm
so that $z \simeq 80\ d^2/\lambda_e$. The hologram was formed at a low magnification of around
3000 X and the reconstruction apparatus used a single-mode He-Ne laser source. The
gold particles were quite clearly reconstructed although with a poorer resolution
than the in-focus electron micrograph.

Some attempts to produce sideband holograms were made by TONOMURA [6.32] and
by MÖLLENSTEDT and WAHL [6.33]. The first of these authors used a thin single
crystal as a beam splitter and this corresponds to an amplitude division of the
wave front. The latter used an electron biprism [6.34] illuminated by a slit source.
The result was essentially a one-dimensional hologram produced by wave front division.

In the work of TOMITA et al. [6.35,36] a Möllenstedt-type biprism was placed
after the back focal plane of the electron microscope objective lens. A pointed
filament was used as an electron source.

The object was placed off the optic axis of the objective lens so that the
scattered wave could be supposed to pass on one side of the centered biprism filament and the reference wave on the other side. Magnesium oxide crystals were used
as test objects. Typical experimental parameters were defocus = 10^{-2} cm; magnification = 10^4; exposure time = 1 min. The reconstruction apparatus used a He-Ne .
laser source and a spatial frequency filter to eliminate the conjugate image and

the straight-through beam. No attempt was made to correct for spherical aberration. The resolution in the reconstructions was estimated as around 20 Å judged from the corner roundness of the crystal images. An estimate of the attainable resolution taking into account the source brightness, image magnification, film sensitivity, spatial coherence, and the film resolution was close to the experimental value. Reconstructions were made using ordinary and bleached (phase) holograms, from the real and the virtual images and from the over- and underfocused holograms. Very strong laser speckle is apparent in all the reconstructed images.

A typical experimental arrangement used in this and subsequent work on sideband holography is shown in Fig.6.10. The object is offset from the axis of the objective lens which produces an image in the plane I. The biprism consists of a fine central filament of the order of 1 μm in diameter isolated from and running between two grounded parallel conducting plates. When a potential is applied to the filament the field distribution between the filament and the plates is such that to a good approximation, the beam is uniformly deflected in opposing senses on either side of the filament. The biprism is placed near the image plane of the objective lens. The interference field between the inclined object wave and the oppositely inclined reference wave is formed in the region H. The following lenses serve only to magnify this image.

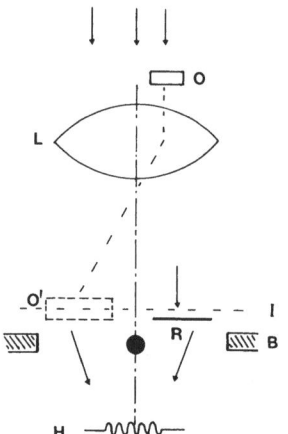

Fig.6.10. A typical experimental arrangment for Fresnel sideband holography in the electron microscopes. The object is set off-axis at O and is imaged in this plane I by the objective lens L. The biprism B deflects the object and reference wave R which superpose and interfere in the plane H

TONOMURA et al. [6.37] have developed a field emission electron gun fitted to a Hitachi HU-12A microscope operating at 100 keV. The two-beam interference region generated by the Möllenstedt-type biprism placed between the objective and the intermediate lens shows up to 3000 fringes. Off-axis holograms of magnesium oxide crystals show carrier fringe spacings which appear to be around 10 Å. The reconstructed image is quite good although no attempt seems to have been made to correct for spherical aberration.

SAXON [6.38] has given an extensive account of work on sideband Fresnel holography using a modified scanning transmission electron microscope fitted with a field emission electron source which has a small emitting tip radius and a brightness several orders of magnitude greater than conventional hairpin filaments. The gun itself was designed to form a real image of the source above the object which is offset from the optic axis; the biprism is placed below the object and produces an overlap of the object and reference waves which have passed on opposite sides of the biprism filament. The hologram plane, which is in this overlap region, is magnified by two lenses and focused onto a fluorescent screen deposited on a fiber optics plate which extracts the image directly, so avoiding the use of photographic plates within the microscope which would be incompatible with maintaining the high vacuum required to operate the field emission source. The image of the emitter tip is the effective illumination source of diameter around 30 Å but stray ac magnetic fields increase this by a factor of ten. The resolution is shown to be diffraction-limited by the hologram size in the reconstruction stage according to the formula

$$\delta = 0.6\lambda_e M z_o / \rho \quad ,$$

where δ is the resolution referred back to the object plane, z_o is the object space defocus, M is the overall magnification, and ρ is the hologram radius. The finite source size smears out the finer outer fringes in the Fresnel diffraction pattern of the object. This reduces the effective radius of the hologram and thereby limits the resolution.

Holograms are recorded with the beam focused a few cm above the biprism the potential on the plate of which is set to produce an interference field of several hundred fringes. The overall magnification is set at around 500 so that the fiber optics plate can resolve the fringes. An exposure time of about 1 min is used. The distance (z_o) from the object to the hologram plane is about 12 cm and in the reconstruction stage the real image is found 460 cm from the hologram.

The apparatus could also be used as a scanning transmission microscope. The magnesium oxide crystals used as test objects were imaged holographically and in the scanning mode. A comparison of the images gave an estimate of around 500 Å for the reconstructed image resolution which was diffraction limited at the reconstruction stage due to the large defocus used.

A second article by SAXON [6.39] deals with the effect of wavefront aberrations on the reconstructed image resolution following the theoretical method and using the terminology of MEIER [6.40-42]. The effect of aberrations can be qualitatively understood by treating the hologram as the exit pupil in the reconstruction process and by comparing the actual imaging wave front to the Gaussian reference sphere centered on a real image point. If the path difference between the real and the ideal wave fronts exceeds a wavelength the resolution at the image point is aberration rather than diffraction limited. In the experiments the hologram resolutions were limited by the coma of the electron lenses. A compensating coma component was introduced into the illuminating wave by using a tilted lens in the reconstruction stage. This compensation improved the resolution of the real image by a factor of three or four down to the diffraction-limited region. The twin image deteriorated because its coma was doubled.

The work described by MUNCH [6.43] extends that of SAXON. A conventional transmission electron microscope was fitted with a differentially pumped field emission gun. A Möllenstedt-type biprism was placed in the selected area aperture plane. The microscope (a Hitachi HU-12) was capable of resolving 3 Å in normal operation. The source image subtended an angle of 10^{-6} rad at the object. Sideband Fresnel holograms of magnesium oxide crystals were recorded at a defocus of about 25 μm. In the reconstruction stage the unwanted sideband and central component were eliminated by a spatial frequency filter. The estimated lateral resolution of the reconstructed images was about 50 Å.

In-line holograms were used to estimate the dependence of the Fresnel fringe resolution on specimen size and defocus. Magnesium oxide crystals, latex, and gold particles were used as objects. The results clearly showed that, in the investigated defocus range of from 10 to 250 μm, the resolution improved as the defocus decreased and as the object size increased. It was found that the conjugate images were very weak in reconstructions from sideband holograms of small objects since even the smallest defocus used satisfied the Fraunhofer condition, (6.9). Further experiments were therefore carried out using the in-line Fraunhofer technique. The test objects were gold particles 10-50 Å in diameter supported on a carbon film about 20 Å thick. The electron holograms were taken at 75 keV with z = 5 μm and $M = 10^5$. In the reconstruction stage 10 Å diameter particles could be detected in the real images which were in focus at 31 cm from the hologram. The conjugate images produced noticeable haloes only around the larger particles. If the holograms were formed at a larger defocus the contrast of the fringes from the smaller particles could not be detected and it follows that in order to preserve sufficient contrast for high-resolution information to be recorded the Fresnel number $N = d^2/\lambda z$ must be large compared to unity. This corresponds to working in the Fresnel or near-field defocus region.

WAHL [6.44,45] gives a good discussion of image plane sideband holography in which the object is imaged in the near-field region. Contrast transfer theory is used to describe the properties of the holograms and the effect of the illumination source size is included in the discussion. The experimental work used a modified Siemens Elmiskop 1 fitted with a pointed cathode, with an adjustable biprism holder and with an image intensifier. Care was taken to reduce stray ac fields. The illumination aperture used was about 10^{-5} rad and the working magnification was 10^4; the object defocus appears to be in the range 1 to 10 μm. Zinc oxide is used as a test object. Light-optical defocus series of the reconstructed image are shown and also the effect of speckle and its reduction by using multiple exposures with different aperture diameters in the reconstruction. A double exposure technique allows the phase and amplitude terms of the complex amplitude transmission of the object to be separated.

A recent article by LAU and POZZI [6.46] describes the use of an electrostatic biprism to form image plane holograms of regions of a nickel single crystal containing a magnetic domain wall. The reconstructions show interference fringes at the wall position as a function of defocus. Holographic interferometry shows that the magnetization directions are not uniform near the wall. No measurements are possible on the wall itself.

Some experimental work on in-line Fraunhofer holography appears in a paper by GALLION et al. [6.47]. A field emission gun developed by TROYON [6.48] was fitted to a differentially pumped Hitachi HU-11C. The test object was a dispersion of gold islands on a carbon substrate. TROYON [6.49] shows holograms taken at 50 keV with a magnification of 2×10^4, a defocus of 45 μm, and an exposure time of 15 s. The reconstructed images show a resolution of particles separated by less than 20 Å.

The possible applications of Fraunhofer holography in high-resolution electron microscopy are discussed by BONNET et al. [6.50]. With an instrumental defocus in the range 1 to 10 μm the reconstructed image quality is stated to be close to that of the focused micrograph. These authors suppose that there is no overlap between the twin images because, in the Fraunhofer limit, diffracted beams corresponding to the lowest harmonic of the sampled object are spatially separated. Our point of view has been that, in the limit of a pointlike object, the interference between the plane background wave and the spherical diffracted wave necessarily generates an in-line zone plate-type hologram. The twin image will be present in the reconstruction but being very diffuse, due to the Fraunhofer condition, will only slightly perburb the in-focus image.

VORONIN et al. [6.51] describe some experiments on in-line Fresnel and Fraunhofer holography using zinc oxide and carbon foil test objects. The effect of the twin image on the reconstructions can be clearly seen since the change in the frequency spectrum shown by the diffractograms suggests that the reconstruction was made from an amplitude hologram. This corresponds to an overall mixed transfer case with the associated aberration doubling [6.52].

The first experiments on single-sideband (SSB) holograms were carried out by THON [6.53] with the aim of demonstrating that this method produces a contrast transfer without zeros. The method has been carefully investigated by DOWNING and SEIGEL [6.54,55]. The phase aberration in the SSB images due to aperture charging could be determined by FRANK's [6.56] method using the fringes generated in the diffractogram of superposed images. The technique was developed with the aim of separating the amplitude and phase components of the images using the complementary

aperture method. The sum and difference images of uranyl acetate stained DNA strands
showed that the method could give an enhancement of the heavy atom stain component
compared to a standard bright-field image. The aberrations were corrected using a
distorted diffraction grating filter [6.57] in one half of the Fourier transform
plane of the light optical reconstruction apparatus. The other half-plane was blocked
out. The electron images were obtained in a Siemens 101 at a magnification of 10^5
using a 100 μm condenser aperture; the spherical aberration of the objective lens
was C_s = 1.35 mm. The SSB method requires a much lower total irradiation than the
dark-field method, although radiation damage may limit its ultimate usefulness in
biological applications.

The wavefront aberrations due to aperture charge may be reduced by heating the
half-plane aperture to 170°C [6.58]. This procedure prevents contamination and
provided the heating current is sufficiently stable ($\Delta i/i < 5 \times 10^{-6}$/min), the SSB
images show no evidence of induced astigmatism.

An intriguing example of a form of Fourier transform electron holography has
been reported by BARTELL [6.59-61]. The apparatus used was essentially a gas phase
electron diffraction unit and the scattering objects were the individual atoms of
a monatomic gas. The electron beam is supposed to be weakly scattered by the atomic
electron cloud but strongly scattered by the atomic nuclei (Rutherford scattering).
This latter component is supposed to act as the reference wave and the hologram is
the average of a large number of elementary "atomic holograms" recorded in the
Fraunhofer diffraction limit which ensures that all the elementary contributions
are centered one on the other. A rotating filter compensates for the falloff of
the Rutherford scattering with diffraction angle. According to the authors the resulting diffraction image can be interpreted as a Fourier transform hologram in
which the reference beam is centered on the object. As a result a light-optical
diffractometer can be used to reconstruct an image which they interpret in terms
of electron cloud projections. Neon and argon atoms were imaged in this way. A
diffraction-limited resolution of 0.08 Å is claimed. Apart from its limited ap-
plicability this scheme seems highly speculative. In particular it is not clear that
the condition $U_r \gg U_0$ is satisfied for all scattering angles so it would seem
necessary to demonstrate that the terms $U_r \otimes U_r$ and $U_0 \otimes U_0$ are not the dominating
terms in (6.13).

Optical modelling applied to electron microscopy and holography has formed the
subject of a recent thesis by ROGERS [6.62,63]. The design of a light-optical lens
system to simulate the electron microscope objective lens is discussed. The lens,
built to a design by Wynne, was incorporated in experiments on sideband holograms
using a Fresnel biprism as a beam splitter. Off-axis reconstructions were made of
a radial grating test object. Attempts at digital reconstructions are described.
Some simulations were made of the effect of the source size on contrast transfer.

6.4 Contrast Transfer and Holography

We have shown in Sect.6.2 how the holographic recording process is based on the
interference between a strong reference wave and the wave scattered by the object.
Similar conditions apply to the application of linear contrast transfer theory in
bright-field microscopy [6.64-67]. Light-optical experiments concerning the contrast transfer interpretation of in-line Fresnel holograms establish that a bright-
field phase or amplitude contrast image obtained under isoplanatic imaging conditions using coherent illumination is identical with an in-line Fresnel hologram
[6.14,52].

In this section we shall develop the bright-field example to show how the contrast transfer and holographic interpretations are complementary descriptions of
the same physical process. We shall distinguish between image treatment and holographic image correction. We shall then show how contrast transfer theory can be
applied to the other holographic methods and how it aids the interpretation of the
reconstruction process. Finally we shall show how the effect of the illumination
source size and chromatic fluctuations can be included in the theory.

Fourier transform holograms [6.24] will not be discussed here because the contrast transfer approach does not add any useful information to the treatment given
in Sect.6.2.4.

6.4.1 In-Line Fresnel Hologram

The in-line mode has been studied meticulously by HANSZEN and ADE [6.14-17,68-73].

In high-resolution electron microscopy essentially the whole range of spatial
frequencies in the scattering distribution associated with the object passes through
the optical system into the image. The scattering of the transmitted electrons can
be described by a complex transmission function $U_0(\underline{r}_0)$ which depends on the position
$\underline{r}_0 = (x_0, y_0)$ within the object plane. This transmission function must usually
account for a small distortion of the outgoing compared to the incident wavefront.
This is the case of a weak phase object, which has the transmission function

$$U_0(\underline{r}_0) = \exp[i\varphi_0(\underline{r}_0)] \simeq 1 + i\varphi_0(\underline{r}_0) \quad \text{for} \quad \varphi_0 \ll 1 \quad .$$

It has been shown in Chap.1 that the isoplanatic imaging condition ensures that
in addition to the uniform background the image amplitude contains the object dependent term

$$U_i(\underline{r}_i) = i \int d\underline{r}'_i \cdot \varphi_0(\underline{r}'_i) G(\underline{r}_i - \underline{r}'_i) \quad ,$$

where $d\underline{r}_i = dx_i \, dy_i$ and $G(\underline{r}'_i)$ is the point spread function associated with the
electron microscope imaging conditions and contains terms arising from the wave
aberrations of the objective lens, notably the spherical aberration and defocus.

The image intensity distribution then takes the form

$$I = 1 + U_i + U_i^* + U_i U_i^* \quad .$$ (6.15)

The last term of (6.15) is the dark-field or speckle term which is ignored when the weak object approximation applies. The terms U_i and U_i^* will be recognized as the twin image components common to all holographic processes.

Contrast transfer theory describes the spatial frequency content of images. This information is contained in the Fourier transform of (6.15) provided the photographic recording of the intensity is a linear process, (6.2). Fourier transformation of the amplitude transmission of the photographic plate will yield, to within multiplicative constants,

$$\tilde{I}(\underline{p}) = \delta(\underline{p}) + \tilde{U}_i^*(\underline{p}) + \tilde{U}_i(\underline{p}) + \int \tilde{U}_i(\underline{p}')\tilde{U}_i^*(\underline{p} - \underline{p}')d\underline{p}' \quad ,$$ (6.16)

where the tilde indicates the Fourier transform, $\underline{p}$ is the spatial frequency vector $\underline{p} = (p,q)$, and the asterisk indicates the complex conjugate.

In the case of a weak phase object axially illuminated by a parallel electron beam (6.16) reduces to

$$\tilde{I}(p) = \delta(p) + i[\tilde{\varphi}_0(p)T(p) - \tilde{\varphi}_0^*(-p)T^*(p)] \quad ,$$ (6.17)

where T, the Fourier transform of the point spread function, is called the complex transfer function. It depends on the wave aberration phase function γ through the relation $T = \exp(-i\gamma)$, where we leave out a multiplicative term describing the aperture size. The wave aberration phase is given essentially by

$$\gamma = \frac{2\pi}{\lambda}\left(-\frac{\Delta\lambda^2 p^2}{2} + \frac{C_s\lambda^4 p^4}{4}\right) = \frac{2\pi}{\lambda} W(p) \quad ,$$

where Δ is the defocus and C_s is the third-order spherical aberration constant of the objective lens. The wave aberration W is a measure of the distortion of the real imaging wavefront at the exit pupil of the objective lens compared to the ideal sphere centered on the Gaussian image point.

Since φ_0 is real the object-dependent part of (6.17) can be written

$$i\tilde{\varphi}_0(\underline{p})[T(\underline{p}) - T^*(\underline{p})] = 2\tilde{\varphi}_0(\underline{p}) \sin\gamma(\underline{p}) \quad .$$ (6.18)

The contributions of the components T and T^* due to the twin images superpose. Their mutual interference produces the familiar contrast transfer function $\mathcal{K}(p)$ = $\sin\gamma$ which is observed in optical diffractograms of high-resolution micrographs of carbon foils [6.74,75].

Clearly ideal imaging corresponds to having a contrast transfer function of unity. This is possible only in the absence of wave aberrations and will produce a diffraction-limited image due to the aperture-dependent function which we have not included in T. In the case of the weak phase object, setting K = 1 implies that the image is equivalent to a Zernike phase contrast image produced by a $\pi/2$ phase shift of the undiffracted beam. It is perhaps timely to draw a distinction between image treatment and holography. We can do this by reference to the definition of the contrast transfer function,

$$\mathcal{K} = T - T^* = \sin\gamma \quad . \tag{6.19}$$

The aim of both image treatment and holography is to achieve the ideal contrast transfer, $\mathcal{K}_I = 1$. Holography attempts to do this by finding some means of subtracting or spatially separating the two image contributions, reduced to T and T^* in (6.19). Image treatment concerns itself with the means of finding a function H such that $\mathcal{K}H = \mathcal{K}_I = 1$. The problems involved in achieving this are dealt with in Chap.1. It is immediately apparent that while almost all the computer and light-optical image treatment procedures of electron micrographs have been carried out on in-line Fresnel holograms this particular type of hologram is ill adapted for holographic reconstruction in that the twin image components can never be separated. The best that can be achieved by in-line holography is a large spatial dispersion of one image component in the reconstruction stage so that it represents a widely spread low-contrast background to the other in-focus image. This can be achieved by in-line Fraunhofer holography which has allowed resolutions down to around 10 Å to be achieved in the reconstructed image.

6.4.2 Fresnel Sideband Hologram

We have already described the physical basis of Fresnel sideband holography and attempts to apply this method to electron microscopy. This technique can also be described in terms of contrast transfer theory [6.18-21,44,76].

If we suppose that the biprism only produces opposite tilts of the reference and object waves which have already passed through the objective lens (Fig.6.10), the hologram intensity $I(r_i)$ is expressed by

$$I = U_r U_r^* + U_r U_i^* + U_r^* U_i + U_i U_i^* \quad , \tag{6.20}$$

where the reference wave is $U_r = \exp(-i\pi k_o \cdot r_i)$ and the object wave

$U_i = \exp(i\pi k_o \cdot r_i) \int dk \, \tilde{U}_o(k) T(k) \exp(i2\pi k \cdot r_i)$, where r_i are the image plane coordinates; k_o is related to the beam tilt α through $|k_o| = \alpha/\lambda$; similarly $|k| = \theta/\lambda$ where θ is the scattering angle. $\tilde{U}_o$ is the Fourier transform of the object transmission function U_o.

The Fourier transform $\tilde{I}$ of the image intensity, (6.20), will describe the spatial frequency content of the image

$$\tilde{I} = \tilde{U}_r \otimes \tilde{U}_r^* + \tilde{U}_r \otimes \tilde{U}_i^* + \tilde{U}_r^* \otimes \tilde{U}_i + \tilde{U}_i \otimes \tilde{U}_i^* \quad , \tag{6.21a}$$

where $\otimes$ symbolizes the convolution integral. The first term on the RHS of (6.21a) is an autocorrelation which arises from the intensity $U_r U_r^*$ of the reference wave; similarly the last term arises from the object-dependent intensity. The second and third terms are the spectral components of the twin images. For a tilted plane reference wave (6.21a) reduces to

$$\tilde{I}(\underline{p}) = \delta(\underline{p}) + \tilde{U}_o^*(\underline{p} + \underline{k}_o)T(\underline{p} + \underline{k}_o) + \tilde{U}_o(\underline{p} - \underline{k}_o)T(\underline{p} - \underline{k}_o)$$

$$+ \ \tilde{U}_o^*(\underline{p})T^*(\underline{p})] \otimes [\tilde{U}_o(\underline{p}) \cdot T(\underline{p})] \quad . \tag{6.21b}$$

In (6.21b) the two autocorrelation terms are centered on the origin (the first term is reduced to a delta function) while the twin image terms are offset, the one being centered on $p = -k_o$, the other on $p = k_o$.

$\tilde{I}(\underline{p})$ is shown schematically in Fig.6.11. The condition for good reconstruction is clearly that the components $\tilde{U}_i$ and $\tilde{U}_i^*$ of the twin images be completely separated from each other and preferably from the weaker central term.

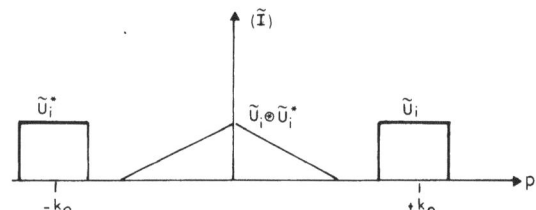

Fig.6.11. A representation of the spatial frequency content of a Fresnel sideband hologram, (6.21b)

The reconstruction process can be considered to consist of three stages:

1) forming the Fourier transform I of the image and selecting one of the twin image terms U_i (or U_i^*) by means of an aperture,

2) multiplying this term by a complex filter function which corrects the wave aberration, e.g., the twin term containing $T(\underline{p} + \underline{k}_o)$ must be multiplied by $T^*(\underline{p} + \underline{k}_o)$,

3) inverse Fourier transformation to produce the corrected image.

One disadvantage of this method is that the separation of the twin image components requires that the carrier fringes of the hologram be at least twice as fine as the finest detail required in the image. If the microscope is capable of resolving such fine detail, image treatment of an in-line image might still be preferable to the reconstructed sideband image. In addition HANSZEN [6.16] remarks

that since the object is offset from the optical axis the off-axis aberrations of
the objective lens will have a severe effect on the imaging. These aberrations are
at present rather poorly known and it is questionable as to whether full correction
would be possible. Furthermore the biprism aberrations would almost certainly have
to be accounted for in high-resolution applications. The best resolutions which
have been achieved experimentally with this method are at present of the order of
20-30 Å.

6.4.3 Single-Sideband Hologram

This method originally proposed by LOHMANN [6.77] and developed by BRYNGDAHL and
LOHMANN [6.78] was historically the first proposal for sideband holography.
It has generated a good deal of interest in view of its possibilities in electron
microscopy [6.16,79-84].

The half-plane aperture, which allows the central beam plus half the object
scattering amplitude to contribute to the hologram, produces the following twin
terms in the intensity distribution of the hologram:

$$\tilde{U}_i(\underline{r}_i) = \int_0^\infty d\underline{k}\ \tilde{U}_0(\underline{k})T(\underline{k}) \cdot \exp(i2\pi\underline{k} \cdot \underline{r}_i)$$

$$U_i^*(\underline{r}_i) = \int_{-\infty}^0 d\underline{k}\ \tilde{U}_0(\underline{k})T^*(\underline{k}) \cdot \exp(i2\pi\underline{k} \cdot \underline{r}_i)$$

provided the object transmission function $U_0(\underline{r})$ is real. The spatial frequency
content of the hologram which contains these twin terms in given by

$$\tilde{I}(\underline{p}) = \delta(\underline{p}) + \tilde{U}_0(\underline{p})T(\underline{p})b_1(\underline{p}) + \tilde{U}_0(\underline{p}) \cdot T^*(\underline{p}) \cdot b_2(\underline{p}) \qquad (6.22)$$

where

$$b_1(\underline{p}) = 0 \quad \text{and} \quad b_2(\underline{p}) = 1 \quad \text{for} \quad \underline{p} > 0$$

$$b_1(\underline{p}) = 1 \quad \text{and} \quad b_2(\underline{p}) = 0 \quad \text{for} \quad \underline{p} < 0 \quad .$$

Consequently the reconstruction process consists of:

1) forming the Fourier transform of the hologram, (6.22),

2) removing one side of the spatial frequency spectrum with a half plane
aperture and multiplying the other side by the complex conjugate of its transfer
function,

3) inverse Fourier transformation.

If $U_0(r_0)$ has both a real and an imaginary component, two images taken with complementary aperture positions are required in order to separate them. The separation
is possible because the diffraction amplitude of the real component remains sym-

metrical for the two aperture positions while the asymmetry of the imaginary component reverses. If the positive spatial frequency region is conserved during reconstruction of both images the spatial frequency content of the final two images is

$$\tilde{I}_1 = \delta(\underline{p}) + T(\underline{p})[i\tilde{\psi}_1(\underline{p}) + \tilde{\psi}_2(\underline{p})]$$

$$\tilde{I}_2 = \delta(\underline{p}) + T^*(\underline{p})[-i\tilde{\psi}_1(\underline{p}) + \tilde{\psi}_2(\underline{p})] \quad ,$$

(6.23)

where the original object transmittance is $U_0 = 1 + i\psi_1 + \psi_2$. It is interesting to note that addition of the final two images gives

$$\tilde{I}_1 + \tilde{I}_2 = \delta(\underline{p}) + 2\psi_1 \sin\gamma + 2\psi_2 \cos\gamma \quad .$$

(6.24)

The second and third terms on the RHS of (6.24) are simply the phase and amplitude contrast transfer functions for conventional bright-field images, so the double image has recreated an in-line Fresnel hologram. Correction of the aberration terms yields the spectra $\tilde{I}_1 T^*$ and $\tilde{I}_2 T$. Reference to the pair of equations (6.23) shows that the sum of the two images then yields ψ_2 while subtraction together with a $\pi/2$ phase shift of the background component gives an image contrast ψ_1 which is equivalent to a phase contrast image.

A major problem in the experimental implementation of this method is the additional aberrations introduced by electrostatic charging of the aperture due to the buildup of an insulating contamination layer. DOWNING and SIEGEL [6.54] have shown how these aberrations can be determined and corrected, and the work of SEIBER [6.58] indicates that contamination can be avoided using a heated aperture.

Despite the initial promise of the results presented by DOWNING and SIEGEL [6.54,55] on the discrimination of heavy and light atomic components of an object it is not clear at present what this method can ultimately contribute to high-resolution imaging.

6.4.4 The Effect of Partial Coherence on Resolution

We shall now give a brief account of the combined effect of 1) a finite sized illumination source (spatial partial coherence) and 2) a finite energy spread of the electron beam (temporal coherence) which can be grouped with instabilities arising from the electrical power supplies into a chromatic defocus spread term.

1) If the object is illuminated by electrons emitted randomly from all points of the source surface the complete image is obtained by summing the elementary image intensities due to each source point. In terms of the contrast transfer formulation this involves integrating (6.17) over the range of spatial frequencies equivalent to the illuminating angles subtended at the object by the source ac-

cording to a weighting factor which accounts for the angular intensity distribution associated with the source.

2) Electron lenses have large chromatic aberrations and as a result defocus fluctuations arise from electrical instabilities in the high-voltage supply and in the objective lens excitation current as well as from the thermal energy spread of the electron beam and from Boersch effect broadening [6.85]. All these terms can be grouped into the chromatic defocus spread $\Delta_c = C_c \Delta W/W$ where C_c represents the chromatic aberration constant of the objective lens and $\Delta W/W$ describes the combined relative instabilities. In this case the observed image will be the intensity sum of the image due to each elementary defocus component. The contrast transfer content of the image is obtained by integrating (6.17) over an appropriate defocus distribution.

The reader is referred to the articles by HANSZEN and TREPTE [6.86a,b], FRANK [6.87], and WADE and FRANK [6.88] for details of the calculations involved in integrating (6.17) over both the source intensity and the defocus spread distributions. An article by FERWERDA [6.89] deals with the effective source concept in electron microscopy.

The essential result for the present purpose is that for axial illumination (in-line holography) it is a very good approximation to represent the modified contrast transfer $\mathcal{K}_m$ as being related to the coherent contrast transfer $(\mathcal{K})$ by the envelope function multiplication

$$\mathcal{K}_m = \mathcal{K} \cdot E_1 \cdot E_2 \; , \tag{6.25}$$

where E_1 is the envelope describing the effect of the source size and E_2 is the envelope arising from the chromatic defocus fluctuations.

In the case of tilted illumination the resulting transfer expression is more complicated. Reference should be made to the articles by McFARLANE and COCHRAN [6.90], McFARLANE [6.91], HOPPE et al. [6.92], and WADE and JENKINS [6.93].

The terms E_1 and E_2 of (6.25) are given explicitly by

$$E_1(\underline{p}) = \exp - [\pi^2 q_0^2 |\underline{\nabla} W(\underline{p})|^2]$$

$$= \exp - [\pi q_0 \underline{p}(-\Delta\lambda + C_s \lambda^3 p^2)]^2 \tag{6.26a}$$

$$E_2(\underline{p}) = \exp - (\pi\lambda p^2 \Delta_0/2)^2 \; . \tag{6.26b}$$

In these expressions q_0 and Δ_0 are the e^{-1} half widths of the source intensity and the defocus distributions, respectively, which are supposed to have a Gaussian form; $\underline{\nabla} W(\underline{p})$ stands for $[\partial W(\underline{p})/\partial p, \partial W(\underline{p})/\partial q]$.

   Essentially the same envelope functions limit the complex transfer terms T in
the sideband holographic methods.

   In the coherent illumination limit, $q_0 = 0$, the source-dependent envelope $E_1 = 1$.
Furthermore whatever the value of the source size q_0, $E_1 = 1$ whenever $\underline{\nabla}W = 0$. This
means that the coherent and partially coherent transfer functions are identical in
spatial frequency regions where the wave aberration $W(\underline{p}) = \text{constant}$. These regions,
found at the spatial frequencies $\underline{p} = 0$ and $|\underline{p}| = (\Delta/C_s\lambda^2)^{\frac{1}{2}}$, can correspond to the
standard optimum transfer zones which can be swept out to higher spatial frequencies
by increasing the instrumental defocus Δ. For a typical angular source size of
around 10^{-3} rad and in the defocus range up to a few thousand Ångström units the
envelope E_1 typically has the form shown in Fig.6.12 which shows that relatively
little attenuation of the curve $\mathcal{K}$ will occur in the lower spatial frequency
region up to the optimum transfer zone after which a sharp cutoff occurs.

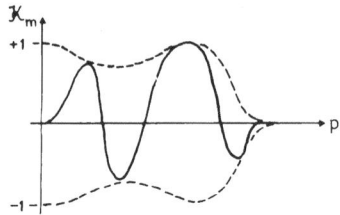

Fig.6.12. The effect of the envelope term E_1
(dashed curve) on the overall transfer function
$\mathcal{K}_m$, (6.25) (with $E_2 = 1$); the figure is not to
scale

   When the defocus term becomes so large as to dominate the spherical aberration
term in $\underline{\nabla}W$ over the entire spatial frequency range of interest the term E_1 reduces
to

$$E_1'(\underline{p}) = \exp - (\pi q_0 \lambda \underline{p} z)^2 \quad .$$

Large values of the argument of the exponential function give small values of E_1'
and consequently a strong attentuation of the contrast transfer. We shall suppose
that satisfaction of the condition

$$\pi q_0 \lambda p z < 1 \quad p = |\underline{p}|$$

ensures that E_1' is a low attenuating factor in the frequency range $p < 1/\pi q_0 \lambda z$
and that reversing the inequality corresponds to complete attenuation of the corresponding spatial frequency range. If d corresponds to a resolvable distance in
the object we find that the inequality implies

$$d > \pi \alpha_0 z \quad ,$$

where $d = 1/p$ and the angular source size $\alpha_0 = q_0\lambda$. The resolvable distances in the object are seen to depend directly on the angular source size and on the defocus. It is interesting to remark that this condition can also be obtained from simple geometric arguments and that it corresponds to the resolution condition given by HAINE and MULVEY [6.10].

The resolution will not be limited by the envelope E_1' when the illumination is completely coherent, $\alpha_0 = 0$, or when the image is perfectly focused, $z = 0$. It was on this basis that image plane sideband holography appeared superficially as being particularly interesting in electron microscopy [6.18-21,44,45,94,95]. However, the limit $E_1 = 1$ for all f can never hold in the electron microscope which has a finite source size and where spherical aberration must be taken into account. The same envelope function limitations apply for the in-line and sideband methods so from this point of view a sideband hologram has no inherent superiority over an in-line hologram.

For a given chromatic defocus spread the envelope E_2 is invariant. Its width is determined by the overall defocus spread which is composed of contributions from:

a) The thermal energy spread which follows a Maxwellian distribution with a maximum at around 0.25 eV and a halfwidth of the order 0.6 eV for a tungsten filament heated to 2900 K.

b) The Boersch effect spread [6.96] which originates in the Coulomb interaction between the electrons in high current density regions of an electron beam [6.97]. The energy distribution is supposed to have a Gaussian distribution whose standard deviation will depend on the total beam current and the beam forming geometry and may readily attain values of 1 eV and upwards [6.98].

c) The high-voltage supplies of modern electron microscopes have short-term relative stabilities of the order $\Delta E/E \simeq 2 \times 10^{-6}$ and the object lens current supplies are stabilized to within a relative fluctuation $\Delta I/I \simeq 1\text{-}2 \times 10^{-6}$.

Experimental measurements of the energy spread of the electron beam show the halfwidth to lie in the range 1.5-2.5 eV for standard operating conditions [6.86a, 98,99]. Δ_0 in (6.26b) will take on a value in the range 2×10^{-6} cm for $\Delta W/W = 10^{-5}$ and $C_c = 2$ mm. Using the same condition for low attenuation as for the source envelope implies that

$$d > \left(\frac{\pi\lambda\Delta_0}{2}\right)^{1/2} \simeq 3.5 \times 10^{-8} \text{ cm}$$

for $\lambda = 3.7 \times 10^{-10}$ cm, giving the chromatic fluctuation-dependent resolution of 3.5 Å. This will only be significantly improved either by a strong reduction of the chromatic aberration of the objective lens or by reducing the energy spread of the beam. There is some hope that this may be possible with field emission sources since energy spreads of around 0.2 eV have been reported for such sources.

6.5 Additional Reading

A number of proposals have been made in favor of other geometries for forming
electron holograms. Since no experimental work has been reported on these they have
not been described here but the interested reader is referred to:
- an article on crysto-holography [6.100] in which a crystal is imaged using one of
  its own diffraction spots as a reference;
- a proposal by POZZI [6.101] to combine a single-crystal specimen support and an
  electron biprism for off-axis holography;
- the interpretation of scanning transmission electron microscope images in terms
  of Gabor projection holograms [6.102], and the remarks of HANSZEN and ADE [6.103,
  104] concerning the spatial variations of the scanned images due to spherical
  aberration;
- GREENAWAY and HUISER [6.105] according to whom off-axis holography can separate
  the elastic and inelastic images.
- The holographic process placed in a general mathematical framework in recent work
  by LANNES [6.106,107]. This allows an analysis of the effect of errors and comparison with related reconstruction techniques.

6.6 Conclusions

The survey of experimental electron holography in Sect.6.3 shows that reconstructed
images from off-line and Fraunhofer in-line holograms have never approached the
quality of direct bright-field micrographs which are themselves Fresnel in-line
holograms. This tends to show that the holographic methods which might ultimatively
prove the most useful must be simple to realize experimentally. The only two methods
which qualify on this basis at present are the in-line and the single-sideband
techniques.

   The effect of the illumination source size and the chromatic defocus spread is
to limit the resolution. No particular holographic scheme is exempt from this limit.
Cold field emission sources may help to reduce the chromatic defocus spread provided
that Boersch effect energy broadening is not particularly significant in the cross-
over regions of the beam from these high brightness sources. A considerable improvement in the practical overall relative stability $\Delta W/W$ is necessary to lower the resolution limit which depends on the square root of the chromatic defocus spread. In
the low defocus region and for source sizes below 5×10^{-4} rad partial spatial coherence should not be a limiting factor. At the large defocus required for Fraunhofer holography it would be essential to verify that the required coherence conditions are satisfied. Since the Fraunhofer method lacks contrast when the particle

size is small [6.43] it may well turn out that "image treatment" of the Fresnel in-line image (hologram) will provide better final images than can be obtained by image reconstruction from other types of hologram.

References

6.1   H. Busch: Ann. Phys. *81*, 974-993 (1926)
6.2   L. de Broglie: Philos. Mag. *47*, 446 (1924)
6.3   D. Gabor: Canberra (1974), Vol. 1, pp. 6-12
6.4   O. Scherzer: Z. Phys. *101*, 593-603 (1936)
6.5   J. Hillier, E.G. Ramberg: J. Appl. Phys. *18*, 48-71 (1947)
6.6   D. Gabor: Nature *161*, 777-778 (1948)
6.7   D. Gabor: Proc. R. Soc. London *A197*, 454-487 (1949)
6.8   D. Gabor: Proc. Phys. Soc. *B64*, 449-469 (1951)
6.9   M.E. Haine, J. Dyson: Nature *166*, 315-316 (1950)
6.10 M.E. Haine, T. Mulvey: J. Opt. Soc. Am. *42*, 763-773 (1952)
6.11 J.B. Develis, G.O. Reynolds: *Theory and Applications of Holography* (Addison-Wesley, Reading, Mass. 1967)
6.12 J.W. Goodman: *Introduction to Fourier Optics*, (McGraw-Hill, San Francisco 1968)
6.13 M. Françon: *Holographie* (Masson, Paris 1969)
6.14 K.J. Hanszen: Optik *32*, 74-90 (1970)
6.15 K.J. Hanszen: Optik *35*, 431-444 (1972)
6.16 K.J. Hanszen: "Contrast Transfer and Image Processing" in *Image Processing and Computer-Aided Design in Electron Optics*, ed. by P.W. Hawkes (Academic Press, London and New York 1973) pp. 16-53
6.17 K.J. Hanszen: Optik *39*, 520-542 (1974)
6.18 I. Weingärtner, W. Mirandé, E. Menzel: Optik *29*, 87-104 (1969)
6.19 I. Weingärtner, W. Mirandé, E. Menzel: Optik *29*, 537-548 (1969)
6.20 I. Weingärtner, W. Mirandé, E. Menzel: Optik *30*, 318-322 (1969)
6.21 I. Weingärtner, W. Mirandé, E. Menzel: Optik *31*, 335-353 (1970)
6.22 E.N. Leith, J. Upatnieks: J. Opt. Soc. Am. *52*, 1123-1130 (1962)
6.23 E.N. Leith, J. Upatnieks: J. Opt. Soc. Am. *53*, 1377-1381 (1963)
6.24 E.N. Leith, J. Upatnieks: J. Opt. Soc. Am. *54*, 1295-1301 (1964)
6.25 P.W. Hawkes: "Partial Coherence in Electron Optics" in *Advances in Optical and Electron Microscopy*, *7*, 101-184 (1978)
6.26 G.L. Rogers: Nature *166*, 237 (1950)
6.27 G.L. Rogers: Proc. R. Soc. Edinburgh *A63*, 193-221 (1950-51)
6.28 J. Dyson: Paris (1950) pp.126-128
6.29 T. Hibi: J. Electron Micros. *4*, 10-15 (1956)
6.30 A. Tonomura, A. Fukuhara, H. Watanabe, T. Komoda: Jpn. J. Appl. Phys. *7*, 295 (1968)
6.31 A. Tonomura, A. Fukuhara, H. Watanabe, T. Komoda: Rome (1968), Vol. 1, 277-278
6.32 A. Tonomura: J. Electron Micros. *18*, 77 (1969)
6.33 G. Möllenstedt, H. Wahl: Naturwiss. *55*, 340-341 (1968)
6.34 G. Möllenstedt, H. Düker: Z. Phys. *145*, 377-397 (1956)
6.35 H. Tomita, T. Matsuda, T. Komoda: Jpn. J. Appl. Phys. *9*, 719 (1970)
6.36 H. Tomita, T. Matsuda, T. Komoda: Jpn. J. Appl. Phys. *11*, 143-149 (1972)
6.37 A. Tonomura, T. Matsuda, T. Komoda: Toronto (1978), Vol. 1, pp.224-225
6.38 G. Saxon: Optik *35*, 195-210 (1972)
6.39 G. Saxon: Optik *35*, 359-375 (1972)
6.40 R.W. Meier: J. Opt. Soc. Am. *55*, 987-992 (1965)
6.41 R.W. Meier: J. Opt. Soc. Am. *56*, 219-223 (1966)
6.42 R.W. Meier: J. Opt. Soc. Am. *57*, 895-900 (1967)

6.43   J. Munch: Optik *43*, 79-99 (1975)
6.44   H. Wahl.: Proc. Kontron Seminar, Munich, pp. 86-113 (1973)
6.45   H. Wahl.: Optik *39*, 585 (1974)
6.46   B. Lau, G. Pozzi: Optik *51*, 287 (1978)
6.47   P. Gallion, M. Troyon, A. Beorchia: Opt. Acta *22*, 731-743 (1975)
6.48   M. Troyon, P. Gallion, A. Laberrigue: Jerusalem (1976), Vol. 1, pp.344-345
6.49   M. Troyon: Thèse, University of Reims (1977)
6.50   N. Bonnet, M. Troyon, P. Gallion: Toronto (1978), Vol. 1, pp.222-223
6.51   Yu.M. Voronin, I.P. Demenchenok, A.V. Mokhnatkin, R.Yu. Khaitlina: Bull. Acad.
       Sci. USSR, Phys. Ser. *36*, 1154-1156 (1972)
6.52   R.H. Wade: Optik *40*, 201-216 (1974)
6.53   F. Thon: Rome (1968), Vol. 1, 127-128
6.54   K.H. Downing, B.M. Siegel: Optik *38*, 21-28 (1973)
6.55   K.H. Downing, B.M. Siegel: Optik *42*, 155-175 (1975)
6.56   J. Frank: Optik *35*, 608-612 (1972)
6.57   A.W. Lohmann, D.P. Paris: Appl. Opt. *7*, 651-655 (1968)
6.58   P. Sieber: Canberra (1974), Vol. 1, 274-275
6.59   L.S. Bartell, C.L. Ritz: Science *185*, 1163-1165 (1974)
6.60   L.S. Bartell: Optik *43*, 373-390 (1975)
6.61   L.S. Bartell: Optik *43*, 403-418 (1975)
6.62   J. Rogers: Ph.D. Thesis, University of London (1978)
6.63   J. Rogers: ICO 11, Madrid (1978), p.235
6.64   O. Scherzer: J. Appl. Phys. *20*, 20-29 (1949)
6.65   K.J. Hanszen, B. Morgenstern, K.J. Rosenbruch: Z. Angew. Phys. *16*, 477-486
       (1964)
6.66   K.J. Hanszen, B. Morgenstern: Z. Angew. Phys. *19*, 215-227 (1965)
6.67   F.A. Lenz: "Transfer of Image Information in the Electron Microscope" in
       Electron Microscopy in Material Science, ed. by U. Valdrè (Academic Press,
       London and New York 1971) pp. 541-569
6.68   K.J. Hanszen: Grenoble (1970), Vol. 1, 21-22
6.69   K.J. Hanszen: "The Optical Transfer Theory of the Elctron Microscope", in
       Advances in Optical and Electron Microscopy, 4, 1-84 (1971)
6.70   G. Ade: Dissertation, PTB-Bericht, APh-3, Braunschweig (1973)
6.71   K.J. Hanszen, G. Ade: PTB-Bericht APh-5, Braunschweig (1974)
6.72   K.J. Hanszen, R. Lauer, G. Ade: Optik *36*, 156-159 (1972)
6.73   K.J. Hanszen, G. Ade, R. Lauer: Optik *35*, 567-590 (1972)
6.74   F. Thon: Z. Naturforsch. *21a*, 476-478 (1966)
6.75   F. Thon: "Phase Contrast Electron Microscopy", in *Electron Microscopy in Ma-
       terial Science*, ed. by U. Valdrè (Academic Press, London and New York 1971)
       pp. 571-625
6.76   R.H. Wade: In *Developments in Electron Microscopy and Analysis*, ed. by J.A.
       Venables (Academic Press, London and New York 1976) pp. 197-200
6.77   A.W. Lohmann: Opt. Acta *3*, 97-99 (1956)
6.78   O. Bryngdahl, A.W. Lohmann: J. Opt. Soc. Am. *58*, 620-624 (1968)
6.79   W. Hoppe, R. Langer, F. Thon: Optik *30*, 538-545 (1970)
6.80   W. Hoppe: Z. Naturforsch. *26a*, 1155-1168 (1971)
6.81   D.L. Misell, R.E. Burge, A.H. Greenaway: Nature *247*, 401-402 (1974)
6.82   D.L. Misell, A.H. Greenaway: J. Phys. D *7*, 832-855 (1974)
6.83   D.L. Misell: J. Phys. D *7*, L69-71 (1974)
6.84   W.O. Saxton: J. Phys. D *7*, L63-64 (1974)
6.85   H. Boersch: Z. Phys. *139*, 115-146 (1954)
6.86   K.J. Hanszen, L. Trepte: (a) Optik *32*, 519-538 (1971); (b) Optik *33*, 182-198
       (1971)
6.87   J. Frank: Optik *38*, 519-536 (1973)
6.88   R.H. Wade, J. Frank: Optik *49*, 87-91 (1977)
6.89   H.A. Ferwerda: Optik *45*, 411-426 (1976)
6.90   S.C. McFarlane, W. Cochran: J. Phys. C *8*, 1311-1321 (1975)
6.91   S.C. McFarlane: J. Phys. C *8*, 2819-2836 (1975)
6.92   W. Hoppe, D. Köstler, D. Typke, N. Hunsmann: Optik *42*, 43-56 (1975)
6.93   R.H. Wade, W.K. Jenkins: Optik *50*, 1-17 (1978)

6.94   L. Rosen: Appl. Phys. Lett. *9*, 337-339 (1966)
6.95   G.B. Brandt: Appl. Opt. *8*, 1421-1429 (1969)
6.96   H. Boersch: Phys. Bl. *23*, 393-404 (1967)
6.97   B. Zimmermann: "Broadened Energy Distributions in Electron Beams", in *Adv. Electronics and Electron Physics*. Vol. 29, ed. by L. Marton (Academic Press, New York and London 1970) pp. 257-312
6.98   R.W. Ditchfield, M.J. Whelan: Optik *48*, 163-172 (1977)
6.99   J. Frank: Optik *44*, 379-391 (1976)
6.100 R.H.T. Bates, R.M. Lewitt: Optik *44*, 1-16 (1975)
6.101 G. Pozzi: Optik *47*, 105-107 (1977)
6.102 L.H. Veneklasen: Optik *44*, 447-468 (1975)
6.103 K.J. Hanszen, G. Ade: Jerusalem (1976), Vol. 1, 446-447
6.104 K.J. Hanszen, G. Ade: PTB-Bericht APh-11, Braunschweig (1976)
6.105 A.H. Greenaway, A.M.J. Huiser: Optik *45*, 295-300 (1976)
6.106 A. Lannes: Toronto (1978), Vol. 1, p. 228
6.107 A. Lannes: Opt. Commun. *20*, 356 (1977)

7. Analog Computer Processing
of Scanning Transmission Electron Microscope Images

M. Isaacson, M. Utlaut and D. Kopf

With 14 Figures

By analog processing, we do not imply optical processing of micrographs, but rather electronic techniques for functional manipulation. These electronic techniques are ideally suited for the STEM since the information from such an instrument is available as a time sequential electrical signal. Moreover, this type of processing cannot easily be performed with a fixed beam instrument (i.e., a conventional transmission electron microscope) [7.1]. It is not our intention to elaborate in detail on the electronic circuitry involved in such analog processing. We prefer, instead to discuss the technique in terms amenable to block diagram descriptions. For the details of analog function circuitry, the reader is referred to some recent comprehensive reviews on the subject [7.2,3].

7.1 Organization

Unlike the other chapters in this volume, this chapter considers mainly the *analog* processing of images. The primary reason for this difference is that most of the digital processing methods discussed so far for use with conventional transmission electron microscope images can be used for scanning transmission electron microscope (STEM) images as well, the main difference being that the intermediate densitometry of electron microscope plates has been eliminated.

This chapter is divided into four sections. First, we discuss the basic signals that are available in modern STEMs and the type of information about the sample that is contained in these various signals. Our main consideration is the utilizations of the current of electrons transmitted through thin specimens. However, the techniques discussed can easily be extrapolated for use with any signal available (e.g., secondary electron signals, Auger electron signals, etc.). In fact, some of the processing techniques are in common use in conventional SEMs [e.g., 7.4-6].

Secondly, we consider how the electrical signals that we actually collect are related to those basic signals and the type of processing that is needed to extract the basic signals from the detected ones. Thirdly, we discuss the types of signal manipulation that can be useful in presenting a more pleasing visual image, or one in which there is apparently more observable information than in the raw signal. Included here is a brief discussion of scan conversion techniques coupled with storage displays which allow us to display continually relatively low noise images for focusing and viewing. Although we have been using such techniques for over five years [7.7], it appears that these systems are rarely used with commercial instruments although the advantages seem to be appreciable.

Finally, we present particular examples to demonstrate some of the applications of analog processing and indicate future directions and optimal ways of coupling analog processing with digital techniques. One point that we hope to make is that analog processing can be used to obtain quantitative information from a specimen. Admittedly, not as many computations can be performed as compared with digital techniques, but the processing is *on-line* and because of the relative inexpense of an analog processing system, such limitations can be tolerable.

7.2 Characteristics of Analog Processing

The analog processor is basically some form of electronic functional amplifier. It allows us to perform linear operations on an electrical signal, such as adding, multiplying, dividing; time sequential operations, such as taking derivatives; and integrals and nonlinear operations, such as taking logarithms or raising an input signal to some power. We consider such processing elements to be *black box* units and speak of these elements in terms of their functions, i.e., adders, dividers, etc. For those interested in the actual construction of such units, the reader is referred to [7.2,3].

The types of image processing operations that can be performed in an analog fashion in the STEM can be broken down into two main categories; those operations whose primary objective is to provide a more pleasing visual image and those operations that allow us to increase the information content in the image as well as provide a means of on-line quantitation of the signal.

Let us begin our discussion by considering those operations which improve the picture quality. In talking about such processing one should bear in mind that since some functional operations are time dependent, they are dependent upon the direction of scanning so that the resultant processed image need not appear the same if the image is scanned in a different direction.

7.2.1 Grey Scale Modification

This is one of the simplest methods of improvement of picture quality and can be divided into three or four categories (and hence three or four separate knobs on the processor panel).

a) *Black Level Subtraction*: This operation has no explicit time dependence and so is independent of the scan direction. It consists of applying an electrical off-set to the video output signal in order to reduce the dc level of the output. For instance, if the signal due to the object of interest happens to appear on a large dc signal background (which is of little interest), one can subtract this background so that the signal of interest lies more in the center of the grey scale. This technique is generally coupled with the next operation.

b) *Differential Amplification*: After the dc offset has been put into the signal, the remainder signal can be linearly amplified above the new *black level*. Thus, small modulations on a large background become more visible and the features of interest are better matched to the grey scale of the final image. The successful use of differential amplification is quite dependent on the signal to noise in-herent in the original signal. That is, since the eye can only distinguish a limited range of grey levels (about ten [7.8]) noise is usually not prominent in the image if it is less than the difference between two visually separated grey levels. If we reduce the black level and then differentially amplify the remainder signal, we increase the size of this noise relative to the grey scale, eventually making it visible. Once the noise is equivalent to several grey levels, the re-sulting image is extremely difficult to interpret. The obvious rule of thumb is that if the original signal modulations due to the specimen are no greater than the statistical noise in the image, contrast expansion alone will result in no image improvement, and is generally undesirable.

c) *Nonlinear Amplification*: A limitation of the above is that if the image sig-nal consists of small modulations on both near-black and near-white levels of the video scale, only one end of the scale at a time can be accentuated through dif-ferential amplification. The other end of the scale is eliminated by the necessary dc offset, thus limiting the information content of the resulting image. One method around this difficulty is the use of nonlinear amplification. Gamma control is a commonly used process for video recording systems. Basically, it processes the sig-nal such that the output signal is proportional to the $1/\gamma$ root of the input sig-nal, where γ is adjustable (i.e., a knob on the processor panel). The γ root oper-ator provides increased contrast at either the black or white end of the grey scale, but at the expense of compressing the contrast at the opposite end. However, un-like linear differential amplification, we need not lose the information completely in one of the extreme regions. Logarithmic amplification is another means of ex-panding one region of the grey scale at the expense of compression of the other regions.

As an example of the utility of logarithmic amplification, consider a signal of the form

$$I(x) = A \sum_{n} \exp[-(x - ns)^2/a^2]\exp[-(x/s)] \tag{7.1}$$

where n = 0, 1, 2, ..., and x is a function of time (i.e., x is the position of the beam as it is swept either across the sample or across the detector). For such a signal, the intensity of the peaks at x = ms is decreasing exponentially as (x/s). Thus, because of the limited range of the video display, one could not simultaneously observe peaks at m = 0 and some large m. By taking the logarithm of this signal we get an output signal,

$$I_{OUT}(x) = \ln(A) + \ln\left\{\sum_n \exp[-(x - ns)^2/a^2]\right\} - (x/s) \tag{7.2a}$$

which becomes at the peaks (x = ms), for s > a, m = 0, 1, 2, ... ,

$$I_{OUT}(x = ms) \approx \ln(A) - m. \tag{7.2b}$$

The peak intensities are now in an arithmetic progression, thus somewhat equalizing the intensity differences so that all peaks can simultaneously be displayed (after suitable black level subtraction).

7.2.2 Filters

A filter operation is easily performed with analog components and the common types are either high-pass or low-pass filters (although more complicated functions can be used). The simplest cases of such filters are differentiators and integrators. It should be noted that these processes are time dependent and, therefore, only operate in the scan direction. Thus, caution is the password in efficiently utilizing them. The filter always has to be some compromise that is determined by the signal-to-noise ratio in the original image and the size of the detail that one wishes to visualize. For instance, an image with very little small-dimensional detail but a poor statistical signal-to-noise ratio can be made relatively noise free using a low-pass filter (an integrator). A signal that is relatively noise free can be somewhat enhanced to show small detail with a high-pass filter (a differentiator).

Differentiation is a useful process if there is a dc level change in the signal due either to the specimen or to slowly varying fluctuations in the incident beam current. For example, consider the signal

$$I_{IN}(x) = Ax + B \cos(\omega x) \quad . \tag{7.3}$$

Ax could represent, for example, a change in the beam current as the beam is scanning across one line in the frame or the signal due to the background of a wedge-shaped specimen. The processed signal becomes

$$I_{OUT}(x) = dI_{IN}/dx = A - \omega B \sin(\omega x) \tag{7.4}$$

and removes the slowly varying component in the image while accentuating the higher frequency components. An example of this is given later in this chapter.

Similarly, the integration process is just a bandwidth limit. The general rule in using a bandwidth limit process on the video signal is that the bandwidth of the video amplifier need not be larger than the modulation frequency of the finest detail one wishes to visualize and in no instance should the time constant associated with this bandwidth be less than the time the beam spends at each picture element (i.e., bandwidth frequency $\leq$ frequency of pixels). Otherwise, one is allowing an unnecessary amount of high-frequency noise into the image. For example, in recording a 512 × 512 element picture with a horizontal line speed of 1/60 s, if one utilizes a bandwidth much greater than 40 KHz, one is allowing excess noise into the image.

7.2.3 Signal Mixing

This type of processing is particularly well suited for signals from the STEM because a variety of different signals can be made available simultaneously in a time-sequential form. This will be discussed in more detail in Sect.7.3. As the complexity of detectors increases, so will the complexity of these techniques. Logically, signal mixing is divided into two types; the type of mixing that is performed for image improvement (i.e., to get the most pleasing visual image) and the type needed to extract signals which depend upon particular properties of the specimen. Since the image improvement mixing is, to a large extent, empirical, it should be functionally separated on the analog processor panel.

Unfortunately, in most commercial analog processors used with STEMs, all the signal mixing takes place with the same set of "knobs" so that it becomes difficult to separate the two functions. An example is the mixing system of the VG Microscopes Ltd. model HB5 STEM described in [7.9]. In this system, one can take up to three input detector signals u, v, and w and feed them through variable gain buffer amplifiers into two independent summing amplifiers to get the signals au+bv+cw and lu+mv+nw. These two additive signals are then divided to produce a final output

$$I_{OUT} = (au + bv + cw)/(lu + mv + nw) \quad . \tag{7.5}$$

One thus has six independent controls to utilize in trying to form a pleasing signal, a not inconsequential task. In the systems we describe, we have tried to separate the image-improvement mixing from the information-extraction mixing.

7.3 Types of Signals Available in the STEM

7.3.1 Basic Signals

In principle, every type of interaction that the incident electron undergoes in passing through the specimen can be used to form an image in the STEM. We shall only consider here the types of information that can be extracted from the transmitted electrons. Secondary electron images, X-ray maps, etc., are covered in the literature [e.g., 7.6,10-13].

One can simply divide the current of electrons transmitted through the specimens used in electron microscopy into three broad classes: 1) elastically scattered electrons (the characteristic scattering angle being of the order of 1 degree for 100 keV incident electrons); 2) inelastically scattered electrons (the characteristic scattering angle being of the order of 0.01 degree for valence shell excitations and fractions of a degree for inner shell excitations for 100 keV incident electrons). Since most of the inelastic excitations are valence shell excitations, the average scattering angle for inelastic scattering is much smaller than for elastic scattering. 3) The last class are those electrons that have not scattered at all in traversing the sample. (The angular distribution of these is just that of the illuminating beam.) For thin enough specimens, we can generally separate these three classes using the simple detector scheme shown in Fig.7.1.

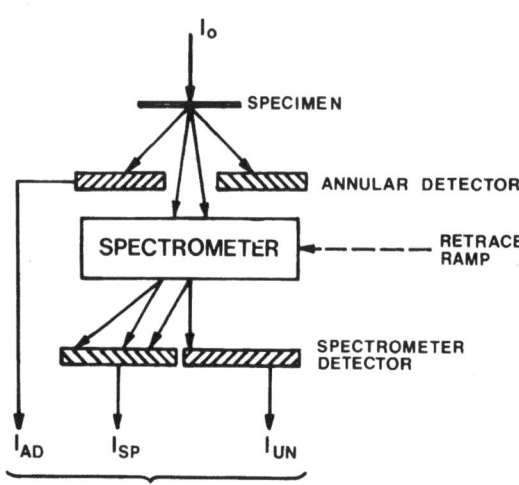

Fig.7.1. General schematic of a simple detector arrangement in the STEM. The three microscope signals that are indicated are defined in the text. They are the input signals into the analog processor

By placing an annular detector at the beam exit side of the specimen, one can collect most of those electrons elastically scattered from the sample at angles greater than the beam convergence angle. If the sample thickness is less than about one-third of a mean free path n for elastic scattering (0.33n), the annular detector signal I_{AD} is proportional to the current of elastically scattered electrons.

If the hole in the annular detector is sufficiently larger than the size of the unscattered beam at the detector plane, both the unscattered electrons and inelastically scattered electrons pass through the hole where they can be separated by a spectrometer into a spectrometer signal I_{SP} and a "no-loss" signal. Again, for thin specimens, I_{SP} is proportional to the inelastically scattered electron current and the no-loss signal is practically the same as the unscattered electron current I_{UN} (small angle elastically scattered electrons can pass through the annular detector hole and appear as a no-loss signal).

We now have three signals running simultaneously out of the microscope. The aim of processing is to extract information about the sample by relating these signals to the three classes of "scattered" currents. It should be noted that there is additional information to be obtained by subdividing the classes of electrons passing through the hole in the annular detector [e.g., 7.14,15]. For instance, using concentric ring detectors within this hole one can extract "phase contrast" information about the sample [7.16] by suitable signal mixing. We shall not consider that aspect here, but rather concentrate on the information available from the detector arrangement described by Fig.7.1. It is at this point that we can think about analog processing. That is, in the case of very thin samples, we can extract the elastic, inelastic, and unscattered currents from the three signals by just multiplying the detector signals by the appropriate proportionality constants. This is a simple analog operation: just route the detector currents into variable gain buffer amplifiers.

One can further perform a normalization process, which has the effect not only of relating the signal to the mass thickness of the sample but also of reducing the effects of beam current fluctuations on the resulting image [7.17]. For the perfect microscope and for specimens where absorption of the incident beam is negligible, we can sum up all the detector signals with an operational amplifier "adder". This resultant sum is just $I_0 \cong I_{AD} + I_{SP} + I_{UN}$, the incident beam current. Thus, this sum allows us simultaneously to monitor the beam current. If we then divide our signals by this sum, beam current fluctuations cancel out and we get a signal that is directly related to the properties of the specimen. That is,

$$I_{SC}/I_0 \cong [1 - \exp(-N\sigma_{SC}T)] \cong N\sigma_{SC}T \qquad (7.6)$$

for thin enough specimens, where N is the number of atoms (or molecules) per unit volume, T is the local specimen thickness, and σ_{SC} is the appropriate scattering cross section. In this case, simple amplifiers take us from the raw detector signal to a calibrated output which directly relates to the scattering power of the specimen [e.g., 7.17-19,23].

7.3.2 Detected Signals

In general, the specimen thickness may be greater than one-third of a mean free path for elastic scattering (for carbon this means a thickness greater than about 400 Å for 100 kV electrons). In that case, the classes of electrons striking each detector are somewhat different. We now have a nonnegligible probability of an incident electron undergoing two or more collisions in traversing the sample. Thus, we can have electrons both elastically and inelastically scattered. This fourth class can be detected either by the spectrometer detector and appear as a contribution to I_{SP} or by the annular detector and appear as a contribution to I_{AD}. Our detector signals are now related to the basic currents in a more complicated fashion. This has been discussed in detail in the literature where it has been shown that the three detector signals are related to the "true currents" as [7.20-23]

$$\frac{I_{AD}}{I_0} = A \frac{I_{IN}}{I_0} + B \frac{I_{EL}}{I_0} \tag{7.7}$$

$$\frac{I_{SP}}{I_0} = C \frac{I_{IN}}{I_0} + D \frac{I_{EL}}{I_0} \tag{7.8}$$

$$\frac{I_{0\ Loss}}{I_0} = \frac{I_{UN}}{I_0} + [1 - (B + D)] \frac{I_{EL}}{I_0} \tag{7.9}$$

where A is the fraction of inelastically scattered electrons which strike the annular detector, B is the fraction of elastically scattered electrons which strike the annular detector, C = constant - A and D = K(n) (1-B) where K(n) is a number that depends somewhat on the thickness (n = $N\sigma_{EL}T$, where σ_{EL} is the total cross section for elastic scattering). The "basic" currents are:

1) I_{EL}, the current of all elastically scattered electrons
2) I_{IN}, the current of electrons only inelastically scattered
3) I_{UN}, the current of electrons not scattered at all.

The point to note is that we can get from our three raw detected signals to the basic signals with a simple matrix inversion and four variables. It turns out that, for most electron microscope specimens, A and C are thickness independent and B and D are only slightly dependent on thickness. Therefore, to a first approximation, we can perform this inversion operation with four variable gain amplifiers (the gain controlled by potentiometers on the front panel of an "efficiency" box); see Fig.7.2. For samples thinner than one mean free path for elastic scattering this simple scheme suffices. For thicker specimens one must be able to vary B and D depending upon the local mass thickness.

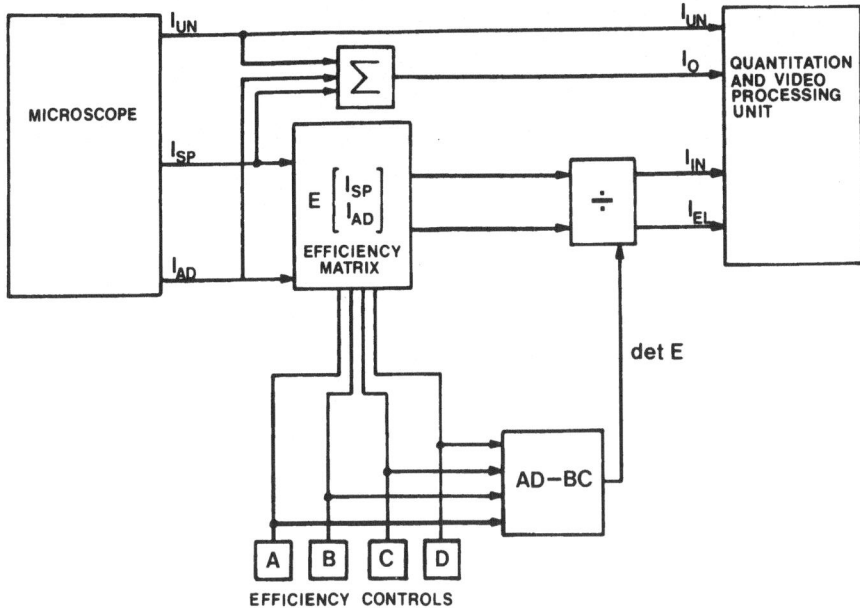

Fig.7.2. First logical section of an analog processor. This section converts the input detector signals into I_0, the incident beam current, and three other signals directly related to properties of the specimen

7.3.3 Extraction of Basic Signals

The implementation of this conversion of raw to true signals is illustrated schematically in Fig.7.2 where we show the "efficiency" controls and the operations that give us the true signals. For example, the current for elastic scattering is

$$I_{EL} = (AI_{SP} - CI_{AD})/(AD - BC) \tag{7.10}$$

and that for inelastic scattering only is

$$I_{IN} = (DI_{AD} - BI_{SP})/(AD - BC) \quad . \tag{7.11}$$

Note that for very thin samples, our processing scheme is considerably simplified. If $n < 1/3$ then $A \approx 0$ and $D \approx 0$ so that we are left with only two variables from which to extract the true signals. In many cases, that is sufficient for the operation of the microscope if one wants to perform the calculations later using digital techniques.

These three basic currents plus their sum (which is eventually the incident beam current) are now available for further processing in the "quantitation and video processing" unit (shown in Fig.7.3).

The rationale for separating out this efficiency correction and conversion to true signals from other processing functions is simple. Their controls depend upon the particular specimen and electron optical geometry. Once they are set for a particular sample (or a region on the sample) they do not have to be adjusted. Further processing is used to obtain a proper combination of these signals to give the appropriate information and to perform operations which allow for a picture that is pleasing to the eye.

Before we conclude this section, a few words should be said about what simple combinations of the true signals are appropriate. An extremely useful combination is that of the ratio of elastic to inelastic scattering which was first put forth in [7.24,25] and sometimes called the "Z contrast signal".

The Z contrast comes about as follows: For any specimen thickness

$$I_{EL}/I_{IN} \approx [\exp(n) - 1]/[1 - \exp(-Rn)] \quad . \tag{7.12}$$

For thickness less than one mean free path ($n \lesssim 1$), this becomes

$$I_{EL}/I_{IN} \approx (1 + n/2)/(1 - Rn/2)R \quad . \tag{7.13}$$

Since σ_{EL}/σ_{IN} is approximately proportional to the atomic number Z and $\sigma_{EL}/\sigma_{IN} = 1/R$, then

$$I_{EL}/I_{IN} \approx \text{const } Z(1 + n/2)/(1 - Rn/2) \quad . \tag{7.14}$$

The correction term is close to unity for carbonlike materials for $n < 1$ and for heavy metals for $n < \frac{1}{3}$. Thus, under these conditions the I_{EL}/I_{IN} signal is proportional to Z. A further simplification in the processing occurs for the very thin samples since then $I_{EL} \approx I_{AD}/B$ and $I_{SP} \approx CI_{IN}$; thus the ratio of the raw detector signals is

$$I_{AD}/I_{SP} \propto I_{EL}/I_{IN} \propto Z \tag{7.15}$$

hence the term "Z contrast signal".

Note that even if the sample is approaching $n \sim 1$, we can still extract a Z contrast signal, by just multiplying the ratio I_{EL}/I_{IN} by the n-dependent correction, n being determined from the unscattered signal as

$$\ln(I_0/I_{UN}) = n(1 + R) \quad . \tag{7.16}$$

There are, of course, many other useful signal combinations, some of which are discussed in Sect.7.5.

Although not as powerful as digital methods in terms of quantitating the image, analog processing techniques are still a very useful aid in trying to understand the image visualized by the microscope operator and as an inexpensive method of extracting small amounts of numerical information concerning the specimen. For example, for very thin specimens ($n < 1/3$), the true computed signals are linearly related to the mass thickness as

$$I_{EL}/I_0 \approx N\sigma_{EL}T, \ I_{IN}/I_0 \approx N\sigma_{IN}T \ . \tag{7.17}$$

For samples of known composition, either signal can be used to determine the local mass thickness quickly, since both signals can be displayed on an oscilloscope trace for measurement (see Fig.7.4). For specimens whose composition is not quite known (for example, biological particles or catalytic particles), the use of both signals can give us information about the composition. From the ratio signal I_{EL}/I_{IN}, we get the effective Z of the particle (or the local area). With this we can then deduce an effective cross section and from the separate signals I_{EL}/I_0 and I_{IN}/I_0 deduce a mass thickness NT. Using this ratio signal is also helpful in quickly detecting the presence of excess salt in a biological specimen preparation [7.19]or to determine whether foreign objects on the specimen are organic in nature.

The extension of the determination of mass thickness to thicker specimens ($n > 1/3$) can be made using the no-loss signal since to a first approximation

$$I_{NO \ LOSS}/I_0 \approx \exp[-NT(\sigma_{EL} + \sigma_{IN})] \ . \tag{7.18}$$

Thus, a logarithmic amplification of this signal gives us a video output signal proportional to NT [7.17]. This is just the electronic analog of the photographic method developed by BAHR and ZEITLER [7.26] for measuring mass in the conventional electron microscope. It has been successfully used to generate images with contours of equal mass thickness by converting the continuous signal $\ln(I_0/I_{NO \ LOSS})$ = $NT(\sigma_{EL} + \sigma_{IN})$ to a step signal having distinct grey levels [7.27].

7.3.4 Normalization

Although the three basic signals we have described could be channeled through the analog processing box at once, one can just as easily use only two detector channels. In fact, in our system we have at present only two video channels. By necessity we have devised a simple scheme to obtain the same types of information as with three channels. Instead of sampling the zero loss signal all the time, we merely measure it during the time it takes for the scanning beam to go from one line to the next (the "retrace" period). In our system that is about 1 ms. During that time, we apply a pulse to the spectrometer so that both the unscattered beam and the energy loss beam strike the spectrometer detector (see Fig.7.1). This signal is essentially the "bright-field" signal. By then integrating this signal during the retrace time, dividing it by the integration time, and sending it to a sample and hold circuit, we have thus generated a signal level that is the average of $I_{SP} + I_{UN}$ of the preceding scanned region. If, in addition to this, we also integrate the annular detector signal, divide by the integration time, and sum it with the preceding $I_{SP} + I_{UN}$, we get a signal that is the average of the incident beam current from the preceding scan line. We can then use this value to "normalize" the signals from the following line. Thus, with only two video channels (and in fact, only two detectors), we can extract the signals I_{EL}/I_0 and I_{IN}/I_0.

This type of normalization is termed "line-by-line" normalization as opposed to "point-by-point" normalization which is essentially instantaneous. The advantage of line-by-line normalization is that only two detectors need be used and one gets a good statistical sample of I_0 since $\bar{I}_0 = T^{-1} \int_0^T I_0(t)dt$. The disadvantage is that if there are fast fluctuations of the incident beam current (faster than the time to scan one line), this type of normalization cannot cancel out such fluctuations. This is not the case for point-by-point normalization in which any fluctuations longer than the time for one pixel are the same for the numerator and the summed denominator signal and are thus cancelled out. The disadvantage of point-by-point normalization is that I_0 is not as statistically well defined since the sampling time (one pixel) is much shorter than for line-by-line normalization. In practice, one uses whichever normalization gives a more noise-free image.

7.4 Instrumental Characteristics

7.4.1 The Analog Processor

We have utilized three different types of analog processor units, each one only slightly different than the other, the differences being generally due to historical circumstances or the personalities of the designers. A logical schematic of such a processor is shown in Figs.7.2 and 7.3.

As we have mentioned before, it is helpful operationally to divide the processing into two sections (or modules). The section shown in Fig.7.2 directly accepts the raw detector signals from the microscope. Using a series of multipliers and adders, it allows us to insert corrections to the raw signals that take into account such things as; 1) detector efficiencies (all detectors do not necessarily have 100% efficiency), 2) electron optical geometrical efficiency of collection (i.e., what fraction of electrons of a particular class actually strike the detector or pass through the spectrometer?), etc. The final output of this "efficiency" module gives us the basic signals I_{EL}, I_{IN}, I_{UN} and I_0 described in Sect.7.3.

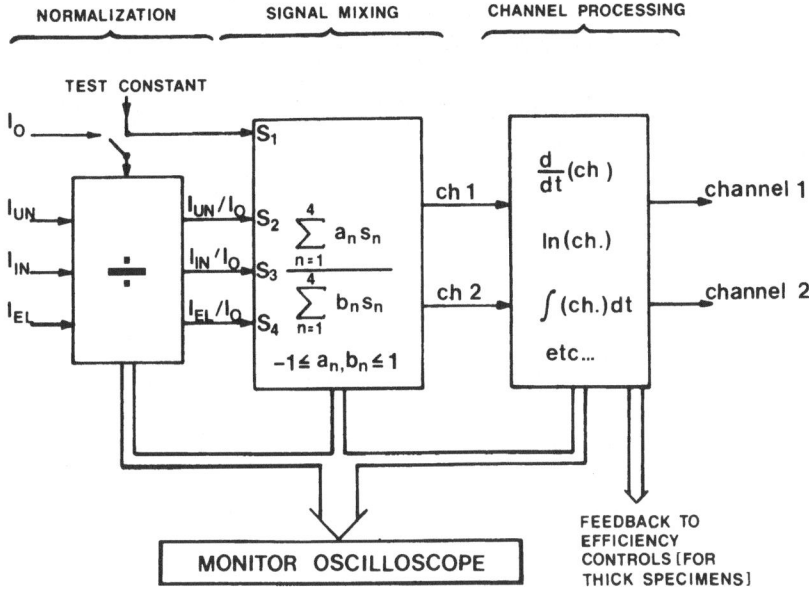

Fig.7.3. Block diagram of the remainder of the analog processor. The three basic signals can either be divided by the incident beam current signal I_0 or by a fixed test signal. These normalized signals are then routed to a mixing section which can form the ratio of the two sums. From this section, two video channels are available for further processing and the final output is fed into the display system for viewing or into a high-resolution CRT for recording

For thin samples, all these corrections are multiplicative, but for thicker samples where there is mixing of the basic signals onto the various detectors, one must solve the set of Eqs.(7.7-9) to extract the basic signals. In other words, these corrections take the form of a matrix with various parameters (up to four to be precise). To first approximations these parameters are determined by four knobs on variable gain amplifiers (A, B, C, and D). Within this efficiency module, one also sums up all the microscope signals [7.17]. If there is negligible absorption in the sample, this gives us the incident beam current, a signal that we have discussed and which is used for normalization purposes to reduce the effects of beam current fluctuations on the image as well as for quantitative purposes. On one of our systems [7.17], the I_{UN} signal from the microscope is the "retrace" signal described in the last section. In that case, the sum box also includes a sample and hold circuit. It is obvious that this efficiency module scheme can be cascaded to accept any number of detectors.

The output of this efficiency section is routed into the "processing" module shown schematically in Fig.7.3. This module is divided into three sections; 1) a normalization section, 2) a signal mixing section, and 3) a "pretty picture" section. The first two sections are predominantly used for extraction of information from the sample, either in terms of a video signal that gives an intensity representation of some known sample parameter (such as mass thickness or Z contrast) or in terms of a number that can be read off an oscilloscope trace.

The output of any one of these three sections can be fanned out to modulate an axis of a dual trace oscilloscope which is arranged to allow us simultaneously to display the intensity on a vertical axis vs beam position as a horizontal axis of the video signal that one is viewing (as an intensity display) plus any other intermediate signal output one chooses. For example, such a CRT display is shown in Fig.7.4, where we show a typical scope trace of the processed video signal I_{EL}/I_0 on the upper trace and the corresponding spectrometer signal I_{SP} on the lower trace along with the bright-field signal obtained from the spectrometer detector by ramping the spectrometer during the retrace so that the bright-field signal falls on the spectrometer detector during the retrace. The detector scheme is exactly as shown in Fig.7.1.

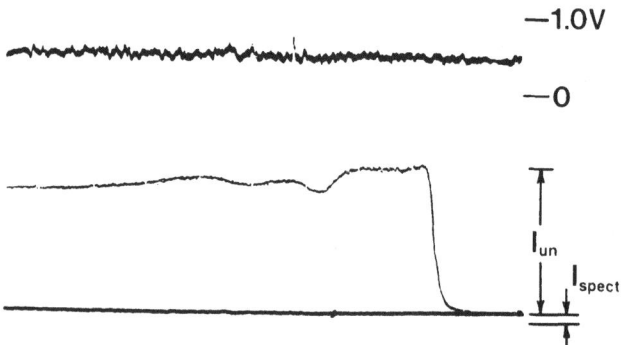

Fig.7.4. Photograph of the monitor oscilloscope trace showing the video signal output (upper trace), the direct signal from the spectrometer detector (I_{SP}), lower trace), and the signal obtained from the spectrometer signal during the retrace period of the horizontal line sweep. The slight dip in this signal is due to a non-uniformity of the detector (the beam is swept across the spectrometer detector during the retrace rather than remaining in a fixed position)

The last section (the channel processing section) is used mainly to create a more pleasing picture (high-pass and low-pass filters are in this section, for instance), or in the case of thick objects, output can be fed back into the efficiency module controls (since the efficiencies then are somewhat thickness dependent).

7.4.2 Display System

The last link of the analog network is the display system, the device into which the final processed video signal is fed to create an image used for focusing the microscope and general viewing. This part is essential for efficient utilization of the information available from the STEM and, therefore, must be discussed as an inherent part of the total processing system [e.g., 7.28].

Usually, the final video signal output of the analog processor is used to modulate the intensity of a cathode ray tube (CRT) scanned synchronously with the beam in the microscope. Focusing and visualization of the image is performed in one of two ways: 1) by viewing a noisy image in real time as the beam in the CRT and the beam in the microscope are scanned at TV rates, or 2) by viewing a fading image on a long persistence phosphor CRT as the beam in the CRT and the beam in the microscope are scanned at much slower rates to increase the signal-to-noise ratio in the image. The final image is then viewed by photographing a high-resolution CRT at some slow frame rate (of the order of 10 s frame time). The disadvantage of either method is that by scanning at slow rates, one does not really know what the image looks like until a photograph is produced and scanning at TV rates results in a noisy image. One could remedy this last fact by substantially increasing the beam current to allow one to obtain noise-free images at TV rates. But this may destroy the sample in some cases and in other cases sufficient beam current may just not be available (even with field emission sources).

As an example, consider the statistics that are available in trying to form a 1024 × 1024 line picture using a beam current of 1 nanoamp (this current would be close to the maximum that could be put into a 5 Å diameter beam using standard field emission sources). If we are interested in imaging a very thin film, then we might expect 5% of the incident beam current to be scattered. Thus, we have 3×10^8 electrons/s striking some detector (assuming 100% collection efficiency) which seems like quite a large value. However, if we are scanning the beam at TV rates (15 frames/s) then each pixel is only 67 ns long. Thus, the signal statistics is due to only 18 electrons per pixel! A very noisy image indeed. If we instead scan the beam at an 8 s frame rate then the statistics in the image is due to more than 2000 electrons per pixel or an increase in signal-to-noise of more than ten. From a signal-to-noise point of view we would then like to scan at slow rates, but viewing a fading CRT screen does not allow us to judge our focusing or image processing without obtaining a photograph and that considerably slows down the interactive process thus defeating the use of analog techniques.

What we need are the capabilities of slow scanning of the beam in the microscope but a continual display of the image at TV rates (the image having the signal-to-noise of the slower scan rate). This is simply done with commerically available analog storage devices and scan converters (or in a more elaborate and expensive fashion with the new generation of digital displays).

A schematic of one of the display systems that we use for accomplishing this is shown in Fig.7.5. This system encompasses both analog storage and scan conversion in addition to the appropriate interface to allow us to transfer data to a mini-computer in digital form for storage onto magnetic tape [7.7,29]. The system operates as follows: Horizontal and vertical ramps generated by a digital controller are applied to the scanning coils of the microscope. The horizontal ramp is usually

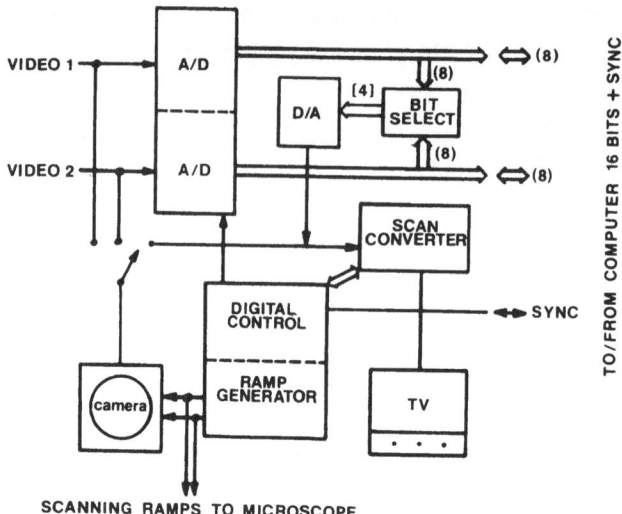

<figure>VIDEO 1 → A/D

VIDEO 2 → A/D

D/A   [4]   BIT SELECT

(8)

(8) ⇔

(8) ⇔

SCAN CONVERTER

DIGITAL CONTROL

RAMP GENERATOR

camera

TV

SYNC

TO/FROM COMPUTER 16 BITS + SYNC

SCANNING RAMPS TO MICROSCOPE</figure>

Fig.7.5. Block diagram of the display system and computer interface. Note that the signal directed to the high-resolution CRT (camera) for photographic recording can come either directly from the analog processor or it can be processed through the display interface. In that case it can be generated through a minicomputer from images stored on magnetic tape

operated at 60 Hz and synchronized to the power mains to minimize raster distortion caused by ambient magnetic fields. Two channels of video signal from the analog processor (shown in Fig.7.4) are digitized into eight bit bytes at the horizontal pixel rate (60 kHz for a 1024 line picture). These bytes can then be transferred and stored as one word in a minicomputer memory (in our case a Data General Corp. NOVA 2), the address being generated at the software level via a synchronous lock to the digital ramp generator.

Either channel of the video signal can also be stored in a high-speed buffer memory until one complete horizontal line is accumulated (about 1/60 s). One then selects any four bits of the eight bit bytes (using the bit range switches on the interface panel shown in Fig.7.6) and reconverts that to an analog signal. That signal is rapidly written onto the storage tube (chip) of a scan converter (Princeton Electronic Products, PEP 400 or PEP 500) during the vertical retrace period of the display monitor, which is then refreshed at TV rates from the same storage tube. The microscope operator thus views a conventional TV image of the current micrograph overwriting the previous image at a 60 line/s rate. The statistics of the image are now that of the microscope frame rate (8.7, 17.2, 32, 64 s, etc.) rather than that of the viewing frame rate thus gaining an order of magnitude in signal-to-noise over actually scanning the beam at the TV rates. The TV monitor is then used for viewing the image while focusing and processing.

We find that such a display system greatly simplifies focusing and processing since the entire picture can be viewed at once without the need for a photographic record. If desired, the digitized output can be stored on magnetic tape or disk at the end of each frame. Although our system only displays four bit TV images, we have found it to be far more efficient for viewing and focusing than trying to focus from a fading phosphor screen in which only a very small portion of the image is visible and then only for a fleeting moment.

One point that should be emphasized here is that this system need not be interfaced to a minicomputer. It can act as a stand-alone package since all that is needed for the display is the high-speed buffer memory used to accumulate the video signal from one horizontal line. In fact, two of our display systems are not interfaced to a minicomputer at all. The principal cost of such a display system is the $4000-$5000 cost of the scan converter (1978 prices); thus it is considerably cheaper than the cost of the new digital displays.

Because of the limited resolution of most TV monitors and the fact that we do not display all the bits of the video signal, a photographic record is not taken off the TV monitor but instead from a high-resolution CRT (2000 lines) which is scanned with the microscope ramps (see Fig.7.5) and intensity modulated with the video signal output directly from the analog computer. The number of horizontal pixels in the photograph is then limited only by the bandwidth of the CRT, the analog processor, and the detectors within the microscope. The number of vertical pixels is of course determined by the number of vertical lines in the raster. We generally find 1024 line rasters to be suitable for obtaining a high-quality photographic record on TRl-X 35 mm film (ASA 400) with a 17.2 s exposure (frame) time.

An example of the use of this display system is given in Fig.7.7 where we show a picture of the TV displayed image in the middle of a microscope scan. The upper portion of the picture is the new, refreshed image that has been written onto the storage tube and the lower portion is the image of the previous scan. The magnification in the microscope was reduced by a factor of ten between scans to illustrate the method. The black bar between the upper and lower portions of the picture marks the current position of the microscope ramps; a new line is written during each vertical retrace of the monitor. As the scan progresses, the bar moves from the top to the bottom of the screen. The entire picture can be displayed for any length of time between scans (i.e., before one starts the next scan), this time being controlled by the delay potentiometer shown on the panel in Fig.7.6. For normal operation, this delay is about 0.3 s.

One final note: if this display system is interfaced to a minicomputer, the interface should be bidirectional. In that way, the computer memory can be scanned through the same display system to generate images on the TV monitor or on the high-resolution CRT. Thus, digitally processed images can be viewed and photographed during the time that the microscope is not being operated.

274

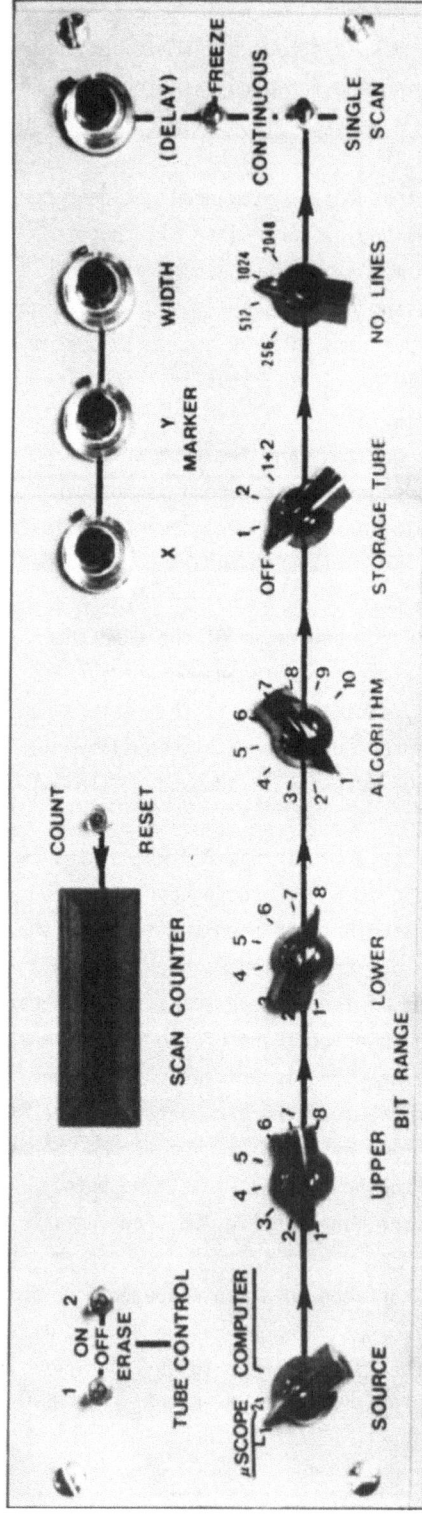

Fig.7.6. A picture of the front panel of the display system interface. The switch labeled NO. LINES refers to the number of horizontal lines per frame. The interface generates lines at a fixed rate of one horizontal line per 1/60 s (the mains frequency in North America). There are capabilities for routing the two input video channels (shown in Fig.7.5) into two separate storage tubes

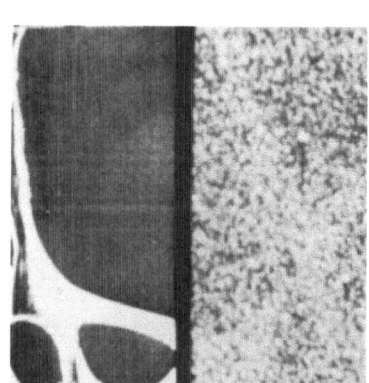

Fig.7.7. A photograph directly from the TV display during one scan. The upper portion is about 2 μm full horizontal scale while the lower portion (the image from the previous scan) is 2000 Å full horizontal scale

7.5 Applications

7.5.1 Basic Operations

The purpose of this section is to present some typical results obtained with the processing systems we have described. They are not meant to demonstrate every aspect, but merely give an indication of the kinds of information extractable with such a system [7.30,31].

The simplest processing that can be done is to take the "Z contrast" or ratio signal (I_{EL}/I_{IN}). For very thin samples we have shown, (7.15), that this ratio is just proportional to the ratio of the annular detector signal to the spectrometer signal (I_{AD}/I_{SP}). An example of this is shown in Fig.7.8 for a specimen consisting of uranium atoms on a thin carbon substrate. We show, (a) the annular detector signal, (b) the spectrometer signal, and (c) the ratio signal. The ratio signal effectively removes most of the intensity modulations due to the varying thickness of the carbon film background, thus making the bright spots (which are due to scattering from uranium atoms) more visible.

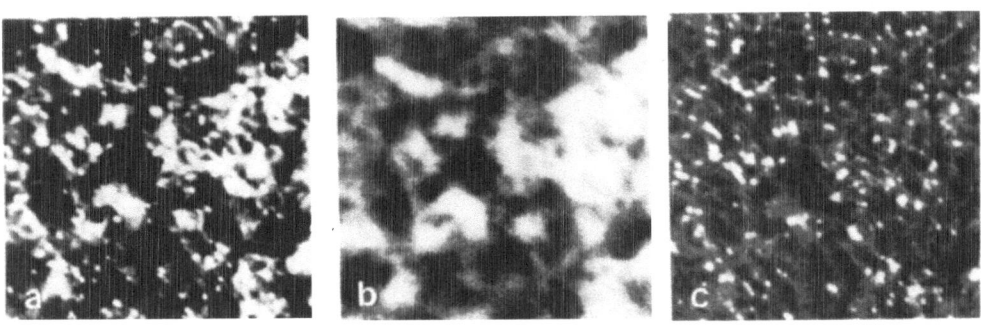

Fig.7.8a-c. Micrographs of a specimen made by solvent evaporation of a 10^{-4} M solution of UO_2Cl_2 onto a 15 Å thick carbon film. All three micrographs are from the same area but using different signals to form the image. The field of view is 225 Å × 225 Å. (a) Signal from the annular detector I_{AD}; (b) signal from the spectrometer detector I_{SP}; (c) ratio signal, I_{EL}/I_{IN} showing reduction in the intensity modulation due to thickness fluctuations of the carbon support

Note the fuzziness of the image made using the spectrometer signal. This is due to the nonlocalization of inelastic scattering [7.32] and is again evident in Fig.7.9b where we show a specimen of platinum and palladium atoms on a 15 A thick carbon substrate. Because the inelastic scattering process (for valence shell excitations) is only localized to about 10 Å, then utilizing such a ratio signal only effectively removes substrate thickness variations in the image to that spatial resolution level [7.24].

In Fig.7.10, we show a succession of images obtained from a thin aluminum foil indicating the variety of contrast available in the STEM. Figures 7.10a-e are images

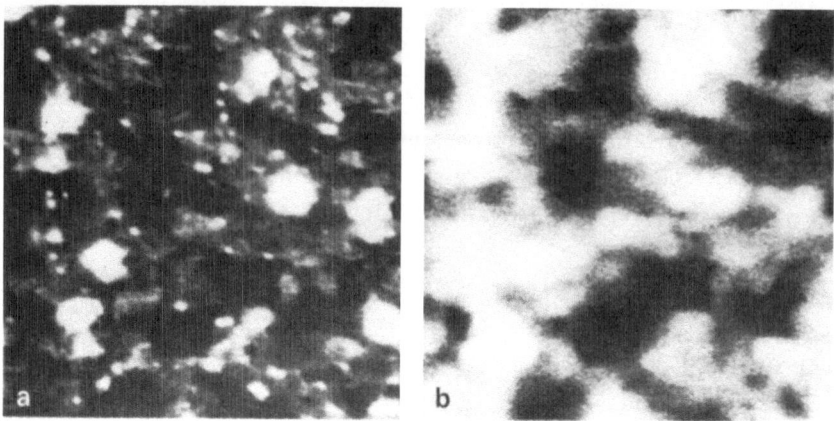

Fig.7.9a,b. Micrographs of a specimen made by resistive evaporation of a wire of 80%Pt-20%Pd onto a 15 Å thick carbon substrate [7.35]. The full horizontal scale is 230 Å. (a) The annular detector signal I_{AD}; (b) the same field of view obtained with the spectrometer signal I_{SP}

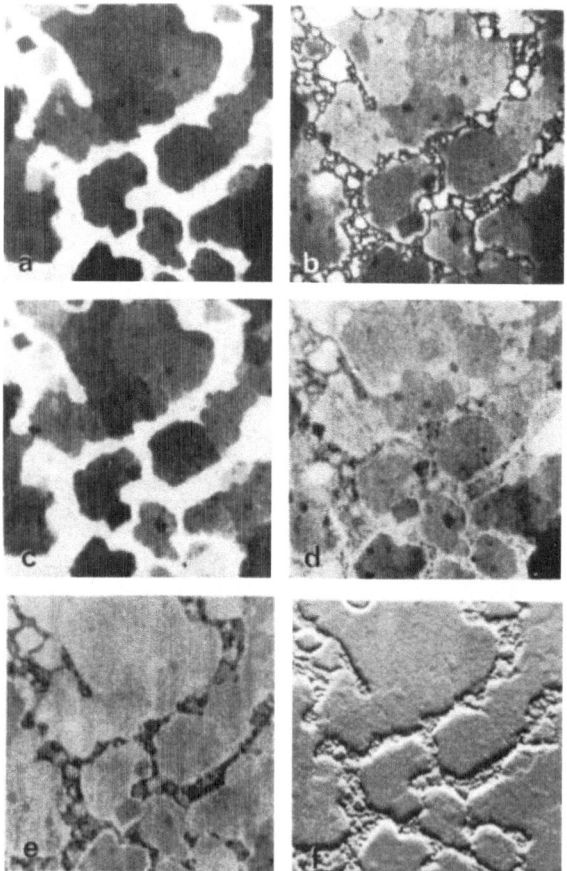

Fig.7.10a-f. Micrographs of a thin evaporated aluminum foil about 300 Å thick. (a)-(e) are images obtained from the spectrometer signal with the detector set to accept various energy loss electrons. (a) 0 ± 0.5 eV loss; (b) 15 ± 0.5 eV loss; (c) 22 ± 0.5 eV loss; (d) 30 ± 0.5 eV loss; (e) 40 ± 0.5 eV loss; (f) shows the signal from (a) after it has been passed through a differentiator. The full scale of each micrograph is 2400 Å

obtained using the spectrometer signal. The spectrometer detector was arranged to accept a given range of energy loss electrons from 0 ± 0.5 eV loss in Fig.7.10a to 40 ± 0.5 eV loss in Fig.7.10e. Because of the high probability of a 15 eV energy loss in aluminum due to plasmon excitation, the spectrometer signal increases whenever one is collecting electrons that have lost an integral multiple of 15 eV in aluminum. Thus, the 15 and 30 eV loss images look similar and the 0 and 22 eV loss images look similar [7.33]. A simple high-pass filtering of the zero-loss image is shown in Fig.7.10f where we have just differentiated the image in Fig.7.10a. Note that the high spatial frequency structure which was not evident in the other images can now be seen quite clearly. The three-dimensional appearance of the image is an artefact of differentiation in only one direction. This appearance has resulted in this mode of processing sometimes being called pseudo three-dimensional processing.

A more elaborate example of the different types of processing on an image is shown in Fig.7.11 where we show four types of signals obtained from a thin section of *Helix aspersa* (snail) sperm tails. Figures 7.11a,c are the unprocessed bright-field and annular detector images, respectively. In Fig.7.11b we have displayed the derivative of the bright-field signal and in Fig.7.11d we display the logarithm of the dark-field signal. Beneath each micrograph is the oscilloscope trace of the video output signal taken along the line indicated by arrows. Note again the apparent shadowing of the differentiated signal in the direction of the horizontal scan.

An example of the use of analog processing to extract more information about properties of the specimen is shown in Fig.7.12. Here we show a field of view of the iron storage protein, ferritin. In Fig.7.12a we show a micrograph of a field of ferritin molecules on a thin carbon substrate taken using the processed elastic signal (corrected for collection efficiency). The protein is white since it scatters more electrons than the carbon film background, but we see no additional evidence of the fact that this globular molecule has an iron-containing core. This can be dramatically shown by using an appropriate linear combination of I_{EL} and I_{IN} as follows: using expressions for elastic and inelastic scattering, we find that $I_{EL} \approx 2.89\ I_{IN}$ for iron and $I_{EL} \approx 0.67\ I_{IN}$ for carbon for the incident electron energy used to take this image. If we form a video signal by subtraction of some of I_{EL} from I_{IN} (i.e., $I_{OUT} = I_{IN} - C\ I_{EL}$), then by choosing $C = 0.35$, we will get $I_{OUT} \approx 0$ for iron but $I_{OUT} > 0$ for carbonlike materials. Thus, we expect a micrograph of ferritin molecules obtained using this signal to exhibit black cores (no signal; indicative of iron) surrounded by bright shells. This is shown in Fig.7.12b where we can now see the various amounts of iron in the cores. This simple technique has allowed the performance of a crude type of elemental quantitation [7.34,35].

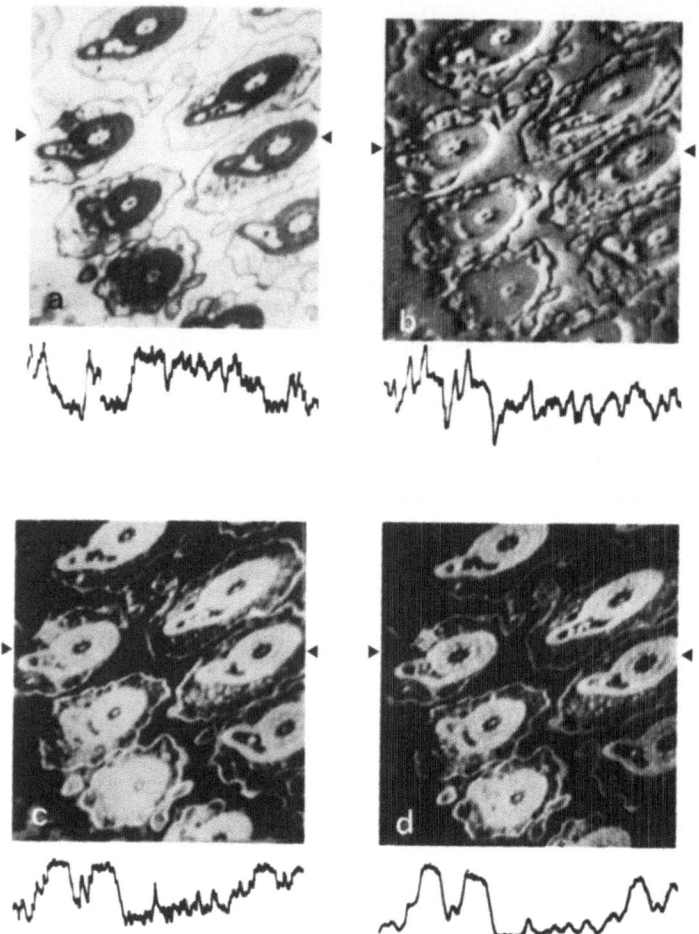

Fig.7.11a-d. Micrographs obtained from a thin section of *Helix aspersa* (snail) sperm tails showing various processed signals. The full scale of each picture is 6.3 µm and the trace below each micrograph is the intensity line scan obtained across the region of the image indicated by the arrows. It corresponds to the same region of the specimen in all four cases. (a) Bright-field signal $I_{NO\ LOSS}$; (b) derivative of $I_{NO\ LOSS}$ image; (c) dark-field image I_{AD}; (d) logarithm of I_{AD} image

7.5.2 Color Conversion Techniques

One example of analog processing that has not been discussed yet is the technique of converting the intensity variations in the image into different colors. This technique is useful for several reasons. First, since the eye can only distinguish ten grey levels [7.8] and in some cases there are more than ten grey levels in the electronic signal, we throw away information by recording the intensity levels directly onto black and white film. While many grey levels may be recordable on the negative, only ten are apparent to the eye on any particular print. This can be alleviated by converting the intensity levels into different colors. Since the

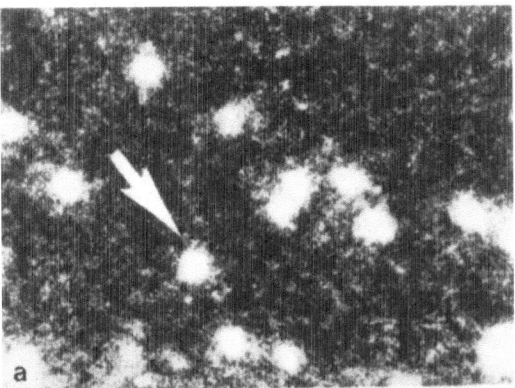

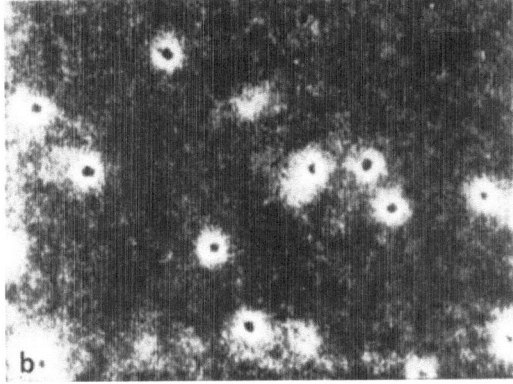

Fig.7.12a,b. Micrographs of ferritin molecules on a thin carbon substrate. The full horizontal scale is about 3800 Å. (a) The annular detector signal; (b) the same field of view obtained using the processed signal $I_{IN} - 0.35\ I_{EL}$. The iron-containing cores now appear black [7.34]

eye can distinguish many more than ten colors (up to 1000), we can, in effect, present more grey levels in a color print. Secondly, since the eye can more easily distinguish different colors, small intensity differences in the signal can be accentuated into distinct colors. Thus, we can get a very effective isointensity color map of the sample which can be more easily interpreted than a black and white image [7.36]. This can be done directly from the electronic signal. Finally, since different signals can be formed with the STEM, it sometimes aids in interpretation to be able to superimpose the different signals in different colors to form a composite color picture rather than separately viewing different pictures. This can be especially useful in looking at energy loss images in which each separate image corresponds to some specific elemental characteristic. This last technique can also be performed off-line from black and white photographs using inexpensive filter techniques.

Some specific applications of these color conversion methods are shown in Figs.7.13 and 7.14. In Fig.7.13a, we show the same image as in Fig.7.11a. Next to it in Fig.7.13b, we show a color-converted image in which the intensity levels have been converted into different colors. Note how the fine details within the very dark regions of the sperm tails in Fig.7.13a become more evident in the color-converted image.

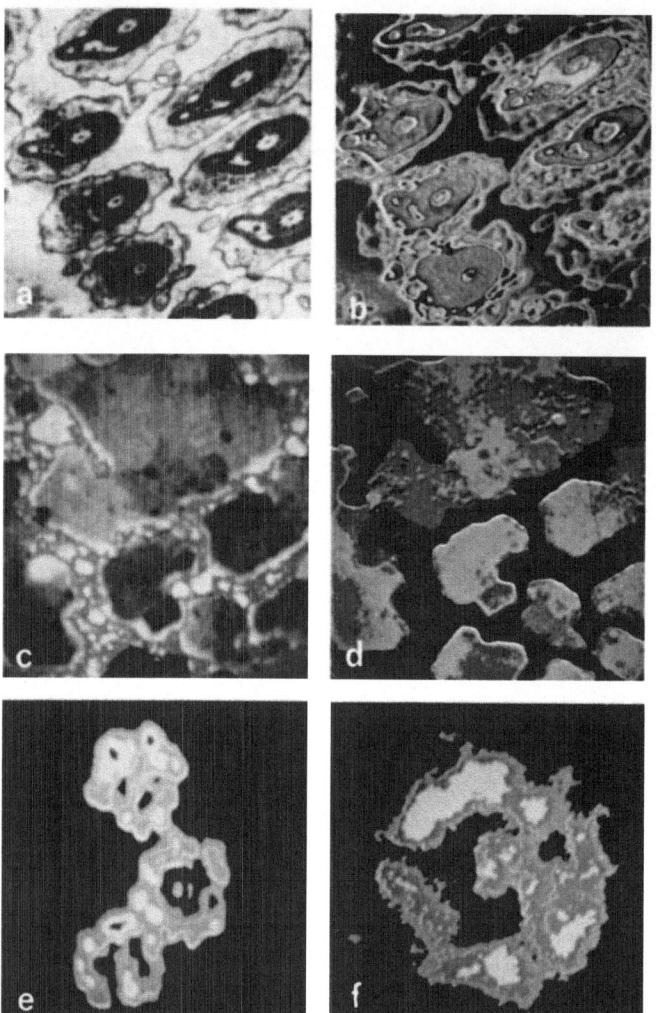

Fig.7.13a-f. Micrographs illustrating the use of color conversion of black and white intensity levels. (a) The same micrograph shown in Fig.7.11a. (b) The color converted image of (a) showing the accentuated fine structure. (c) The same field of view of an aluminum film·that is shown in Fig.7.10 but obtained by superimposing the 15 eV image (Fig.7.10b) in green and the 22 eV image (Fig.7.10c) in red. (d) The same field as in (c) but converting the intensities in the 0 eV image (Fig.7.10a) into different colors. That is, it is an isointensity map of the film. (e) A color converted isointensity map of the image of uranyl acetate-stained mouse tumor nucleosomes obtained using the annular detector signal I_{AD}. The full scale is 335 Å. (f) A high-magnification view of another nucleosome. The full scale is 135 Å (Sample courtesy of S. Usala)

In Fig.7.13c, we show a two-color superposition of two of the images of Fig.7.10. The 15 eV loss image (Fig.7.10b) is projected through a green filter and the 22 eV loss image (Fig.7.10c) is projected through a red filter. The composite image is then photographed. Regions where there are equal intensities in the red image and the green image are superimposed to form yellow. Regions in which either a 15 or a

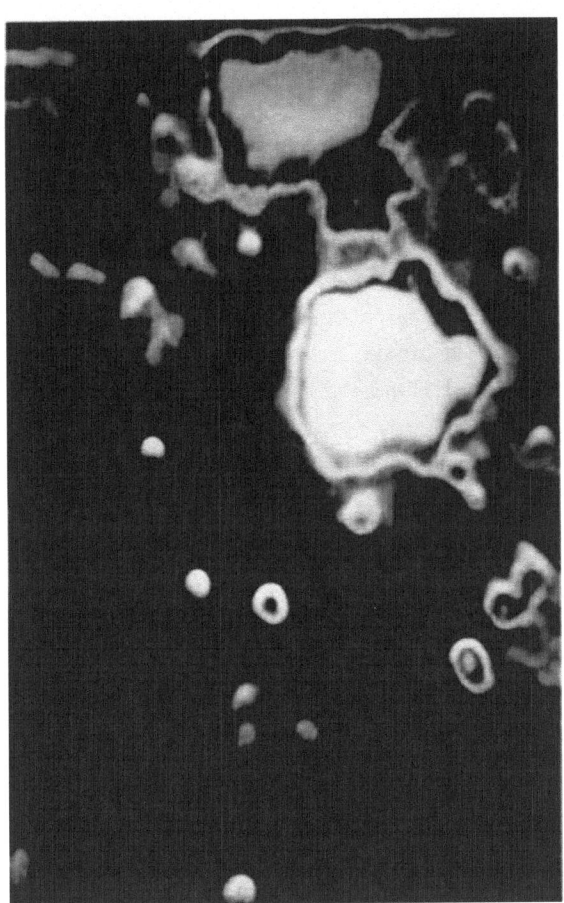

Fig.7.14. A color conversion micrograph from the annular detector signal of a sample of platinum and palladium atoms (the same sample used for Fig.7.9). The color conversion scheme is given in the text. Full horizontal scale is 95 Å

22 eV loss predominates show up as either green or red, respectively, in the composite image. Since the 15 eV loss is the plasmon excitation of pure aluminum and the 22 eV loss is the plasmon excitation of aluminum oxide, this superposition results in an elemental map. Regions that are yellow are equal amounts of Al and Al_2O_3. Red regions are predominantly Al_2O_3 and green regions predominantly Al. Thus, the color superposition has greatly enhanced the information extractable from one photograph. Although this superposition was done from black and white pictures, it can also be accomplished with our display system using the two different video channel outputs to modulate two different colors of a color TV monitor, resulting in a final image being the composite of the two channels [7.7].

Using the same sample we can also further demonstrate the use of color conversion to form isointensity maps. In Fig.7.13d we show the bright-field image of the aluminum film (shown in Fig.7.10a) converted into color by mapping different intensity levels into color. In this figure, yellow regions correspond to the thickest region of the film and dark blue regions, the thinest. Thus, in the upper portion of this micrograph, we see fine detail corresponding to small thickness changes which was barely visible in Fig.7.10 or in the color composite of Fig.7.13c.

A further example of isointensity mapping is shown in Figs.7.13e,f. Here we show two different magnifications of uranyl-stained mouse tumor nucleosomes (positive staining) converted into color by mapping the highly scattering regions into white and the least scattering regions (the substrate) into black. This isointensity map then shows us the regions of equal staining concentration.

A final example of isointensity mapping is shown in Fig.7.14. Here we show a color mapping of the annular detector signal image obtained from the same sample as in Fig.7.9, a sample of platinum and palladium atoms. In this case, the isointensity map has the added advantage of helping us to distinguish individual platinum atoms from palladium atoms.

The color conversion scheme is such that black is the thinnest region of the carbon film, dark blue is a slightly thicker region. Bright blue spots are individual palladium atoms, yellow spots are individual platinum atoms, yellow spots with red centers are two platinum atoms, and the larger objects that are white, red and bluish white are the multiatom clusters. Thus, this isointensity color mapping technique can be quite useful in single-atom microscopy where more than one species of atom are present. A recent example of the use of this mapping technique for high-resolution microscopy is given in reference [7.36].

7.6 Conclusion

We have tried to delineate the basic elements of analog image processing as it pertains to the STEM. Some examples of processing have been presented and we have shown the utility of color conversion of black and white intensity levels to ex-tract the most information from an image. Given the space limitations, we have not intended to present examples of every type of operation, but only selected ones to give the reader a feeling for these things.

While it is true that analog processing to produce a more pleasing visual image is somewhat empirical, we have tried to show that quantitative information about an image can be obtained using analog techniques.

References

7.1   K.-H. Herrmann, D. Krahl, A. Kübler, K.-H. Müller, V. Rindfleisch: In *Electron Microscopy in Material Science*, ed. by U. Valdrè (Academic Press, New York and London 1971) pp. 237-272

7.2 J.G. Graeme: *Application of Operational Amplifiers* (McGraw-Hill, New York) 1973)
7.3 Y.J. Wong, W.E. Ott: *Function Circuit Design and Applications* (McGraw-Hill, New York 1976)
7.4 C.E. Fiori, H. Yakowitz, D.E. Newbury: Scanning Electron Microsc. *74*, 167-174 (1974)
7.5 D.E. Newbury: Scanning Electron Microsc. *75*, 727-736 (1975)
7.6 C.W. Oatley: *The Scanning Electron Microscope* (Cambridge University Press 1972)
7.7 V. Beck: Rev. Sci. Instrum. *44*, 1064-1066 (1973)
7.8 A. Rose: *Vision; Human and Electronic* (Plenum Press, 1974)
7.9 A.V. Jones, K.C.A. Smith: Scanning Electron Microsc. *78/I*, 13-26 (1978)
7.10 C.C. Chang: Surf. Sci. *25*, 53-74 (1971)
7.11 P. Duncumb: J. Phys. *E2*, 553-560 (1969)
7.12 R.L. Gerlach, N.C. MacDonald: Scanning Electron Microsc. *76/I*, 199-206 (1976)
7.13 P.F. Knane, G. Larrabee: *Characteristics of Solid Surfaces* (Plenum Press, N.Y. 1974) Chap. 20
7.14 N.H. Dekkers, H. DeLang: Optik *41*, 552-456 (1974)
7.15 H. Rose: Ultramicroscopy *2*, 251-267 (1977)
7.16 H. Rose: Optik *39*, 416-436 (1974)
7.17 M. Isaacson, J. Langmore, J. Wall: Scanning Electron Microsc. *74*, 19-26 (1974)
7.18 A. Engel: Ultramicroscopy *3*, 273-281 (1978)
7.19 J.P. Langmore: Jerusalem (1976) Vol. I, pp. 31-36
7.20 A.V. Crewe, T. Groves: J. Appl. Phys. *45*, 3662-3672 (1974)
7.21 S. Golladay: Toronto (1978) Vol. II, pp. 106-107
7.22 T. Groves: Ultramicroscopy *1*, 15-31 (1975)
7.23 M.K. Lamvik: Phd Thesis, Dept. of Biophysics, Univ. of Chicago (1976) unpublished
7.24 A.V. Crewe, J.P. Langmore, M. Isaacson: In *Physical Aspects of Electron Microscopy and Microbeam Analysis*, ed. by B.M. Siegel, D.R. Beaman (Wiley, New York 1975) pp . 47-62
7.25 A.V. Crewe, J. Wall, J. Langmore: Science *168*, 1338-1340 (1970)
7.26 G.F. Bahr, E.H. Zeitler: In *Quantitative Electron Microscopy*, ed. by G.F. Bahr, E.H. Zeitler (Williams and Wilkens, Baltimore 1964) pp. 217-239
7.27 L. Reimer, P. Hagemann: Ultramicroscopy *2*, 297-301 (1977)
7.28 C.J.D. Catto, K.C.A. Smith: Scanning Electron Microsc. *72*, 41-48 (1972)
7.29 J.W. Wiggins, M. Beer, S.D. Rose, M. Cole, A.A. Waldrop, J. Zubin, J.W. Platner, L. Marzilli, C.H. Chang, L. Kapili: Scanning Electron Microsc. *76*, ·295-300 (1976)
7.30 M. Isaacson, J. Langmore, N.W. Parker, D. Kopf, M. Utlaut: Ultramicroscopy *1*, 359-376 (1976)
7.31 M. Isaacson, D. Kopf, N.W. Parker, M. Utlaut: *Proc. 34th Annual Meeting EMSA, Miami Beach*, ed. by G.W. Bailey (Claitor, Baton Rouge 1976) pp. 584-585
7.32 M. Isaacson, J.P. Langmore, H. Rose: Optik *41*, 92-96 (1974)
7.33 A.V. Crewe, J. Wall: Optik *30*, 461-474 (1970)
7.34 M. Isaacson: In *Proc. Specialist Workshop on Analytical Electron Microscopy*, Ithaca (1976), Cornell Univ. Materials Science Center Report No. 2763, pp. 81-89
7.35 M. Isaacson: Jerusalem (1976) Vol. I, pp. 26-30
7.36 M. Isaacson, A.V. Crewe, D. Kopf, M. Ohtsuki, M. Utlaut: Science (to be published)

Appendix: Publication Details of International and European Congresses on Electron Microscopy

Delft (1949): Proceedings of the Conference on Electron Microscopy, Delft 4-8 July 1949; editorial committee: A.L. Houwink, J.B. Le Poole, W.A. Le Rütte (printed by Hoogland, Delft, 1950)

Paris (1950): Comptes Rendus du Premier Congrès International de Microscopie Electronique, Paris 14-22 Septembre, 1950 (Editions de la Revue d'Optique théorique et instrumentale, Paris, 1953)

London (1954): The Proceedings of the Third International Conference on Electron Microscopy, London, 1954; edited by R. Ross (Royal Microscopical Society, London, 1956)

Stockholm (1956): Electron Microscopy: Proceedings of the Stockholm Conference, September 1956; edited by F.J. Sjöstrand, J. Rhodin (Almqvist and Wiksells, Stockholm, 1957)

Berlin (1958): Vierter Internationaler Kongress für Elektronenmikroskopie, Berlin, 10.-17. September, 1958, Verhandlungen; edited by W. Bargmann, G. Möllenstedt, H. Niehrs, D. Peters, E. Ruska, C. Wolpers (Springer, Berlin, Heidelberg, New York, 1960) 2 Vols.

Delft (1960): The Proceedings of the European Regional Conference on Electron Microscopy, Delft, 1960; edited by A.L. Houwink, B.J. Spit (Nederlandse Vereniging voor Electronenmicroscopie, Delft, n.d.) 2 Vols.

Philadelphia (1962): Electron Microscopy, Fifth International Congress for Electron Microscopy held in Philadelphia, Pennsylvania, August 29th to September 5th, 1962; edited by S.S. Breese (Academic Press, New York and London, 1962) 2 Vols.

Prague (1964): Electron Microscopy 1964, Proceedings of the Third European Regional Conference, Prague; edited by M. Titlbach (Publishing House of the Czechoslovak Academy of Sciences, Prague, 1964) 2 Vols.

Kyoto (1966): Electron Microscopy 1966, Sixth International Congress for Electron Microscopy, Kyoto; edited by R. Uyeda (Maruzen, Kyoto, 1966) 2 Vols.

Rome (1968): Electron Microscopy 1968, Pre-congress Abstracts of Papers presented at the Fourth Regional Conference, Rome; edited by D.S. Bocciarelli (Tipografia Poliglotta Vaticana, Rome, 1968) 2 Vols.

Grenoble (1970): Microscopie Electronique 1970, résumés des communications présentées au septième congrès international, Grenoble; edited by P. Favard (Société de Microscopie Electronique, Paris, 1970) 3 Vols.

Manchester (1972): Electron Microscopy, 1972, Proceedings of the Fifth European Congress on Electron Microscopy, Manchester (Institute of Physics, Bristol, 1972)

Canberra (1974): Electron Microscopy 1974, Abstracts of Papers presented to The Eighth International Congress on Electron Microscopy, Canberra; edited by J.V. Sanders, D.J. Goodchild (Australian Academy of Science, Canberra, 1974) 2 Vols.

Jerusalem (1976): Electron Microscopy 1976, Proceedings of the Sixth European Congress on Electron Microscopy, Jerusalem; edited by D.G. Brandon, Vol.I, and Y. Ben-Shaul, Vol.II (Tal International, Jerusalem, 1976), 2 Vols.

Toronto (1978): Electron Microscopy, 1978, Papers presented at the Ninth International Congress on Electron Microscopy, Toronto; edited by J.M. Sturgess (Microscopical Society of Canada, Toronto, 1978) 3 Vols.

Additional References with Titles

Chapter 1

U. Aebi: Image processing of electron micrographs of ordered biomacromolecular assemblies. Toronto (1978) Vol.III, pp.81-86

E.V. Ageev, I.F. Anaskin, P.A. Stoyanov: Experimental analysis of image correction methods for electron microscopy. Izv. Akad. Nauk SSSR (Ser. Fiz) 41, 1447-1451 (1977) [English transl.: Bull. Acad. Sci. USSR (Phys. Ser.) 41 (7), 116-119 (1977)]

W. Baumeister (ed.): *Electron Microscopy in Molecular Dimensions. State of the Art and Strategies for the Future* (Springer, Berlin, Heidelberg, New York 1980) In preparation

P. Bonhomme, A. Beorchia: A light-optical diffractometer for electron microscopical images operating in line. Toronto (1978) Vol.I, pp.86-87

N. Bonnet, A. Beorchia, P. Bonhomme: Fonctions de transfert en microscopie électronique avec éclairage conique annulaire. J. Microsc. Spectrosc. Electron. *3*, 497-511 (1978)

J.M. Cowley, A.Y. Au: Bright-field image contrast and resolution in STEM and CTEM. Toronto (1978) Vol.I, pp.172-173

K.H. Downing: Possibilities of heavy atom discrimination using single-sideband techniques. Ultramicroscopy *4*, 13-31 (1979)

J. Frank, B. Shimkin: A new image processing software system for structural analysis and contrast enhancement. Toronto (1978) Vol.I, pp.210-211

L.A. Freeman, A. Howie, M.M.J. Treacy: Bright field and hollow cone dark field electron microscopy of palladium catalysts. J. Microsc. *111*, 165-178 (1977)

T.G. Frey: The structures of glutamine synthetase and cytochrome c oxidase — studies by electron microscopy and image analysis. Toronto (1978) Vol.III, pp.107-119

V.P. Gilev: A simple method of optical filtration. Ultramicroscopy *4*, 323-336 (1979)

T.A. Grishina: Diffraction effect in a transmission electron microscope: violation of spatial invariance and the appearance of pseudostructures on the image. Izv. Akad. Nauk SSSR (Ser. Fiz.) 41, 904-908 (1977) [English transl.: Bull. Acad. Sci. USSR (Phys. Ser.) 41 (5), 57-60 (1977)]

R. Guckenberger, W. Hoppe: On-line electron-optical correlation computing in the CTEM. Toronto (1978) Vol. I, pp.88-89

K.-J. Hanszen, G. Ade: Phase contrast transfer with different imaging modes in electron microscopy. Optik *51*, 119-126 (1978)

P.W. Hawkes: "Some unsolved problems and promising methods in electron image processing", in *Electron Microscopy in Molecular Dimensions. State of the Art and Strategies for the Future*, ed. by W. Baumeister (Springer, Berlin, Heidelberg, New York 1980) In preparation

K.-H. Herrmann, E. Reuber, P. Schiske: A simple way for producing holographic filters suitable for image improvement. Toronto (1978) Vol.I, pp.226-227

K.-H. Herrmann, D. Krahl, H.-P. Rust: A TV system for image recording and processing in conventional transmission electron microscopy. Ultramicroscopy *3*, 227-235 (1978)

R.W. Horne: Special specimen preparation methods for image processing in transmission electron microscopy: a review. J. Microsc. *113*, 241-256 (1978)

E.J. Kirkland, B.M. Siegel: Error sensitivity as a limit to resolution in computer image processing of electron micrographs. Optik *53*, 181-196 (1979)

N.A. Kiselev: Reconstruction of the structure of enzymes from their images. Toronto (1978) Vol.III, pp.94-106

W. Krakow: Applications of electronically controlled illumination in the conventional transmission electron microscope. Ultramicroscopy *3*, 291-301 (1978)

O.L. Krivanek: EM contrast transfer functions for tilted illumination imaging. Toronto (1978) Vol.I, pp.168-169

W. Kunath: Signal-to-noise enhancement in bright field images by incoherent superposition. Toronto (1978) Vol.I, pp.222-223

W. Kunath: Signal-to-noise enhancement by superposition of bright-field images obtained under different illumination tilts. Ultramicroscopy *4*, 3-7 (1979)

Y. Murata: Studies of radiation damage mechanism—by optical diffraction analysis and high resolution image. Toronto (1978) Vol.III, pp.49-60

H.-P. Rust, D. Krahl, K.-H. Herrmann: A digital storage and processing system which permits a posteriori image accumulation. Toronto (1978) Vol.I, pp.90-91

M.H. Savoji: Le bruit du grain de film en microscopie électronique et l'estimation de sa densité spectrale, J. Microsc. Spectrosc. Electron *4*, 175-188 (1979)

W.O. Saxton, D.J. Smith: "Bright-Field Hollow Cone Illumination—Theory and Experiment", in *Electron Microscopy and Analysis 1979*, ed. by T. Mulvey (Institute of Physics, Bristol 1979)

P. Schiske: Imaging of tilted extended thin phase-objects. Toronto (1978) Vol.I, pp.216-217

M. Troyon: The influence of beam intensity on the contrast transfer function of a field emission electron microscope. Optik *52*, 401-411 (1978/79)

N. Uyeda, K. Kirkland, Y. Fujiyoshi, B. Siegel: Atomic resolution from 500 kV electron micrographs by computer image processing. Toronto (1978) Vol.I, pp.220-221

Yu.M. Voronin, R.Yu. Khaitlina: Electron microscope with optical diffractometer for immediate image analysis. Izv. Akad. Nauk SSSR (Ser. Fiz.) *41*, 871-875 (1977) [English transl.: Bull. Acad. Sci. USSR (Phys. Ser.) *41* (5), 25-28 (1977)]

Yu.M. Voronin, A.V. Mokhnatkin: Production of focused images of phase specimens through a light-optics correction of defocusing and spherical aberration. Izv. Akad. Nauk SSSR (Ser. Fiz.) *41*, 917-921 (1977) [English transl.: *41* (5), 68-71 (1977)]

A.A. Vyazigin: Linear filtering of electron microscope images. Opt. Spektrosk. *45*, 821-822 (1978) [English transl.: Opt. Spectrosc. *45*, 714-716 (1978)]

A.A. Vyazigin: Problems in the reconstruction of electron microscope images. Opt. Spektrosk. *45*, 1008-1011 (1978) [English transl.: Opt. Spectrosc. *45*, 819-821 (1978)]

T.A. Welton: A computational critique of an algorithm for image enhancement in bright field electron microscopy. Adv. Electron. Electron Phys. *48*, 37-101 (1979)

F. Zemlin, K. Weiss, P. Schiske, W. Kunath, K.-H. Hermann: Coma-free alignment of high resolution electron microscopes with the aid of optical diffractograms. Ultramicroscopy *3*, 49-60 (1978)

F. Zemlin: Image synthesis from electron micrographs taken at different defocus. Ultramicroscopy *3*, 261-263 (1978)

F. Zemlin: A practical procedure for alignment of a high resolution electron microscope. Ultramicroscopy *4*, 241-245 (1979)

Chapter 2

R.E. Burge, M.A. Fiddy: Object reconstruction in electron microscopy from the real part of the scattered wave, Optik *54*, 21-26 (1979). (This area has, however, already been covered more fully elsewhere, cf. Sects.2.2 and 2.4)

N.H. Dekkers: Object wave reconstruction in STEM. Optik *53*, 131-142 (1979)

M.A. Fiddy, A.H. Greenaway: Object reconstruction from intensity data. Nature *276*, 421 (1978)

W.O. Saxton: "SEMPER—a Portable Image—Processing System Applied to Electron Microscopy", in *Machine-aided Image Analysis, 1978*, ed. by W.E. Gardner (Institute of Physics, Bristol 1979) pp.78-87

W.O. Saxton, A.J. Pitt, M. Horner: Digital image processing: The SEMPER system. Ultramicroscopy *4*, 343-354 (1979)

J.C.H. Spence: Practical phase determination of inner dynamical reflections in STEM. Scanning Electron Microscopy/1978, Vol.I (SEM, AMF O'Hare, 1978) pp.61-68

Chapter 4

W. Hoppe, J. Gassmann: "Comments on the 3-D reconstruction of periodic structures", in *Electronic Microscopy in Molecular Dimensions. State of the Art and Strategies for the Future*, ed. by W. Baumeister (Springer, Berlin, Heidelberg, New York 1980) In preparation

Chapter 5

N.R. Arnot, W.O. Saxton: Improvement of cross-correlation peaks by image filtering. Optik *53*, 271-279 (1979)

J. Frank: Image analysis in electron microscopy. J. Microsc. *117*, 25-38 (1979)

J. Frank, W. Goldfarb: "Methods for averaging of single molecules and lattice fragments", in *Electron Microscopy in Molecular Dimensions. State of the Art and Strategies for the Future*, ed. by W. Baumeister (Springer, Berlin, Heidelberg, New York 1980) In preparation

J. Frank, W. Goldfarb, D. Eisenberg, T.S. Baker: Addendum to Reconstruction of glutamine synthetase using computer averaging. Ultramicroscopy *4*, 247 (1979)

Z. Kam: "The reconstruction of structure from electron micrographs of randomly oriented particles", in *Electron Microscopy in Molecular Dimensions. State of the Art and Strategies for the Future*, ed by W. Baumeister (Springer, Berlin, Heidelberg, New York 1980) In preparation

M. Kessel, J. Frank, W. Goldfarb: "Low-dose electron microscopy of individual biological macromolecules", in *Electron Microscopy in Molecular Dimensions. State of the Art and Strategies for the Future*, ed by W. Baumeister (Springer, Berlin, Heidelberg, New York 1980) In preparation

W. Saxton: Computer techniques for image processing in electron microscopy, Adv. Electron. Electron Phys. Suppl. 10 (1978)

W. Saxton: "Matching and averaging fragmented lattices", in *Electron Microscopy in Molecular Dimensions. State of the Art and Strategies for the Future*, ed. by W. Baumeister (Springer, Berlin, Heidelberg, New York 1980) In preparation

H.P. Zingsheim, D.C. Neugebauer, F.J. Barrantes, J. Frank: "Image averaging of membrane-bound acetylcholine receptor from *Torpedo marmorata*", in *Electron Microscopy in Molecular Dimensions. State of the Art and Strategies for the Future*, ed. by W. Baumeister (Springer, Berlin, Heidelberg, New York 1980) In preparation

Chapter 6

N. Bonnet, M. Troyon, P. Gallion: Possible applications of Fraunhofer holography in high resolution electron microscopy. Toronto (1978) Vol.I, pp.222-223

A. Tonomura, T. Matsuda, T. Komoda: Off-axis electron holography by field emission electron microscope. Toronto (1978) Vol.I, pp.224-225

A. Tonomura, T. Matsuda, J. Endo: Spherical-aberration correction of an electron lens by holography. Jpn. J. Appl. Phys. *18*, 1373-1377 (1979)
Image holograms of fine gold particles were formed in an electron microscope equipped with a field emission gun and a biprism, at a magnification of 140 000 X (chosen to be equal to the wavelength ratio). The spherical aberration was corrected in a subsequent optical stage; the C_s required, 240 m, was

so large that the reconstructed image had to be reduced by a factor of six using the correction lens.

E. Zeitler: Electron holography. Proc. 37th Annual Meeting EMSA, San Antonio, ed. by G.W. Bailey (Claitor, Baton Rouge 1979) pp.376-379. This is a short but fairly complete review of electron holography (written in Zeitler's inimitable style!)

Subject Index

Digital Pattern Recognition

Editor: K. S. Fu
1976. 54 figures, 4 tables. XI, 206 pages
(Communication and Cybernetics,
Volume 10)
ISBN 3-540-07511-9

Contents:
K. S. Fu: Introduction. – *T. M. Cover,
T. J. Wagner:* Topics in Statistical Pattern Recognition. – *E. Diday, J. C. Simon:* Clustering Analysis. – *K. S. Fu:* Syntactic (Linguistic) Pattern Recognition. – *A. Rosenfeld,
J. S. Weszka:* Picture Recognition. – *J. J. Wolf:* Speech Recognition and Understanding. – Additional References with Titles. – Subject Index.

Digital Picture Analysis

Editor: A. Rosenfeld
1976. 114 figures, 47 tables. XIII, 351 pages
(Topics in Applied Physics, Volume 11)
ISBN 3-540-07579-8

Contents:
A. Rosenfeld: Introduction. – *R. M. Haralick:* Automatic Remote Sensor Image Processing. – *C. A. A. Harlow, S. J. Dwyer III,
G. Lodwick:* On Radiographic Image Analysis. – *R. L. McIlwain, Jr.:* Image Processing in High Energy Physics. – *K. Preston, Jr.:* Digital Picture Analysis in Cytology. – *J. R. Ullmann:* Picture Analysis in Character Recognition.

T. Pavlidis
Structural Pattern Recognition

1977. 173 figures, 13 tables. XII, 302 pages
(Springer Series in Electrophysics, Volume 1)
ISBN 3-540-08463-0

Contents:
Mathematical Techniques for Curve Fitting. – Graphs and Grids. – Fundamentals of Picture Segmentation. – Advanced Segmentation Techniques. – Scene Analysis. – Analytical Description of Region Boundaries. – Syntactic Analysis of Region Boundaries and Other Curves. – Shape Description by Region Analysis. – Classification, Description and Syntactic Analysis.

Syntactic Pattern Recognition, Applications

Editor: K. S. Fu
1977. 135 figures, 19 tables. XI, 270 pages
(Communication and Cybernetics,
Volume 14)
ISBN 3-540-07841-X

Contents:
K. S. Fu: Introduction to Syntactic Pattern Recognition. – *S. I. Horowitz:* Peak Recognition in Waveforms. – *J. E. Albus:* Electrocardiogram Interpretation Using a Stochastic Finite State Model. – *R. DeMori:* Syntactic Recognition of Speech Patterns. – *W. W. Stallings:* Chinese Character Recognition. – *Th. Pavlidis, H.-Y. F. Feng:* Shape Discrimination. – *R. H. Anderson:* Two-Dimensional Mathematical Notation. – *B. Moayer, K. S. Fu:* Fingerprint Classification. – *J. M. Brayer, P. H. Swain,
K. S. Fu:* Modeling of Earth Resources Satellite Data. – *T. Vámos:* Industrial Objects and Machine Parts Recognition.

Springer-Verlag
Berlin
Heidelberg
New York

Optical Data Processing

Applications
Editor: D. Casasent
1978. 170 figures, 2 tables. XIII, 286 pages
(Topics in Applied Physics, Volume 23)
ISBN 3-540-08453-3

Contents:
D. Casasent, H. J. Caulfield: Basic Concepts. –
B. J. Thompson: Optical Transforms and Coherent Processing Systems With Insights from Crystallography. – *P. S. Considine,*
R. A. Gonsalves: Optical Image Enhancement and Image Restoration. – *E. N. Leith:* Synthetic Aperture Radar. – *N. Balasubramanian:* Optical Processing in Photogrammetry. –
N. Abramson: Nondestructive Testing and Metrology. – *H. J. Caulfield:* Biomedical Applications of Coherent Optics. – *D. Casasent:* Optical Signal Processing.

Picture Processing and Digital Filtering

Editor: T. S. Huang
2nd corrected and updated edition. 1979.
113 figures, 7 tables. XIII, 297 pages
(Topics in Applied Physics, Volume 6)
ISBN 3-540-09339-7

Contents:
T. S. Huang: Introduction. – *H. C. Andrews:* Two-Dimensional Transforms. –
J. G. Fiasconaro: Two-Dimensional Nonrecursive Filters. – *R. R. Read, J. L. Shanks, S. Treitel:* Two-Dimensional Recursive Filtering. –
B. R. Frieden: Image Enhancement and Restoration. – *F. C. Billingsley:* Noise Considerations in Digital Image Processing Hardware. –
T. S. Huang: Recent Advances in Picture Processing and Digital Filtering. – Subject Index.

L. Reimer, G. Pfefferkorn
Raster-Elektronenmikroskopie

2., neubearbeitete und erweiterte Auflage.
1977. 146 Abbildungen. XI, 282 Seiten
ISBN 3-540-08154-2

Inhaltsübersicht:
Einleitung. – Wechselwirkung Elektron-Materie. – Elektronenoptik, Aufbau und Funktion des Raster-Elektronenmikroskopes. – Abbildung mit Sekundär-, Rückstreuelektronen und Probenströmen. – Raster-Transmissions-Elektronenmikroskopie. – Elementanalyse und Abbildung mit emittierten Quanten und Augerelektronen. – Auswertemethoden rasterelektronenmikroskopischer Aufnahmen. – Präparation.

X-Ray Optics Applications to Solids

Editor: H. J. Queisser
1977. 133 figures, 17 tables. XI, 227 pages
(Topics in Applied Physics, Volume 22)
ISBN 3-540-08462-2

Contents:
H. J. Queisser: Introduction: Structure and Structuring of Solids. – *M. Yoshimatsu,*
S. Kozaki: High Brilliance X-Ray Sources. –
E. Spiller, R. Feder: X-Ray Lithography. –
U. Bonse, W. Graeff: X-Ray and Neutron Interferometry. – *A. Authier:* Section Topography. – *W. Hartmann:* Live Topography.

Springer-Verlag
Berlin
Heidelberg
New York